A. J. Baker, James D. Freels
Error Freed CFD Mathematics

Also of interest

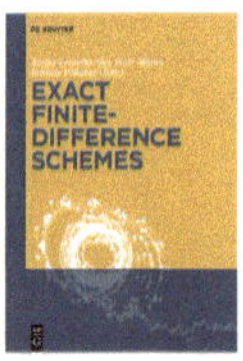

Exact Finite-Difference Schemes
Sergey Lemeshevsky, Piotr Matus
and Dmitriy Poliakov (Eds.), 2016
ISBN 978-3-11-048964-4
e-ISBN (PDF) 978-3-11-049132-6

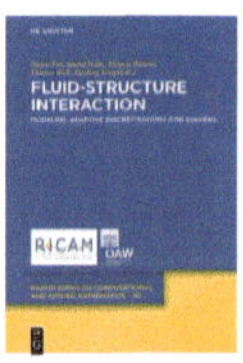

Fluid-Structure Interaction.
Modeling, Adaptive Discretisations and Solvers
Stefan Frei, Bärbel Holm, Thomas Richter, Thomas Wick
and Huidong Yang (Eds.), 2017
ISBN 978-3-11-049527-0
e-ISBN (PDF) 978-3-11-049425-9

Computational Physics.
With Worked Out Examples in FORTRAN® and MATLAB®
2nd Edition
Michael Bestehorn, 2023
ISBN 978-3-11-078236-3
e-ISBN (PDF) 978-3-11-078252-3

Theoretical and Computational Chemistry.
Gauge Theories for Chemical Physics and Excited State Methods
Emanuele Curotto, planned 2026
ISBN 978-3-11-161008-5
e-ISBN (PDF) 978-3-11-161020-7

A. J. Baker, James D. Freels

Error Freed CFD Mathematics

Stability, Monotonicity, Finite Element Theory

DE GRUYTER

Authors
A. J. Baker
Knoxville, TN 37996-2030
USA
ajbaker@utk.edu

James D. Freels
Oak Ridge, TN 37830
USA
jdfreels@proton.me

ISBN 978-3-11-914776-7
e-ISBN (PDF) 978-3-11-220643-0
e-ISBN (EPUB) 978-3-11-220646-1

Library of Congress Control Number: 2025940567

Bibliographic information published by the Deutsche Nationalbibliothek
The Deutsche Nationalbibliothek lists this publication in the Deutsche Nationalbibliografie; detailed bibliographic data are available on the internet at http://dnb.dnb.de.

Cover image: James D. Freels
Typesetting: Integra Software Services Pvt. Ltd.
Printing and binding: CPI books GmbH, Leck

www.degruyterbrill.com
Questions about General Product Safety Regulation:
productsafety@degruyterbrill.com

MIX
Papier | Fördert gute Waldnutzung
FSC
www.fsc.org
FSC® C083411

Dedication

This monograph is dedicated to the memory of the lead author's graduate academic mentor, University Distinguished Teaching Professor Irving H. Shames. His innovative approach to teaching engineering as a science rather than as tenuously related topics in mechanics is archived in his acclaimed McGraw-Hill *engineering science* textbook series, with premier issue *The Mechanics of Fluids* (Shames, 1962).

Accepting the early 1960s offer to move his Pratt Institute engineering science academic program to the Buffalo NY campus of the newly minted State University of New York Graduate School system enabled my exposure to, hence acquired dexterity with the mathematics, differential equations, vector calculus, continuum mechanics, tensor field theory and calculus of variations rigor that is herein identified to *finally enable* (!) computational fluid dynamics to join theoretical fluid mechanics as a mathematically rigorous discipline in the mechanics of continua.

The renewed collegial collaboration leading to this monograph, as enabled by both authors now being retired, follows Professor Shames leadership in the formal archiving of academic advancements in the engineering sciences.

A. J. Baker, PhD
Professor Emeritus
University of Tennessee

James D. Freels, PhD
Senior Research Staff (retired)
Oak Ridge National Laboratory

https://doi.org/10.1515/9783112206430-202

Contents

Prologue

The persistence of approximation error mechanisms *intrinsic* to legacy CFD practices prompted an op-ed in *Aerospace America* that stated, without equivocation, Masters (2014):

> . . . recently perceived ***stagnation*** of CFD capability advancement . . . exists.

boldface emphasis is by the author.

Reflecting on the mathematical rigor with which theoretical fluid mechanics is archived in advanced textbooks, this observation raises the question: "Are difference algebra theorized, '*phake*' physics stabilized, Reynolds stress tensor model closed CFD algorithms/codes the best that the community can bring to today's remotely accessible, massively multi-processor parallel, NVIDIA-accelerated supercomputers with limitless memory and blazing speed?"

The answer lies in responses to two *very* fundamental questions.

> Are you solving the *right* equations?
> Are you solving these equations *right*?

It is rigorously quantified herein that both answers are a resounding ***NO*** for ***all*** current practice open-source/government/commercial CFD codes.

Transitioning the response for

> . . . *right* equations?

to ***YES*** involves fundamental ***alteration*** of the textbook appearance of fluid mechanics' differential equation systems with ***truncation error annihilation*** (TEA) theory-derived *cubically nonlinear* velocity tensor product, Reynolds number (Re) and state variable member-dependent continuum calculus *functionals*. Thus enabled is total ***annihilation*** of *these* legacy practice CFD error mechanisms:

(i) spatial discretization generated algebraic *instability*;
(ii) space-time discretization-induced scalar transport phase *aliasing*;
(iii) *the need for* stabilized shock capture, shock mesh-fitting algorithms;
(iv) weak linear algebra admitted non-converged solutions;
(v) incompressible divergence-freeness *inadequate* enforcement, pressure.

Transitioning the response for

> . . . equations *right*?

to ***YES*** replaces discrete algebra theorization with *continuous* Galerkin *finite element* (FE) bases vector calculus TEA theory infused weak formulations. Coupled with mathematically rigorous derivations of ***for all Re pertinent*** analytical clo-

https://doi.org/10.1515/9783112206430-001

sure of spatially filtered Navier-Stokes (NS) differential equation systems in concert enables:

(vi) modern approximation theory and vector calculus replacement of discrete algebraic schemes;
(vii) rigorously derived global/local *error estimates*, rigorous error *quantification*;
(viii) *stable* tridiagonal stencil equivalent fourth-order *monotone* predictions on ***any*** mesh;
(ix) shock monotone *continuum interpolation* on ***any*** mesh;
(x) ∀Re pertinent spatially filtered NS closure prediction of laminar profile $O(1)$ velocity incipient transition, separation, and turbulent profile reattachment;
(xi) FE basis completeness degree enabled ***higher-order*** asymptotic convergence;
(xii) matrix calculus derived entries for the *quadratically convergent* Newton iteration algorithm Jacobian.

To conclude, the implementation of herein identified, derived, and validated ***error freed*** **CFD** ***mathematics*** enables CFD to ***at last*** (!) join with theoretical fluid mechanics as a well-posed, mathematically *rigorous* discipline in **continuum mechanics**.

Chapter 1
Why this monograph?

The mathematical nonlinearity inherent to the continuum conservation principles of fluid mechanics has long posed challenges to their rigorous characterization. In the *Physics of Fluids* (2022) commemorative issue recognizing fluid mechanics textbook author Professor Frank White, the article "On mathematics of physics of fluids maturation" outlined the historical sequence of analytical successes as follows:

> The Blasius 1903 dissertation on laminar boundary layer matched asymptotic expansion;
> Prandtl's advanced topics 1942 text, *A Brief Introduction to Fluid Dynamics* ([4], in German);
> Its NACA-funded 1952 translation *Essentials of Fluid Dynamics* [5] catalyzes a rash of advanced texts including seven editions of Schlichting's *Boundary Layer Theory* [6];
> The mid-1960s introduction of, then domination by *computational fluid dynamics* (CFD).

The evolution of scientific computing capabilities over the past six-plus decades has quantitatively validated Moore's law, as the *truly primitive* computing environments of the 1960s transitioned into remotely accessible, massively multi-processor parallel, NVIDIA-accelerated supercomputers with limitless memory and blazing speed. This maturation prompted CFD professionals to focus on porting their well-trusted 1980s–1990s theorized codes to these incredibly capable computing environments. Time frame advances in grid generation and data visualization, e.g., the 1970s Calcomp multi-color pen x-y plotter evolution to Tecplot, Ensight, and FieldView dynamic color graphics packages that enable multi-terabyte CFD data files to be converted into 3D color perspective graphics and animations directly supported this action.

These artificially/numerically stabilized, turbulence modeling closed second-order accurate CFD algorithm legacy codes have now generated the sum total of all known fluid mechanics' first-order insights. Despite this accomplishment, their discretized approximation generated error mechanisms, existent at the dawning of CFD a century ago, persist today to thwart their ability to predict incisive *physics of fluids fidelity* information as the *significance* of these non-quantifiable features is the same order as the error mechanisms intrinsic to legacy practice.

In his reflections on the status of CFD, Jameson [2] states:

> . . . worldwide, commercial and government CFD codes are based on algorithms developed in the 80s and 90s . . .
> . . codes can handle complex geometry . . . limited to 2*nd* order accuracy . .
> . . current CFD codes cannot handle turbulence without modeling . . .

Inadequacies cited for the NPARC Alliance US government-commercial aerospace industry collaborative WIND-US code include [1]:

https://doi.org/10.1515/9783112206430-002

> . . . laminar-turbulent boundary layer transition . . .
> . . . shock-turbulent boundary layer interaction . . .
> . . . scalar transport aliasing, turbulent Prandtl/Schmidt number *models* . . .
> . . . compressible mixing . . .

The forward-thinking NASA contractor report, “CFD Vision 2030 Study: A Path to Revolutionary Computational Aerosciences,” cites these key limitations [7]:

> . . . single most *critical* . . . turbulent . . . transition . . . separation . . .
> . . . lack of *theoretically based* hybrid RANS-LES . . . capability . . .
> . . . key impediments . . . complexity of the *models* . . .

The mathematical rigor with which theoretical fluid mechanics is presented in advanced textbooks is truly noteworthy. And it is precisely ***these*** Navier-Stokes (NS) partial differential equation (PDE) systems that have been space/time discretized to generate legacy CFD algorithms/codes that, *without* exception (!), are *algebraically unstable*, hence pathologically divergent in finite time/iterations.

This persistence of these legacy CFD practice intrinsic error mechanisms prompted an op-ed in *Aerospace America* which stated, Masters (2014), [3].

> . . . recently perceived ***stagnation*** of CFD capability advancement . . . exists.

boldface emphasis by the author. Reflecting on the mathematical rigor with which theoretical fluid mechanics is archived in advanced textbooks, this observation raises the question,

> Are difference-algebra theorized, *phake* physics stabilized, Reynolds-stresstensor model closed CFD algorithms the best the community can bring to today’s high performance peta-scale scientific computing environment?

The answer lies in the responses to two absolutely fundamental queries:

> Are you solving the right equations?
> Are you solving these equations right?

The authors of this monograph, now both retired following ~90 years accumulated CFD professional practice, assert that the answer to each of these questions is a resounding ***NO*** for all current practice open-source/government/commercial CFD codes.

Alterations to ***YES*** are monograph identified, rigorously derived and validated by implementing the *engineering sciences* rigor of nonlinear partial differential equations, vector calculus, continuum mechanics, tensor field theory, the calculus of variations, and energy methods combined with modern approximation theory that in concert identify the analytical resolutions that *annihilate every* (!) *error mechanism* intrinsic to legacy practice CFD. Each is derived discrete numerics compatible with “tridiagonal stencil” equivalent CFD second-order difference algebra practice, which when used to alter legacy CFD codes will ***unstagnate*** their inability to generate fourth-order accurate *monotone* distributions that quantify *physics of fluids* first- *and* higher-order *effects fidelity* on ***any*** (!) mesh.

For readers with a serious interest in mathematics, the weak formulation continuous Galerkin finite element (FE) method with higher-order basis functions implemented algorithms used here to validate these assertions has, in fact, generated the mathematical foundation necessary to enable computational fluid mechanics to *finally* join theoretical fluid mechanics as a rigorous discipline in continuum mechanics.

References

[1] Georgiadis, N.J., Yoder, D.A., Towne, C.S., Engblom, W.A., Bhagwandin, V.A., Power, G.D., Lankford, D.W., Nelson, C.C. (2009). "Wind-US Code Physical Modeling Improvements to Complement Hypersonic Testing and Evaluation," NASA/TM-2009- 215615, Tech Rpt AIAA:2009-193

[2] Jameson, A. (2012) *Reflections on Four Decades of CFD – A Personal Perspective*, http://dept.ku.edu/~cfdku/JRV/Jameson.pdf

[3] Masters, J. (2014). "A year of enhancements, plus a groundbreaking vision document," *Aerospace America*, 52(11).

[4] Prandtl, L. (1942). *A Brief Introduction to Fluid Dynamics*, Vieweg und Sohn, Germany.

[5] Prandtl, L. (1952). *Essentials of Fluid Dynamics*, NACA translation, Hafner, NY.

[6] Schlichting, H. (1979). *Boundary Layer Theory*, McGraw-Hill, NY.

[7] Slotnick, J., Khodadoust, A., Alonso, J., Darmofal, D., Gropp, W., Lurie, E., Mavriplis, D., (2014). "CFD Vision 2030 Study: A Path to Revolutionary Computational Aerosciences," NASA Tech Report. CR-2014-218178

Chapter 2
A brief pertinent CFD chronology

Libretto: The inaugural time-explicit Navier-Stokes finite difference (FD) CFD algorithm identified the "CFL" time-step constraint a century ago.

A half-century later, the Cartesian grid stencil restriction for FD CFD algorithms was removed by Poisson equation-generated curvilinear coordinate transformations.

Thus initiated was the replacement of FD recursion stencils with finite volume (FV) flux vector formulations, which are the current practice in most CFD codes.

FV time frame coincident development of the discrete method of weighted residuals (MWR) introduced calculus as an alternative to discrete algebra theorization.

MWR was then replaced with the weak formulation *continuous* Galerkin finite element (FE) trial space bases theorization, which is monograph *validated* to be *optimal* among all alternatives.

Keywords: Navier-Stokes, CFD, FD, FV, FE, MWR, weak formulation, continuous Galerkin

2.1 Evolution, legacy CFD theorizations

The pioneering time-explicit *finite difference* (FD) CFD algorithm of Courant, Friedrichs, and Loewy ([14], in German) identified the "CFL number" explicit time step constraint. Following WW II, CFD theorizations associated with the New York University Courant Institute include Lax [28], Lax and Wendroff [29], and Richtmyer and Morton [35]. FD-theorized CFD algorithms derived by technical staff members at the U.S. Los Alamos Scientific Laboratory (LASL) include Amsden and Harlow [1] and Harlow [25]. These developments were time-frame coincident with the Imperial College (London) development of the famous "SIMPLE" family of FD CFD algorithms, Gosman et al. [22]. The algebraic instability intrinsic to ***every one of these*** NS CFD formulations was *universally* moderated by algorithm insertion of one of a wide variety of *artificial/numerical diffusion* stabilizations. This timeframe CFD FD stencil algorithm repository is Roache [36].

2.2 Legacy CFD NS potential flow shock capturing algorithms

The 1970s witnessed a rash of CFD FD algorithms seeking steady-state prediction of an isentropic transonic shock formulated on the compressible NS mass conservation equation (DM) altered to its *potential theory* elliptic boundary value (EBV) PDE. The

https://doi.org/10.1515/9783112206430-003

resultant code algebraic stability *challenge* involved alteration of the FD stencil in flowfield reaches with supersonic velocity to account for the DM altered EBV PDE transition to hyperbolic.

The inaugural NS mixed elliptic-hyperbolic *perturbation* potential PDE FD algorithm utilized a "typed" central difference FD stencil, altered to upwind in the regions of supersonic velocity, Murman and Cole [32]. The successor *full* potential DM PDE FD algorithm employed a "rotated difference scheme" to align the footprint of the FD stencil with the predicted trajectory of the shock, Jameson [26].

As these typed-FD stencil transonic potential flow schemes proved computationally cumbersome, the focus turned to strictly central difference potential DM PDE algorithm *shock-capturing* schemes stabilized by the insertion of *numerical diffusion*. Due to the modest compute resources of this era, these segregated Jacobian "reverse engineered" algorithms first solved the linearized potential Laplacian DM for velocity using a "dissipation altered" density, followed by an update to the density, Jameson [26], Hafez et al [23]. Notwithstanding, code algebraic instability required "*upwinding*" of the DM laplacian central difference stencil, which amounted to insertion of numerical diffusion stabilization, Shankar et al. [40].

A range of weak formulation dissipative (hence non-Galerkin) finite element (FE) linear ($p = 1$) basis algorithms for NS transonic potential DM followed, Chan et al. [11], Glowinski et al. [20], Hafez and Murman [23]. The latter development observed that the *artificial compressibility* formulation of Eberle [17] was functionally identical to the Jameson [26] *dissipation-altered density* FD scheme for the definitions

$$\text{DM:} \quad -\frac{\partial}{\partial x}\left[\tilde{\rho}\frac{\partial \varphi}{\partial x}\right] - \frac{\partial}{\partial y}\left[\tilde{\rho}\frac{\partial \varphi}{\partial y}\right] = 0$$

$$\tilde{\rho} \equiv \rho + \frac{\mu\rho}{a^2}\left[u\frac{\partial u}{\partial x}\Delta x + v\frac{\partial v}{\partial y}\Delta y\right], \quad \mu = \max(0,\ 1 - \mathrm{Ma}^{-2})$$

$$\int \text{DP:} \quad \frac{\rho}{\rho_\infty} = \left[1 - \frac{(\gamma - 1)\nabla\varphi\cdot\nabla\varphi}{2a_\infty^2}\right]^{1/(\gamma-1)}$$

for a the sound speed with the dissipative density coefficient μ defined Mach number $\mathrm{Ma}^h(\mathbf{x})$ distribution dependent. The dissipative density definition for this weak formulation, numerically stabilized non-Galerkin FE $p = 1$ basis segregated Jacobian algorithm was then altered to dependence on the Mach number DOF $\{\mathrm{Ma}(\mathbf{x})\}$ distribution

$$\tilde{\rho} = \rho - \mu\rho_s\Delta s, \quad \rho_s\Delta s \equiv \left[\frac{u}{q}\frac{\partial\rho}{\partial x}\Delta x + \frac{v}{q}\frac{\partial\rho}{\partial y}\Delta y\right], \quad q = \left(u^2 + v^2\right)^{1/2}$$

$$\mu = \max(0,\ 1 - \{\mathrm{Ma}^{-2}\})$$

to moderate the excess numerical diffusion of the prior FD algorithm [24].

Thus summarized is the "build it and burn it" approach to transonic shock-capturing CFD algorithm theorizations through the early '80s. That this profusion of formulations cannot be linearly independent then prompted the search for a rigorous framework to correlate their mathematical essence. The maturation of this search identified the time Taylor series (TS) correlation theory, Baker and Kim [8], which is detailed in Chapter 4 as the inaugural step to rigorous derivation of *error freed* CFD *mathematics*.

As steady-state iteration is computationally analogous to unsteady time marching, this TS theory indeed enabled the correlation of these 70s-80s-era transonic potential NS shock-capturing FD/FE algorithms as decisions made on the $O(\Delta t)$ truncation error β term in the time TS-augmented unsteady $\mathrm{D}^m\mathrm{M}$

$$\text{TS:}\ \frac{\rho(t+\Delta t)\ -\ \rho(t)}{\Delta t}\ +\ \frac{\partial \rho u_j}{\partial x_j}\ -\ \frac{\beta\Delta t}{2}\frac{\partial}{\partial x_j}\left[u_j u_k\frac{\partial \rho}{\partial x_k}\right]\ =\ 0\ +\ O(\Delta t^2)$$

Replacing $\Delta t/2$ with the local time scale $h/|\mathbf{u}|$, absorbing the ½ into the {Ma} DOF distribution now denoted as the TS {β} coefficient, and taking the limit $\Delta t \Rightarrow \varepsilon > 0$ enabled the correlation of the range of historical FD/FE theorized transonic NS full potential algorithms as individual decisions made on the time TS β-term *modified* potential flow $\mathrm{D}^m\mathrm{M}$

$$\mathrm{D}^m\mathrm{M}\text{:}\ \frac{\partial \rho}{\partial t}\ +\ \frac{\partial}{\partial x_j}\left[\rho u_j -\ \{\beta\}h\frac{u_j u_k}{|\mathbf{u}|}\frac{\partial \rho}{\partial x_k}\right]\ =\ 0\ +\ O(h^p)$$

2.3 Legacy CFD NS aerodynamics algorithms

Technical staff at the NASA Ames Research Center (ARC) initiated compressible aerodynamics NS CFD FD algorithms, specifically the time-explicit algorithm, MacCormack [31], and the time-*implicit* approximate factorization (AF) *numerical linear algebra*-designed algorithm, Beam and Warming [9]. The FD stencil grid regularity requirement was subsequently eliminated via Poisson PDE curvilinear coordinate transformations, Thompson et al. [34], which altered CFD theorization to *finite volume* (FV) flux vector algorithms. The imposition of the divergence theorem on the continuum NS PDE system was subsequently discretized into the union of non-overlapping FV cells, which generated the requirement to form NS state variable *cell face efflux* summation over each FV cell, the discrete approximations of which reverted to difference algebra scheming.

The integral FV formulation thereby engendered the replacement of *numerical diffusion*-stabilized FD operators with stabilized flux vector algorithms formulated on models of the isentropic Euler PDE system altered to *hyperbolic conservation law* form. These CFD stabilizations sought algebraically stable generation of *essentially non-oscillatory* (ENO), then *weakly* ENO (WENO), and then *total variation diminishing*

(TVD) NS state variable degree-of-freedom (DOF) distributions via piecewise discontinuous "Godunov genre" constructions, Godunov [21].

Universally sought was FV algorithm generation of *almost monotone* NS state variable DOF distributions. Resultant was the creation of a variety of difference algebra-implemented flux vector/eigenvector/flux difference split dissipative/dispersive predictor/corrector efflux schemes with limiters and switches. Those of van Leer [44] and Roe [33] are NS/RaNS algorithm stabilization options in select current CFD codes. The timeframe repository for these FD/FV/ENO/WENO/TVD CFD algorithms is Tannehill et al. [43].

Without exception, these stabilized flux vector schemes sought *mitigation* of the algebraic *instability* intrinsic to NS spatially discretized CFD algorithms rather than identification of the root cause, hence its *annihilation*. With few exceptions, current open source/commercial/ government CFD codes function on these 80s–90s stabilized FV flux vector algorithms.

Weak formulation restatements of flux vector split stabilizations then followed as *discontinuous* Galerkin (DG) FE *h,p* basis algorithms, Bey and Oden [10], Cockburn and Shu [12], Cockburn et al. [13]. These Euler PDE *isentropic* CFD algorithms then challenged FV flux vector split algorithm-stabilized *shock capture* with mesh realignment *shock fitting* strategies incorporating the Rankine-Hugoneou *shock jump* relationship.

2.4 Evolution of integral calculus expressed CFD algorithms

Select 1970s–1980s FD/FV CFD algorithms then became correlated in terms of specific decisions made in the spatially discrete *method of weighted residuals* (MWR), Finlayson [18]. In an MWR algorithm, the NS discrete *approximation error* is constrained by requiring local integrals containing "weights" to vanish. Within the class, selecting the weights as Dirac delta functions recovered *collocation* and reproduced classic FD stencils. An FV formulation was recovered by selecting weights the set of all constants. Generalizing weights to local polynomials led to the discrete *pseudo-variational* continuous Galerkin FE basis formulation, named after the non-discrete, non-CFD theory of Galerkin [19]. Finally, defining weights as the PDE differential operator string itself reproduced a *least squares* algorithm.

Inaugural NS analog integral theorizations are reported by Oden (1969, 1972). Continuous Galerkin MWR FE linear basis CFD algorithm *validations* for two-dimensional compressible and incompressible NS PDE systems are Baker [2–5] and Baker and Zelazny [6]. A least-squares MWR boundary layer algorithm that retained a symmetric global matrix is Lynn [30].

Galerkin MWR FE linear basis algorithm asymptotic convergence theory for the laminar incompressible boundary layer PDE system was *validated* in the NS velocity PDE intrinsic *energy norm*, Popinski and Baker [37] and Soliman and Baker [41]. Extension to turbulent boundary layer asymptotic convergence theory validation is documented in Soliman and Baker [42], who also reported Galerkin MWR algorithm *error extremization* in comparison with the same-order accurate Crank-Nicolson [15] FD algorithm. The time-period MWR FE linear basis archive is Baker [7].

2.5 Taylor series correlation of legacy NS CFD stabilizations

Time TS alteration of an NS PDE system was first reported by Donea [16]. Its generalization, Baker and Kim [8], led to the correlation of a dozen *independently* derived FD, FV, FE, and Godunov genre CFD *stabilizations*. Shortly thereafter, classic mechanics-rooted symmetric quadratic forms calculus of variations extremization, coupled with concepts of completeness and data smoothness, became recognized as the *continuum mechanics* foundation underlying discrete MWR.

Thus, a research agenda was initiated to replace CFD discrete theorization with weak formulation *continuous* Galerkin constructions that retained *continuum* vector calculus through to identification of the terminal algebraic matrix statement. The transition to TS spatial manipulations on a model NS PDE enabled rigorous derivation of the spatial discretization-generated Reynolds number-dependent *cubically nonlinear* velocity tensor product $O(h^2)$ truncation error (TE) *continuum* functional, Kolesnikov and Baker [27], h the measure of the mesh. Innumerable NS/RaNS PDE continuous Galerkin CFD predictions have quantified that this $O(h^2)$ TE term is ***the root cause*** (!) for NS PDE spatial discretization induced algebraic *instability*.

Extension of TS manipulations to Fourier wave number space led to derivation of the *cubically nonlinear* continuum functional responsible for NS space-time discretization-induced scalar Fourier spectral content transport *aliasing* error, Sahu and Baker [38]. The resultant *truncation error annihilation* (TEA) theory altered the compressible NS/RaNS PDE system's weak formulation enabling *continuous* Galerkin FE *h,p* basis algorithms to secure shock continuum *monotone interpolation*, ***the*** mathematically rational alternative to numerically stabilized shock capture and/or shock fitting algorithms. Resultant is validation of continuous Galerkin CFD algorithm *physics of fluids fidelity* prediction of first- and higher-order effects in transonic shock characterization and shock-turbulent boundary layer interaction diagnostics.

Monograph Part B content validates the mathematically rigorous derivation of *analytical closure* for incompressible and compressible space-filtered NS PDE systems' weak formulation Galerkin CFD algorithm, Sekachev et al. [39]. As the word *turbulent* is never mentioned during derivation, the resultant ***F***iltered ***a***nalytically ***N***avier-***S***tokes (FaNS theory) PDE system *continuous* Galerkin FE basis implemented algorithms are pertinent to simulations at *arbitrary* Reynolds number (Re) specifications. Validations are detailed for Reynolds numbers ranging $100 \leq \mathrm{Re} \leq 12{,}000$, for which the latter lead to validation of TEA theory infused, *error freed* CFD *mathematics* prediction of space filtered laminar profile resolved scale velocity insipient transition, separation, turbulent profile reattachment then relaminarization.

References

[1] Amsden, A., Harlow, F. (1970). "The SMAC Method: A Numerical Technique for Calculating Incompressible Fluid Flow," Univ. Calif. LASL Tech. Rpt. LA-4370

[2] Baker, A.J. (1971). A Finite Element Computational Theory for Mechanics and Thermodynamics of a Viscous Compressible Multi-Species Fluid, Research Rpt. 9500-920200, Bell Aerospace Corp., Buffalo NY.

[3] Baker, A.J. (1973). "Finite Element Solution Algorithm for Viscous Incompressible Fluid Dynamics," *J Num Mtd Engineering*, V. 6.

[4] Baker, A.J. (1974a) "A Finite Element Algorithm for the Navier-Stokes Equations," NASA Contractor Report CR-2391.

[5] Baker, A.J. (1974b). "Finite Element Solution Theory for Three-Dimensional Boundary Flows," *Comp Mtd Appl Mech Engr*, 4.

[6] Baker, A.J., Zelazny, S.W. (1975). "Predictions in Environmental Hydrodynamics Using the Finite Element Method, I. Theory, II. Applications," *AIAA Journal*, 13(1).

[7] Baker, A.J. (1983). *Finite Element Computational Fluid Mechanics*, Hemisphere, NY.

[8] Baker, A.J., Kim, J.W. (1987). "A Taylor Weak Statement Algorithm for Hyperbolic Conservation Laws," *J Num Mtd Fluids*, 7.

[9] Beam, R., Warming, R. (1976). "An Implicit Finite-Difference Algorithm for Hyperbolic Systems in Conservation Law Form," *J Comp Physics*, 22.

[10] Bey, K.S., Oden, J.T., Patra, A. (1996). "A parallel hp-adaptive discontinuous Galerkin method for hyperbolic conservation laws," *Appl Num Math*, 20.

[11] Chan, S.T.K., Brashears, J.R., Young, V.Y.C., (1975). "Finite element analysis of transonic flow by the method of weighted residuals," Tech Rpt AIAA75:xxx

[12] Cockburn, B., Shu, C.W. (1998). "The local discontinuous Galerkin method for time dependent convection-diffusion systems," SIAM *J Num Analysis*, 35.

[13] Cockburn, B., Karniadakis, E., Shu, C.W. (2000). "The Development of Discontinuous Galerkin Methods." in *Discontinuous Galerkin Methods*, Lecture Notes Comp Science Engineering, Springer, Berlin.

[14] Courant, R., Friedrichs, K., Lewy, H. (1928). "Uber die Partiellen Differenzengleichungen der Mathematischen Physik," *Mathematische Annalen*, 100.

[15] Crank, J., Nicolson, P. (1947). "A Practical Method for Numerical Evaluation of Solutions of Partial Differential Equations of Heat-Conduction Type," *Cambridge Phil Soc*, 43.

[16] Donea, J. (1984). "A Taylor-Galerkin algorithm for hyperbolic conservation laws," *J Num Mtd Engr*, 20.

[17] Eberle, A. (1977). "Eine Methods Finiter Elements Zur Breshnung der Transsonischen Potential-Stromung um Profile," Tech Rept MBB Dericht Nr. UEE, 132(0).

[18] Finlayson, B.A. (1972). *The Method of Weighted Residuals and Variational Principles*, Academic Press, NY.

[19] Galerkin, B.G. (1915). "Series Occurring in Some Problems of Elastic Stability of Rods and Plates," *Engr Bulletin*, 19.

[20] Glowinski, R., Periaux, J., Pironneau, O.,(1976). "Transonic flow simulation by the finite element method via optimal control, " Proc. 2nd Int. Symposium in Flow Problems, Italy

[21] Godunov, S.K. (1959). "Finite-Difference Method for Numerical Computation of Discontinuous Solution of the Equations of Fluid Dynamics," *Mat Sb*, 47.

[22] Gosman, A., Pun, W., Runchal, A., Spalding, D., Wolfshtein, M. (1969). *Heat and Mass Transfer in Recirculating Flows*, Academic, London.

[23] Hafez, M., Murman, C. (1979). "Artificial compressibility methods for numerical solution of the transonic full potential equation," *AIAA Journal*, V.17.

[24] Habashi, W., Hafez, M. (1981). “Finite element solution of transonic flow problems,” *AIAA Journal*, V. 20.

[25] Harlow, F. (1971). “Fluid Dynamics – a LASL Monograph,” Univ. Calif., LASL Tech Rpt LA-4700.

[26] Jameson, A., (1975). “Transonic potential flow calculations using conservation form,” Proc. AIAA 2nd CFD conference.

[27] Kolesnikov, A., Baker, A.J. (2001). “An efficient high order Taylor weak statement for the Navier-Stokes equations,” *J Comp Physics*, 173.

[28] Lax, P. (1954). “Weak Solutions of Nonlinear Hyperbolic Equations and their Numerical Computation,” *Comm Pure Appl Math*, 7.

[29] Lax, P., Wendroff, B. (1960). “Systems of Conservation Laws,” *Comm Pure Applied Math*, 13.

[30] Lynn, P.O. (1974). “Least squares finite element analysis of laminar boundary layers,” *J Num Metd Engineering*, V.8.

[31] MacCormack, R. (1969). “The Effect of Viscosity in Hypervelocity Impact Cratering,” Tech Rpt. AIAA:69–354

[32] Murman, E., Cole, J.D. (1971). “Calculation of plane steady transonic flow,” *AIAA J*, V.9.

[33] Roe, P.L. (1981). “Approximate Riemann Solvers, Parameter Vectors and Difference Schemes,” *J Comp Physics*, 43.

[34] Thompson, J.F., Warzi, Z.U.A., Mastin, C.W. (1985). *Numerical Grid Generation, Formulations & Applications*, Elsevier, NY.

[35] Richtmyer, R.D., Morton, K.W. (1967). *Difference Methods of Initial-Value Problems*, 2nd Ed., Interscience, NY.

[36] Roache, P.J. (1972). *Computational Fluid Mechanics*, Hermosa, Albuquerque.

[37] Popinski, Z., Baker, A.J. (1976). “An Implicit Finite Element Algorithm for the Boundary Layer Equations,” *J Comp Physics*, 21.

[38] Sahu, S., Baker, A.J. (2007). “A Modified Conservation Principles Theory Leading to an Optimal Galerkin CFD Algorithm,” *J Num Mtd Fluids*, 55(8).

[39] Sekachev, M.A., Baker, A.J., Wong, K.L. (2016a,b,c). “Totally Analytical Closure of Space Filtered Navier-Stokes for Arbitrary Reynolds Number, Part I. Theory, Resolutions,” *J Num Heat Trans, Part B*, 70(4).

[40] Shankar, N.L., Malone, J.B., Tassa, Y., (1981). “An implicit conservative algorithm for steady and unsteady three-dimensional transonic potential flow,” Tech Rpt. AIAA:81-1016

[41] Soliman, M.O., Baker, A.J. (1981a). “Accuracy and convergence of a finite element algorithm for laminar boundary layer flow,” *J Computers Fluids*, 9.

[42] Soliman, M.O., Baker, A.J. (1981b). “Accuracy and convergence of a finite element algorithm for turbulent boundary layer flow,” *Comp Mtd Appl, Mech Engr*, 28.

[43] Tannehill, J.C., Anderson, D.A., Pletcher, R.H. (1997). *Computational Fluid Dynamics and Heat Transfer*, 2nd Ed, Taylor & Francis, Washington DC.

[44] van Leer, B. (1979). “Towards the Ultimate Conservative Difference Scheme, V. A Second Order Sequel to Godunov’s Method,” *J Comp Physics*, 32.

[45] Oden, J. T. (1969). “Finite Element Applications in Fluid Dynamics,” ASCE *J. Eng Mech Div*, 9.

[46] Oden, J. T., (1972). *Finite Elements of Nonlinear Continua*, McGraw-Hill, NY.

Chapter 3
Fluid mechanics conservation principles, PDE notation

Libretto: Compressible, incompressible fluid mechanics continuum conservation principles are expressed by Navier-Stokes (NS) partial differential equation (PDE) systems

NS PDE system alterations to their isentropic (Euler), time-averaged (RaNS), and space-filtered (LES) PDE statements are detailed in vector and tensor notation.

A brief exposure to heuristic "physics-based" closure modeling for RaNS/LES PDE systems documents the significant limitations of current practice formulations.

Keywords: Navier-Stokes, PDE, isentropic, time-averaged, space-filtered alterations, Reynolds stress tensor

3.1 Textbook appearance of Navier-Stokes (NS) PDE systems, monograph notation

The state variable members for the compressible fluid NS PDE system are density ρ, linear momentum vector $\mathbf{m} \equiv \rho\mathbf{u}$ and total internal energy E. Herein an NS state variable is expressed as a column matrix of vector form $\{q(\mathbf{x},t)\} \equiv \{\rho,\mathbf{m},E\}^T$. The members of the compressible NS PDE system are denoted DM, **DP** (bold face) and DE. Their dimensional vector forms are

$$\text{DM: } \frac{\partial \rho}{\partial t} + \nabla \cdot \mathbf{m} = 0$$

$$\textbf{DP}\text{: } \frac{\partial \mathbf{m}}{\partial t} + \nabla \cdot \mathbf{um} + \nabla p - 2\nabla \bullet \left(\mu\mathbf{S} - \frac{2\mu}{3}\mathbf{S}\delta \right) - \rho\mathbf{g} = \mathbf{0}$$

$$\text{DE: } \frac{\partial E}{\partial t} + \nabla \cdot \mathbf{u}(E + p) - \nabla \bullet (k\nabla T) = 0$$

wherein D(•) is classic mechanics notation for the *substantial* (Eulerian reference frame) time derivative which permeates all NS PDE systems.

As NS $\{q(\mathbf{x},t)\}$ state variable members are linearly independent, the first-order DM is a pure initial value statement in time. Conversely, **DP** and DE are initial-elliptic boundary value (EBV) statements wherein velocity $\mathbf{u} \equiv \mathbf{m}/\rho$ is computed as "*data*." The pressure p in **DP** is predicted algebraically, usually from a perfect gas equation of state, and the gravity body force vector $\rho\mathbf{g}$ is usually discarded. The fluid specific dependencies in **DP** and DE reside in the temperature dependent absolute viscosity $\mu = \rho\nu$ and thermal conductivity k. The dyadic form of the linear Stokes

https://doi.org/10.1515/9783112206430-004

symmetric strain rate tensor is $\mathbf{S} \equiv (\nabla u + \nabla^T u)/2$ and the Kronecker delta signifies the dyadic dilatation term $\mathbf{S}\delta$.

The preferred practice in fluid mechanics is to non-dimensionalize NS PDE systems to expose the key non-dimensional groups that parameterize a specific application. DM is unaltered by this operation. For selected reference length scale L, velocity magnitude U, time scale L/U and the farfield stagnation reference states ρ_∞, p_∞, T_∞, compressible non-D NS **DP** and DE PDE statements expressed in vector notation

$$\mathbf{DP}: \frac{\partial \mathbf{m}}{\partial t} + \nabla \cdot \mathbf{um} + \mathrm{Eu}\nabla p - \frac{1}{\mathrm{Re}}\nabla \cdot 2\mu^* \left(\mathbf{S} - \frac{2}{3}\mathbf{S}\delta\right) = \mathbf{0}$$

$$\mathrm{DE}: \frac{\partial E}{\partial t} + \nabla \cdot \mathbf{u}(E + p) - \frac{1}{\mathrm{Re\,Pr}}\nabla \bullet (\kappa^* \nabla T) = 0$$

now contain the Euler, Reynolds and Prandtl non-D groups, the definitions for which are $\mathrm{Eu} \equiv \rho_\infty \mathrm{U}^2/p_\infty$, $\mathrm{Re} \equiv \rho_\infty \mathrm{UL}/\mu_\infty$, and $\mathrm{Pr} \equiv c_p\mu_\infty/k_\infty$. Therein $\mu^* \equiv \mu(T)/\mu_\infty$, and $\kappa^* \equiv \kappa(T)/\kappa_\infty$ denote temperature dependent non-D absolute viscosity and thermal diffusivity distributions for reference values defined by the farfield definitions μ_∞ and κ_∞.

The significance of the Eu and Pr groups is of the order unity, herein symbolized as $O(1)$, while typical velocity and length reference scale selections generate the significance of inverse Re to be $O(10\pm^{-6})$. This significance disparity substantiates the *gas dynamics* simplification which discards the impact of fluid viscosity and thermal diffusivity. This defines the *isentropic* Euler PDE system, which is then cast in *hyperbolic conservation law* form for $\{q(\mathbf{x}, t)\} \equiv \{\rho, \mathbf{u}\}^T$

$$\mathrm{DM}: \frac{\partial \rho}{\partial t} + \nabla \cdot \rho\mathbf{u} = 0$$

$$\mathbf{DP}: \frac{\partial \mathbf{u}}{\partial t} + \mathbf{u} \cdot \nabla \mathbf{u} + \frac{\mathrm{Eu}}{\rho}\nabla p = \mathbf{0}$$

As briefly introduced in Chapter 2, linear models of Euler **DP** were the basis for Godunov [13] genre difference algebra derived flux vector/predictor-corrector *stabilized* shock capturing schemes that replaced the direct insertion of artificial/numerical stabilization. The companion weak formulation *discontinuous* Galerkin (DG) stabilized hyperbolic conservation law form algorithms employ shock fitting mesh realignment strategies, cf. Cockburn et al. [4].

For an incompressible fluid and selecting the Boussinesq assumption for the buoyancy body force $\rho\mathbf{g}$, Gebhart et al. [10], the classic vector appearance of NS non-D DM:**DP**:DE PDE system for state variable $\{q(\mathbf{x}, t)\} \equiv \{\boldsymbol{u}, P, \Theta\}^T$ is

$$\mathrm{DM}: \nabla \cdot \mathbf{u} = 0$$

$$\mathbf{DP}: \frac{\partial \mathbf{u}}{\partial t} + \mathbf{u} \cdot \nabla \mathbf{u} + \nabla P - \frac{1}{\mathrm{Re}}\nabla^2 \mathbf{u} + \frac{\mathrm{Gr}}{\mathrm{Re}^2}\Theta\, \widehat{\mathbf{g}} = \mathbf{0}$$

$$\text{DE: } \frac{\partial \Theta}{\partial t} + \mathbf{u} \cdot \nabla \Theta - \frac{1}{\text{Re Pr}} \nabla^2 \Theta = 0$$

The substitution of DM into **DP** identifies the *kinematic* pressure $P \equiv p/\rho_\infty$ as the state variable member for reference density ρ_∞. The non-D DE state variable is potential temperature with definition $\Theta \equiv (T - T_{\min})/(T_{\max} - T_{\min})$. Reynolds and Grashof non-D group definitions are $\text{Re} \equiv \text{UL}/\nu$, $\text{Gr} \equiv g\beta\Delta\text{TL}^3/\nu^2$ wherein $\Delta T \equiv (T_{\max} - T_{\min})$ for compressibility factor definition $\beta = 1/T_{\text{abs}}$.

DM has now transitioned to a *differential constraint* which requires *all* incompressible fluid velocity vectors $\boldsymbol{u}$ to be *divergence-free* (!). The range of iterative/non-iterative computable strategies for approximating DM requisite "divergence-freeness" constraint are archived [23]. Finally, the standard assumption is that kinematic viscosity ν and thermal conductivity k do not depend on temperature.

3.2 Textbook time averaged Navier-Stokes (RaNS) PDE systems

The literature refers to time averaged NS PDE system CFD algorithms as *Reynolds-averaged Navier-Stokes* (RaNS) formulations. For an assumed *fully turbulent* flowfield the NS unsteady velocity vector **u(x**,*t*) is *resolved* into the steady, i.e., time averaged, component that is assumed to be resolvable on a spatial discretization of mesh measure *h*, and the time-dependent fluctuations thereabout.

Switching to tensor notation, the CFD predictable component of this RaNS resolution for an incompressible fluid

$$\mathbf{u}(\mathbf{x}, t) \Rightarrow u_i(x_k, t) \equiv \bar{u}_i(x_k, T) + u_i'(x_k, t)$$

traditionally signified by an overbar

$$\bar{u}_i(x_k, T) \equiv \frac{\lim}{T \to \infty} \frac{1}{T} \int_t^{t+T} u_i(x_k, \tau)\, d\tau$$

explicitly assumes that a time interval *T*, extending to infinity, exists such that the time integrated sample $\bar{u}_i(x_k, T)/T$ possesses physical significance. Note that this time interval *T* is *never* (!) specified in RaNS code practice and/or in the reporting of results. As a RaNS code is often used to predict *unsteady* incompressible flow fields, this *explicitly* assumes that the time interval *T* of the CFD prediction *is* the integration time step size $\Delta t \ll T \to \infty$ (!)

By definition the time average of velocity fluctuations about $\bar{u}_i(x_k, T)$ vanishes

$$\overline{u_i'}(x_k, T) \equiv \frac{\lim}{T \to \infty} \frac{1}{T} \int_t^{t+T} u_i'(x_k, \tau)\, d\tau = 0$$

The time average of the RaNS **DP** PDE nonlinear convection term is

$$\overline{u_j u_i} = \bar{u}_j \bar{u}_i + \overline{u'_j u'_i}$$

with the lead term assumed to be the nonlinear convection term in $\mathbf{D\overline{P}}$. The second term is the *a priori* unknown *Reynolds stress tensor* which for an incompressible fluid is traditionally symbolized as $\tau_{ij} = \overline{u'_i u'_j}$, or $\tau_{ij} = \overline{\rho u''_i u''_j}$ for the compressible fluid $\mathbf{D\overline{P}}$ [7].

Fluid mechanics is a *continuum* mechanics discipline, hence also is turbulence. Classical tensor field theory requires that the correlation between the second order stress tensor and the fluid mechanics strain (rate) tensor is a fourth order tensor C_{ijkl}. In linear elasticity, another continuum mechanics discipline, a series of symmetry operations transition the tensor C_{ijkl} into the three scalar constants named Youngs modulus (*E*), the shear modulus (*G*) and the Poisson ratio (ν).

As a fluid cannot withstand tension, heuristically argued "physics based" RaNS stress tensor $\tau_{ij} = \overline{u'_i u'_j}$ models explicitly *assume* that C_{ijkl} is a *scalar*, named the *eddy viscosity*, and it is linearly correlated with the time averaged linear Stokes strain rate *isotropic* tensor $\bar{S}_{ij} \equiv (\partial \bar{u}_i / \partial x_j + \partial \bar{u}_j / \partial x_i)/2$. The resultant stress tensor definition for an incompressible fluid

$$\overline{u'_i u'_j} \cong \tau_{ij} \equiv -2\nu^T \left[\bar{S}_{ij} - \frac{2}{3} \bar{S}_{kk} \delta_{ij} \right]$$

thus involves the *distribution* $\nu^T(x_i)$ which is named the *kinematic* eddy viscosity

The representative Reynolds stress tensor closure model for a compressible fluid is

$$\overline{\rho u''_i u''_j} \cong \tau_{ij} \equiv -2\mu^T \left[\bar{S}_{ij} - \frac{2}{3} \left(\bar{S}_{kk} + \bar{\rho} k \right) \delta_{ij} \right]$$

wherein $\mu^T(x_j)$ is the *absolute* eddy viscosity distribution with k the *turbulence kinetic energy*, which is defined equal to one-half of the square root of the *trace* of the time averaged Stokes strain rate tensor $\bar{S}_{kk} \equiv \left(\overline{u'_1 u'_1} + \overline{u'_2 u'_2} + \overline{u'_3 u'_3} \right)$.

These RaNS tensor linear isotropic closure models are literature validated to adequately characterize the *first-order* effects existent in shear dominated turbulent velocity vector fields. In distinction, these isotropic tensor models fail to predict the higher-order *physics of fluids* effects that are induced by a flowfield bounding geometry that is not aerodynamic. One specific example of this error is the velocity quadruple vortex existent in the transverse plane of the steady three-dimensional velocity vector distribution in a square cross-section duct.

Classic tensor field theory guides the derivation of higher-order and/or *anisotropic* nonlinear Reynolds stress-Stokes strain rate correlations [18]. One such tensor field theory derived *anisotropic* nonlinear quadratic strain rate tensor model

$$\overline{u'_i u'_j} \cong \tau_{ij} = k\alpha_{ij} - \nu^T \bar{S}_{ij} - \frac{C_2 (\nu^T)^2}{C_4\, k} \bar{S}_{ik} \bar{S}_{kj} + \ldots$$

is validated to predict the higher-order effects existent in the square cross-section duct, Baker [1, Ch. 8], for coefficient set definitions $C_\alpha \equiv \{0.94, 0.067, 0.56, 0.068\}$, [16], and diagonal anisotropic tensor $\alpha_{ij} \equiv \text{diag}[C_1, C_3, C_3]$, [22].

Of note, the influence of the nonlinear Stokes strain rate tensor product therein is overwhelmed by that of the linear *anisotropic* tensor. This anisotropic tensor was also validated as critical to the prediction of the mean and fluctuating velocity *distribution* resolutions in the merging of upper – lower surface turbulent boundary layers immediately downstream of a sharp trailing edge airfoil [2]. A nonlinear quadratic Reynolds stress tensor closure model is validated to generate first and higher-order effects characterization of turbulent boundary layer separation in a three dimensional wing-body juncture region, [5].

Characterization of the balance of laminar and turbulent *diffusion processes* in RaNS stress tensor closures is clearly quantified by definition of the *turbulent* Reynolds number

$$\text{Re}^T \equiv \frac{\mu^T}{\mu_\infty} \; or \; \frac{\nu^T}{\nu_\infty}$$

the ratio of non-D eddy viscosity distribution to the fluid reference kinematic viscosity. For the Stokes strain rate tensor $\bar{S}_{ij} \equiv (\partial \bar{u}_i / \partial x_j + \partial \bar{u}_j / \partial x_i)/2$ correlation, the non-D magnitude of the $\text{Re}^T(x_j)$ distribution for incompressible $\text{D}\bar{\mathbf{P}}$

$$\frac{1}{\text{Re}} \nabla^2 \mathbf{u} + \tau_{ij} \Rightarrow \frac{2}{\text{Re}} \frac{\partial}{\partial x_j} \left(\left(1 + \text{Re}^T\right) \left[\bar{S}_{ij} - \frac{2}{3} \bar{S}_{kk} \delta_{ij} \right] \right)$$

ranges unity to $O(10^4)$. This extremum can increase an order of magnitude ± for the compressible $\text{D}\bar{\mathbf{P}}$ closure non-D $\text{Re}^T(x_j)$ distribution

$$\frac{1}{\text{Re}} \nabla \cdot 2\mu^* \left(\mathbf{S} - \frac{2}{3} \mathbf{S}\delta \right) + \tau_{ij} \Rightarrow \frac{2}{\text{Re}} \frac{\partial}{\partial x_j} \left(\left(\mu^* + \text{Re}^T\right) \left[\bar{S}_{ij} - \frac{2}{3} \left(\bar{S}_{kk} + \bar{\rho} \bar{k} \right) \delta_{ij} \right] \right)$$

wherein $\mu^* \equiv \mu(T)/\mu_\infty$ is the distribution of non-D $O(1)$ temperature-dependent absolute viscosity.

3.3 Textbook spatially filtered Navier-Stokes (LES) PDE systems

CFD algorithms written on spatially filtered NS PDE systems are literature referenced as *Large Eddy Simulation* (LES) formulations. Herein, the NS velocity vector is resolved into the *time-dependent* mesh measure *h resolved scale* velocity with that left over by definition the *unresolved* scale

$$u_i(x_k,t) \equiv \overline{u_i}(x_k,t) + u'_i(x_k,t)$$

Convolution (*) of NS velocity $u_i(x_k,t)$ with a spatial filter function $g_\delta(x_s)$ of *measure* (diameter) δ defines the *resolved scale* velocity vector to be

$$\bar{u}_i(x_k,t) \equiv g_\delta(x_s)^* u_i(x_k,t)$$

which is the output of legacy LES CFD codes.

The filter functions $g_\delta(x_s)$ common in the legacy LES literature are the "box," an FD discrete cell, the FD box in transform space named the "sharp spectral" filter, and a Gaussian. The selection of a specific filter exerts absolutely *no* influence on the convolution of operands in NS PDE systems. The space-filtered DM and **DP** PDE system for an incompressible fluid is

$$\text{DM:} \frac{\overline{\partial u_j}}{\partial x_j} = \frac{\partial \bar{u}_j}{\partial x_j} = 0$$

$$\textbf{DP:} \frac{\partial \bar{u}_i}{\partial t} + \frac{\partial}{\partial x_j}\left[\overline{u_j u_i} + \frac{\bar{p}}{\rho_0}\delta_{ij} - \frac{1}{\text{Re}}\left(\frac{\partial \bar{u}_i}{\partial x_j} + \frac{\partial \bar{u}_j}{\partial x_i}\right)\right] + \frac{\text{Gr}}{\text{Re}^2}\bar{\Theta}\,\hat{g}_i = 0$$

and for a compressible fluid

$$\text{DM:} \frac{\partial \bar{\rho}}{\partial t} + \frac{\partial\, \bar{m}_j}{\partial x_j} = 0$$

$$\textbf{DP:} \frac{\partial \bar{m}_i}{\partial t} + \frac{\partial}{\partial x_j}\left[\overline{u_j m_i} + \text{Eu}\,\bar{p}\,\delta_{ij} - \frac{\mu^*}{\text{Re}}\left(\frac{\partial \bar{u}_i}{\partial x_j} + \frac{\partial \bar{u}_j}{\partial x_i} - \frac{2}{3}\frac{\partial \bar{u}_k}{\partial x_k}\delta_{ij}\right)\right] = 0$$

The theoretical requirements for spatial filtering to be *rigorous* are that (1) differentiation and convolution *commute*, and (2) the linearity constraint for λ any scalar constant

$$1)\ \frac{\overline{\partial u_i}}{\partial x_j} = \frac{\partial \bar{u}_i}{\partial x_j},\ \frac{\overline{\partial u_i}}{\partial t} = \frac{\partial \bar{u}_i}{\partial t},\ 1 \le (i,j) \le n, \text{ and } 2)\ \overline{u_i + \lambda v_i} = \bar{u}_i + \lambda \bar{v}_i$$

The commutation requirement is severely challenged when space-filtered NS PDE domain $\Omega \subset \mathbf{R}^n$ is bounded by surfaces $\partial\Omega \subset \mathbf{R}^{n-1}$ on which the NS BC is no-slip as any finite measure δ filter will penetrate this surface. One LES legacy practice resolution that circumvents this complication is to employ a variable measure filter $\delta(x_j)$ which is defined to vanish on NS no-slip BC surfaces $\partial\Omega$ [3, 9, 20]. The disadvantages of this *near-wall resolution* (NWR) practice are numerous including violation of commutation rigor [6], the injection of an $O(h^2)$ error mechanism [8, 12], and that the compute resource demand for an LES NWR algorithm/code scales as $O(\mathrm{Re}^{2.5})$ which approximates that of a *Direct Numerical Simulation* code.

The requirement for closure of a space-filtered NS PDE system results from convolution of the nonlinear convection terms. Substituting *resolved* and *unresolved* scale definitions, and including a space-filtered state variable scalar member q, the *a priori* unknown *tensor* and *vector quadruples* that become embedded in the resolved scale $O(1)$ *space-filtered* PDE system

$$\overline{u_j u_i} \equiv \overline{(\bar{u}_j + u'_j)(\bar{u}_i + u'_i)} = \overline{\bar{u}_j \bar{u}_i} + \overline{\bar{u}_j u'_i} + \overline{u'_j \bar{u}_i} + \overline{u'_j u'_i}$$

$$\overline{u_j m_i} \equiv \overline{(\bar{u}_j + u'_j)(\bar{m}_i + m'_i)} = \overline{\bar{u}_j \bar{m}_i} + \overline{\bar{u}_j m'_i} + \overline{u'_j \bar{m}_i} + \overline{u'_j m'_i}$$

$$\overline{u_j q} \equiv \overline{(\bar{u}_j + u'_j)(\bar{q} + q')} = \overline{\bar{u}_j \bar{q}} + \overline{\bar{u}_j q'} + \overline{u'_j \bar{q}} + \overline{u'_j q'}$$

do not correlate with any member of the resolved scale state variable.

The inaugural closure model for incompressible NS space-filtered **DP** is the FD *box-filter*, a *mixing length* genre eddy viscosity model with *single* correlation constant C_S, Smagorinsky [21]. This simple closure model generated inaccurate predictions, as could be well anticipated (!), leading to the replacement of the "Smagorinsky constant" C_S with *distributions,* e.g., the kinetic energy of the subgrid scale concept, Yoshizawa and Horiuti [25], dynamic-subgrid scale solution-adapted on-the-fly mesh refinements, Germano et al. (1991), [17], and the mixed scale Sagaut, [20].

The diversity of these heuristic "physics-based" *subgrid scale* (SGS) tensor models are fully archived in the LES literature, cf. [3, 9, 12, 14, 15, 19, 20, 24].

References

[1] Baker, A.J. (2014). "*Optimal* MODIFIED CONTINUOUS Galerkin CFD," John Wiley, London.

[2] Baker, A.J., Manhardt, P.M. (1978). "Numerical Prediction of Mean and Fluctuating Velocities for Airfoil Trailing Edge Jet-Flap Flows," *AIAA Journal*, 16(7).

[3] Berselli, L.C., Iliescu, T., Layton, W.J. (2006). *Mathematics of Large Eddy Simulation of Turbulent Flows*, Springer, Heidelberg.

[4] Cockburn, B., Karniadakis, E., Shu, C.W. (2000). The Development of Discontinuous Galerkin Methods. in *Discontinuous Galerkin Methods*, Lecture Notes Comp Science Engineering, Springer, Berlin.

[5] Dawson, S., Gonzalez, D. (2020). "Novel tools help uncover new insights into complex flows," *Aerospace America*, 58(11).

[6] Dunca, A., John, V., Layton, W.J. (2003). "The commutation error of the space averaged Navier-Stokes equations on a bounded domain," *J Math Fluid Mech*, 5.

[7] Favre, A. (1965). "Equations des Gaz Turbulents Compressibles: 1. Formes Generales," *J Mechanique*, 4.

[8] Furby, C., Tabor, G. (1997). "Mathematical and Physical Constraints on Large-Eddy Simulation," *J Theo Comp Fluid Dyn*, 9.

[9] Garnier, E., Adams, N., Sagaut, P. (2009). *Large Eddy Simulation for Compressible Flows*, Springer, Berlin.
[10] Gebhart, B., Jaluria, Y., Mahajan, R.L., Sammakia, B. (1988). *Buoyancy-Induced Flow and Transport*, Hemisphere, NY.
[11] Germano, M., Piomelli, U., Moin, P. and Cabot, W. (1991). “A dynamics subgrid scale eddy viscosity model,” *Physics of Fluids* A, V.3 and alter the number system.
[12] Ghosal, S., Moin, P. (1995). “The basic equations for the large eddy simulation of turbulent flows in complex geometry,” *J Comp Physics*, 118.
[13] Godunov, S.K. (1959). “Finite-Difference Method for Numerical Computation of Discontinuous Solution of the Equations of Fluid Dynamics,” *Mat Sb*, 47.
[14] Grinstein, F.F., Fureby, C. (2004). “From Canonical to Complex Flows: Recent Progress on Monotonically Integrated LES,” *Comp Sci Eng*, 6.
[15] Grinstein, F.F., Margolin, L.G., Rider, W.F. (Eds.) (2007). *Implicit Large Eddy Simulation: Computing Turbulent Fluid Dynamics*, Cambridge University Press, England.
[16] Launder, B.E., (1975). “Stress-transport closures: Into the third generation,” Proc. 1st Symp Turb. Shear Flows, Springer-Verlag, NY.
[17] Lilly, D.K. (1992). “A proposed modification of the Germano subgrid scale closure method,” *Phys Fluids*, A,V.4.
[18] Lumley, J.L. (1970). “Towards a turbulent constitutive relation,” *J Fluid Mech*, 41.
[19] Pope, S.B. (2000). *Turbulent Flows*, Cambridge University Press, NY.
[20] Sagaut, P. (2006). *Large Eddy Simulation for Incompressible Flows*, 3rd ed., Springer, Berlin.
[21] Smagorinsky, J. (1963). “General Circulation Experiments with the Primitive Equations, I. The Basic Experiments,” *Monthly Weather Revue*, 91.
[22] Wilcox, D.C. (2006). *Turbulence Modeling for*, CFD, DCW Industries, La Canada, CA.
[23] Williams, P.T., Baker, A.J. (1996). “Incompressible computational fluid dynamics and the continuity constraint method for the 3-D Navier-Stokes equations,” *J Num Heat Trans, Part B.*, 29 (entire issue).
[24] Yang, Z. (2015). “Large-Eddy Simulation: Past, Present and Future,” *Chinese Journal of Aeronautics*, 28(1).
[25] Yoshizawa, A., Horiuti, K. (1985). “A Statistically-Derived Subgrid Scale Kinetic Energy Model for the Large-eddy Simulation of Turbulent flow,” *J Phys Soc Japan*, 54, 2834.

Part A: Space-time discretization error annihilation, algebraic instability

Chapter 4
Legacy CFD difference algebra derived stabilizations, $O(h^2)$ truncation error annihilation (TEA) theory, NS *error freed* PDE *mathematics*, monotonicity, stability

Libretto: CFD algorithms cast on the textbook forms of Navier-Stokes (NS) partial differential equation (PDE) systems are *invariably* (!) algebraically *unstable*.

The legacy CFD remedy is insertion of a variety of finite difference derived numerical diffusion/stabilization schemes determined optimal in some sense.

Fourier modal analysis quantifies the *root cause* of algebraic instability is the spatial discretization generated dispersive $O(h^2)$ truncation error (TE), h the measure of the mesh.

Taylor series (TS) manipulations *analytically* identify that this $O(h^2)$ TE term, omitted in NS CFD algorithms, is a *cubically nonlinear* Reynolds number and state variable member dependent velocity vector tensor product calculus functional.

Reversed sign insertion of this $O(h^2)$ *truncation error annihilation* (TEA) functional into any textbook NS PDE *annihilates* (!) the spatial discretization generated algebraic *instability*.

Thereby, *any* tridiagonal stencil equivalent algorithm for a TEA theory altered NS PDE generates *stable* TS fourth-order accurate distributions *monotone* to iterative convergence digit on *any* mesh

Keywords: Navier-Stokes, CFD, instability, $O(h^2)$ discretization error, TEA theory, $O(h^4)$ accurate, shock continuum interpolation, monotone DOF

4.1 CFD algorithm $O(h^2)$ TE instability, Fourier modal analysis

CFD algorithms cast on textbook presentations of Navier-Stokes (NS) partial differential equation (PDE) systems are invariably *algebraically unstable*. Rigorous identification of the causal mechanism accrues to Fourier modal analysis of the 1D model of the NS PDE substantial time derivative **DP**

$$\mathrm{D}\mathbf{P}\text{: } \frac{\partial q}{\partial t} + \frac{\partial f}{\partial x} \Rightarrow q_t + f_x = 0$$

for scalar variable q and kinetic flux vector $f \equiv uq$. In the Fourier modal resolution of the exact solution to D**P**

https://doi.org/10.1515/9783112206430-005

$$q(x,t) = q(x,t_0)\mathrm{e}^{\mathrm{i}\kappa(x-ut)}$$

$q(x,t_0)$ is the initial condition (IC), e is the exponential operator with $\mathrm{i} \equiv \sqrt{-1}$ the imaginary unit multiplier on Fourier mode of wavelength λ, hence *wavenumber* $\kappa \equiv 2\pi/\lambda$. This resolution states that the Fourier modal content in the model **DP** IC is *exactly* preserved for all time as u is the propagation speed of each mode present in the IC.

Uniform x axis discretization generates the geometric node set Xj. The Fourier representation of the *approximate* solution to thus spatially discretized **DP**, herein universally signified by superscript h

$$\mathrm{D}^h\mathbf{P}\text{: } q^h(j\Delta x, t) = q(x,t_0)\mathrm{e}^{\mathrm{i}\kappa(j\Delta x - \tilde{\mathrm{U}}t)}$$

requires prediction of the spatially discrete speed $\tilde{\mathrm{U}}(\Delta x) \equiv \tilde{\mathrm{U}}_{real}(\Delta x) + \mathrm{i}\tilde{\mathrm{U}}_{imag}(\Delta x)$, a complex function distinct from the u in **DP**. Deferring details to Chapter 5, the weak formulation *continuous* Galerkin (GWS^h) finite element (FE) linear ($p = 1$) basis generated $\mathbf{DP}^h$ element matrix *assembled* (denoted S_e) at generic mesh node X_j is the $O(h^2)$ accurate tridiagonal stencil

$$\mathrm{S}_e(\mathrm{GWS}^h)\text{: } \frac{1}{6}\frac{d}{dt}\left[Q_{j-1} + 4Q_j + Q_{j+1}\right] + \frac{\mathrm{U}}{2\Delta x}\left[-Q_{j-1} + Q_{j+1}\right] = 0 + O(h^2)$$

wherein $Q_{j,j\pm1}$ are the approximation *degrees of freedom* (DOF) co-located at nodes $\mathrm{X}_{j,j\pm1}$

Substituting Fourier modal discrete representation $q^h(j\Delta x, t)$ into the $\mathrm{S}_e(\mathrm{GWS}^h)$ stencil, then projecting all $j \pm 1$ indexed terms to node X_j generates an algebraic statement of the form $[\bullet]q(x,t_0)\mathrm{e}^{\mathrm{i}\kappa(j\Delta x - \tilde{\mathrm{U}}t)} = 0$. As $q(x,t_0)$ and the exponential operator are obviously not zero, the vanishing of $[\bullet]$ predicts the GWS^h FE $p = 1$ basis Fourier mode complex propagation speed is

$$\tilde{\mathrm{U}}_{real} = \mathrm{U}\left[1 - \frac{1}{180}(\kappa\Delta x)^4 + O(\kappa\Delta x)^6\right], \tilde{\mathrm{U}}_{imag} = \mathrm{U}\left[O(\kappa\Delta x)^5\right]$$

where the integer exponents denote the *order* of wavenumber dependence for each truncation error (TE) term.

Inserting this solution into $q^h(j\Delta x, t)$ and noting that $\tilde{\mathrm{U}}$ in actual computations is u, the Fourier modal representation of NS model **DP** GWS^h FE $p = 1$ basis solution is

$$\mathrm{GWS}^h : q^h(j\Delta x, t) = q(x,t_0)\mathrm{e}^{\mathrm{i}\kappa(j\Delta x - u(1+O(\kappa\Delta x)^4)t)}\mathrm{e}^{-\kappa u O(\kappa\Delta x)^5 t}$$

Comparison with the analytical solution $q(x,t) = q(x,t_0)\mathrm{e}^{\mathrm{i}\kappa(x-ut)}$ confirms that the x axis discretization of uniform measure $h \equiv \Delta x$ injects an $O(\kappa\Delta x)^4$ wavenumber dependent *error* into GWS^h algorithm Fourier mode propagation speed, *also* an $O(\kappa\Delta x)^5$ *numerical diffusion* error as the exponent on this term is real.

Recalling that $\lambda \equiv 2\pi/\kappa$, the sketch in Figure 4.1 clearly visualizes that a Fourier mode of wavelength $\lambda = 2\Delta x$ is not resolvable on a mesh of measure Δx as all of the nodal DOF $Q_{j,j\pm1}$ are identically zero! Thereby, mesh scale wavelength Fourier content present in $q(x,t)$ that is not resolvable by $q^h(j\Delta x,t)$ will be *dispersed* into mesh resolvable *error* of longer wavelengths $\lambda = (3, 4, 5, \ldots)\Delta x$.

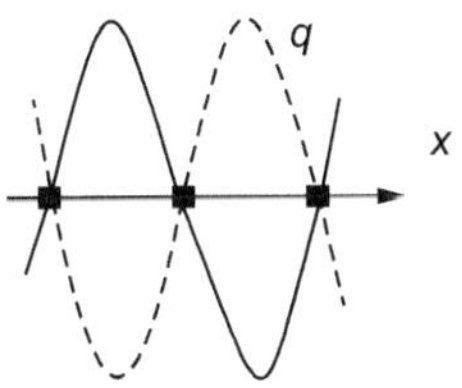

Figure 4.1: Mesh scale Δx Fourier content resolution issue.

Thereby clearly identified is that spatial discretization generated $O(\kappa\Delta x)^4$ TE Fourier mode speed propagation error is *dispersive,* a mechanism manifested as $2\Delta x$ node-to-node oscillations in a $q^h(j\Delta x,t)$ DOF distribution. As NS/RaNS/LES PDE systems are explicitly *nonlinear*, the accumulation of this $O(h^2)$ TE dispersion error renders the CFD code algebraic process *pathologically divergent* in finite time/iterations unless moderated. This algebraic *instability* mechanism is latent in *every* (!) current open source/commercial/government code for which the CFD algorithm is cast on the textbook appearance of NS/RaNS/LES PDE systems.

4.2 Functional correlation of legacy CFD derived stabilizations

Legacy CFD algebraic instability mechanism *moderations* have been typically derived on models of NS substantial derivative **DP**. For generic state variable member q and flux vector $f \equiv uq$, the model unsteady NS **DP** ODE in differential subscript notation is

$$\mathbf{DP}: \frac{\partial q}{\partial t} + \frac{\partial f}{\partial x} \Rightarrow q_t + f_x = 0$$

For the $O(\Delta t^4)$ explicit time Taylor series, with superscripts n, n + 1 denoting time stations

$$\text{TS: } q^{n+1} = q^n + \Delta t q_t^n + \frac{1}{2}\Delta t^2 q_{tt}^n + \frac{1}{6}\Delta t^3 q_{ttt}^n + O(\Delta t^4)$$

and assuming space/time differentiation is interchangeable (Euclidean space), the *non-unique* restatement of this time TS second derivative term upon substitutions from model **DP** is

$$q_{tt} = (q_t)_t = -(f_x)_t = -(f_t)_x = -(f_q q_t)_x = (f_q f_q q_x)_x$$

Similarly, for the third time TS derivative term

$$q_{ttt} = \cdots = (f_q f_q q_{tx})_x = -(f_q f_q (f_q q_x)_x)_x$$

Inserting the coefficient set $\{\alpha, \beta, \gamma, \mu\}$ to account for the *non-uniqueness* and absorbing all signs therein, forming the limit $\Delta t \Rightarrow \varepsilon > 0$ on thus altered time TS divided through by Δt

$$\text{TS:} \quad \frac{q^{n+1} - q^n}{\Delta t} + f_x^n - \frac{\Delta t}{2}\left(\alpha u q_t + \beta u^2 q_x\right)_x^n - \frac{\Delta t^2}{6}\left(\gamma u^2 q_{tx} + \mu u^3 q_{xx}\right)_x^n + O(\Delta t^3)$$

admits retaining the higher order TE terms which generates the time TS-*m*odified $\mathrm{D}^m\mathbf{P}(q)$

$$\mathrm{D}^m\mathbf{P}: \quad \frac{\partial q}{\partial t} + \frac{\partial u q}{\partial x} - \frac{\Delta t}{2}\frac{\partial}{\partial x}\left(\alpha u \frac{\partial q}{\partial t} + \beta u^2 \frac{\partial q}{\partial x}\right) - \frac{\Delta t^2}{6}\frac{\partial}{\partial x}\left(\gamma u^2 \frac{\partial^2 q}{\partial x \partial t} + \mu u^3 \frac{\partial^2 q}{\partial x^2}\right) = 0 + O(\Delta t^3)$$

Identification of $\mathrm{D}^m\mathbf{P}$ enabled establishment of functional correlation for a wide range of *independently* derived legacy CFD stabilizations in one dimension (1-D) (Table 4.1). Therein $\mathrm{D}^m\mathbf{P}$ time step Δt is replaced with local time scale $h/|u|$, which in 1-D is $u/|u| \equiv \hat{u} = \pm 1$, the "sign(u)" terms in the β term column in Table 4.1.

Table 4.1: $\mathrm{D}^m\mathbf{P}$ $(\alpha, \beta, \gamma, \mu)$ correlations of Euler **DP** model independently derived CFD algorithm stabilizations, $n = 1$, adapted from Baker and Kim [1].

Algorithm Name	**θ**	**α**	**β**	**γ**	**μ**
$\mathrm{D}^m\mathbf{P}$	all	arbitrary	arbitrary	arbitrary	arbitrary
Galerkin	all	0	0	0	0
Donor cell FD	0	0	uC	$1/C^2$	0
Lax-Wendroff FD	0	0	sign(u)	0	0
Taylor-Galerkin (TG)	0	0	1	1	0
Crank-Nicolson (CN)	0.5	0	0.5	0	0
Characteristic Galerkin	0	0	1	0	1
Swansea TG	0	0	1	0	0
Walbin	0	sign(u)	2sign(u)	0	0
Dendy	0	*h*sign(u)	*h*sign(u)	0	0
Raymond-Garder	0.5	2v*sign(u)C	2v*sign(u)C	0	0
Hughes SUPG (steady)	–	0	sign(u)	0	0
Euler Petrov-Galerkin	0	0	0	(1-v)	0
SUPG Petrov-Galerkin	0.5	sign(u)	v sign(u)	-v/2	0
Warming-Beam FD	0	0	1	0	−3(1-C)
Van Leer MUSCL	1	0	sign(u)	0	−3
Galerkin least squares	all	2θ	2θ	0	0

Note: v* = 1/sqrt(15), *h* is mesh measure, C < v < 1.0 throughout, C is Courant number.

Classic finite difference (FD) numerical diffusion (donor cell) stabilizations correlated include Courant et al. [3], Lax-Wendroff [11], and Crank-Nicolson [4]. Numerically dissipative, hence non-Galerkin weak formulations correlated are Wahlbin [15], Dendy [5], Dendy-Fairweather [6], Raymond-Garder [12] and streamline upwind Petrov-Galerkin (SUPG), Brooks-Hughes [2]. TS correlated Godunov [8] genre ENO/WENO/TVD FD flux vector schemes are van Leer [14] and Warming-Beam [16]. Extension of this TS correlation to n-D $\mathrm{D}^m\mathbf{P}(q)$ only involves transition to tensor notation and identification of which way the "wind is blowing," i.e., the *upwind* direction.

The β term column entries in Table 4.1 confirm that *every* correlated legacy CFD stabilization scheme contains this *cubically nonlinear* differential functional. For thus time TS β term altered $\mathrm{D}^m\mathbf{P}$

$$\mathrm{D}^m\mathbf{P}\colon\ \frac{\partial q}{\partial t} + \frac{\partial uq}{\partial x} - \frac{\beta \Delta t}{2}\frac{\partial}{\partial x}\left(u^2 \frac{\partial q}{\partial x}\right) = 0$$

the continuous Galerkin (GWS^h) FE $p = 1$ basis element matrix *assembled* at generic node X_j is again a tridiagonal stencil. Replacing Δt with local time scale $h/|u|$, the analytical solution for the Fourier mode complex phase speed for this $\mathrm{D}^m\mathbf{P}$ algorithm

$$\tilde{U}_{real} = u\left[1 + \left(-\frac{1}{180} + \frac{\beta^2}{12}\right)(\kappa\Delta x)^4 + O(\kappa\Delta x)^6\right]$$

$$\tilde{U}_{imag} = |u|\left[-\frac{\beta}{12}(\kappa\Delta x)^3 + O(\kappa\Delta x)^5\right]$$

admits an *optimal* (!) decision. Specifically, the wavenumber *dependence* of the $\mathrm{D}^m\mathbf{P}$ GWS^h Fourier mode propagation speed can be reduced two orders by defining $\beta \equiv (15)^{-1/2}$. This definition then renders the Fourier artificial diffusion TE term two orders *more* significant at $O(\kappa\Delta x)^3$ with coefficient magnitude $\beta/12 \approx 0.02$.

4.3 Algebraic instability, $O(h^2)$ numerical diffusion infused legacy CFD

Extension of the time TS β term algorithm to the $n = 1$ Euler ODE system enables GWS^h FE $p = 1$ basis prediction of the unsteady evolution of a compressible fluid in the Riemann shock tube, Figure 4.2. Illustrated flowfield features generated upon rupture of the diaphragm are a downstream traveling shock wave, the trailing propagation of a contact surface and an upstream traveling expansion fan.

The state variable members for the compressible $n = 1$ Euler ODE system are density ρ, momentum $m = \rho u$ and total internal energy $E = \rho e$. Replacing Δt with $h/|u|$, the time TS β term *m*odified Euler ODE system is

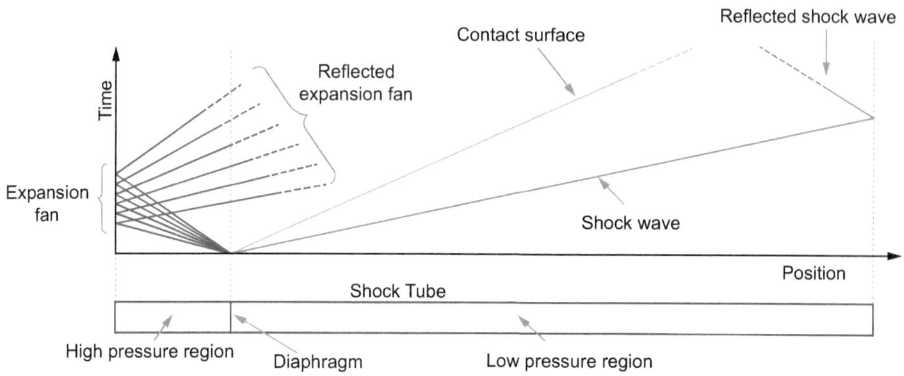

Figure 4.2: Riemann shock tube unsteady compressible flowfield feature evolution, https://en.wikipedia.org/wiki/Shock_tube.

$$\mathrm{D}^m\mathrm{M}: \ \frac{\partial \rho}{\partial t} + \frac{\partial}{\partial x}\left[m - \beta_\rho h\widehat{u}\,u\frac{\partial \rho}{\partial x}\right] = 0$$

$$\mathrm{D}^m\mathbf{P}: \ \frac{\partial m}{\partial t} + \frac{\partial}{\partial x}\left[um + p - \beta_m h\widehat{u}\,u\frac{\partial m}{\partial x}\right] = 0$$

$$\mathrm{D}^m\mathrm{E}: \ \frac{\partial E}{\partial t} + \frac{\partial}{\partial x}\left[uE + up - \beta_E h\,\widehat{u}\,u\frac{\partial E}{\partial x}\right] = 0$$

The thermodynamic pressure $p \equiv (\gamma - 1)e$ is computed algebraically for γ the specific heat ratio. Velocity (speed) unit vectors $\widehat{u} = \pm 1$ correlate with the sign(u) entries in Table 4.1 β column.

Riemann shock tube NS state variable non-D initial condition (IC) that became the 1970–80s era "CFD standard" are $p = 1.0 = \rho$ on $0 \le x < 0.5$, $p = 0.1$, $\rho = 0.125$ on $0.5 \le x \le 1.0$, and $u = 0$ throughout. This definition seriously challenged CFD algorithm prediction *fidelity* of NS total internal energy E, hence the temperature distribution.

For derived *optimal* $\beta = 1/\sqrt{15}$ and uniform M = 100 x axis discretization, following ~1400 time steps nΔt to "CFD standard" output station nΔt = 0.14161s $\equiv$ T, the temperature DOF distribution predicted by the $\beta\mathrm{D}^m\mathrm{M}{:}\mathrm{D}^m\mathbf{P}{:}\mathrm{D}^m\mathrm{E}$ ODE GWS^h algorithm is graphed in Figure 4.3a. The time T exact solution is superposed thereon as the union of essentially vertical and horizontal straight lines. Readily ascertained is that the level of numerical diffusion generated by optimal $\beta = 1/\sqrt{15}$ has *totally compromised* the *fidelity* of the algorithm predicted temperature DOF distribution throughout.

The β coefficients in the $\beta\ \mathrm{D}^m\mathrm{M}{:}\mathrm{D}^m\mathbf{P}{:}\mathrm{D}^m\mathrm{E}$ ODE system have been defined state variable member dependent, as legacy CFD stabilization optimization relied on coefficient *tweaking*, as Table 4.1 well illustrated. Replacing the constant $\beta = 1/\sqrt{15}$ with Euler state variable member-dependent $\{\beta_q\} = \beta\{1/4, 1/3, 1/2\}^T$ leads to a significant improvement in temperature DOF distribution *fidelity* at nΔt = 0.14161s, Figure 4.3b). The shock is sharply resolved over four DOF but the distribution is *non-monotone* due

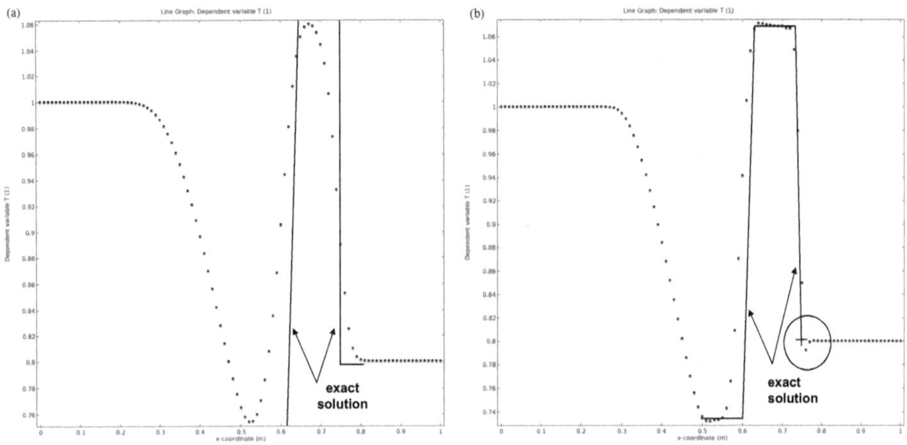

Figure 4.3: Temperature DOF distributions, Riemann shock tube, compressible Euler time TS β altered D^mM:D^m**P**:D^mE GWS^h + θTS FE $p=1$ basis algorithm, M = 100 uniform mesh, nΔt = 0.14161s: a) for *"optimal"* $\beta=1/\sqrt{15}$, b) for alteration to $\{\beta_q\}=\beta\{1/4,1/3,1/2\}$.

to the negative DOF at the shock foot, thereon encircled. The high temperature DOF distribution plateau between the shock and contact surface approaches planarity but it is not horizontal. The low temperature DOF distribution plateau surface separating the expansion fan from initiation of the contact surface is excessively diffused over half its span, an algorithm liability not responsive to additional $\{\beta_q\}$ tweaking.

The Godunov-genre FD flux vector split MUSCL CFD algorithm, van Leer [14], generates quality resolution of the shock tube expansion fan-contact surface plateau. This was enabled by inclusion of the time TS identified "μ" term into the Euler β altered D^m**P**

$$D^m\mathbf{P}: \frac{\partial q}{\partial t}+\frac{\partial uq}{\partial x}-\frac{\beta\Delta t}{2}\frac{\partial}{\partial x}\left(u^2\frac{\partial q}{\partial x}\right)-\frac{\mu\Delta t^2}{6}\frac{\partial}{\partial x}\left(u^3\frac{\partial^2 q}{\partial x^2}\right)=0$$

The third-order derivative in this *fourth-order* nonlinearity required a higher than second-order differencing scheme which ultimately led to the design of algorithm flux vector limiters with switches. Alternative Godunov-genre (β, μ) D^m**P** flux difference split scheme stabilizations are Engquist and Osher [7] and Roe [13], and the latter is a stabilization option in select current commercial and government CFD codes.

4.4 Derivation, NS DP $O(h^2)$ truncation error *continuum* nonlinear functional, $n=1$

No genuine fluid-thermal mechanics process is physics of fluids *fidelity* characterized by an Euler PDE system solution. Furthermore, seven decades of legacy NS/RaNS CFD experience quantifies that code algebraic instability is Reynolds number dependent! Is it thus not *patently logical* (?!) that CFD algorithm stability theorization address a model NS ODE, then the full NS PDE system with extensions to RaNS and LES PDE systems that secure closure via Reynolds stress tensor models?

For generic state variable scalar $q(x)$, flux vector $f(q, x)$ and kinematic viscosity $\nu(x)$, the steady NS model ODE

$$f(q,x)\frac{dq}{dx} - \nu(x)\frac{d^2q}{dx^2} = 0$$

enables *analytical* prediction of the nonlinear $O(h^2)$ TE functional that will be validated as the ***root cause*** for algebraic instability. For an x axis uniform discretization of measure h, hence the geometric node set X_j with coincident DOF Q_j, and evaluating $f(q, x)$ and $\nu(x)$ at X_j, the second-order accurate FD *and* GWS^h FE $p = 1$ basis element matrix assembled $S_e(\mathbf{DP}^h)$ algorithms generate the identical tridiagonal stencil

$$\mathrm{FD/FE}: f_j\left(a_1Q_{j-1} + a_2Q_j + a_3Q_{j+1}\right) - \nu_j\left(a_4Q_{j-1} + a_5Q_j + a_6Q_{j+1}\right) = 0 + O\left(h^{\mathrm{k}}\right)$$

For uniform mesh with lead TE order k = 2, the definition of the stencil coefficient set is well known to be $a_i = (-1/2\ h, 0, 1/2\ h, h^{-2}, -2\ h^{-2}, h^{-2})$.

Alteration of a tridiagonal stencil lead TE term to orders k > 2 is possible while maintaining preservation of the coefficient consistency set general requirement $(a_1 + a_2 + a_3) = 0 = (a_4 + a_5 + a_6)$, $(a_3 + a_1) = 0 = (a_6 - a_4)$, $(a_3 - a_1) = h^{-1}$ and $(a_6 + a_4) = 2\ h^{-2}$. Forming a spatial TS about X_j for each term in the FD/FE tridiagonal stencil, then inserting DOF subscripts to denote the order of spatial derivative, the tridiagonal stencil for this NS model ODE can be recast as the $O(h^\infty)$ accurate *difference-differential* (DD) equation [10]:

$$\begin{aligned}\mathrm{DD:}\ f&\left[\begin{array}{l}(a_1+a_2+a_3)Q + (a_3-a_1)hQ_x + (a_3+a_1)\frac{h^2}{2}Q_{xx}\\ +\,(a_3-a_1)\sum\limits_{n=3,2}^{\infty}\frac{h^n}{n!}Q_{(n)} + (a_3+a_1)\sum\limits_{n=4,2}^{\infty}\frac{h^n}{n!}Q_{(n)}\end{array}\right]\\ -\nu&\left[\begin{array}{l}(a_4+a_5+a_6)Q + (a_6-a_4)hQ_x + (a_6+a_4)\frac{h^2}{2}Q_{xx}\\ +\,(a_6-a_4)\sum\limits_{n=3,2}^{\infty}\frac{h^n}{n!}Q_{(n)} + (a_6+a_4)\sum\limits_{n=4,2}^{\infty}\frac{h^n}{n!}Q_{(n)}\end{array}\right] \equiv 0 + O(h^\infty)\end{aligned}$$

Therein, the summation subscripts $n = 3, 2$ and $n = 4, 2$ denote *all* odd, even integers in increments of two. The integer subscript (n) on each DOF Q denotes the order of spatial differentiation.

That this DD is *infinite* (!) order accurate enables *exact* replacement of any higher-order derivative appearing in a stencil without insertion of an unknown coefficient. Selecting the first three DOF terms *only* in each square bracket of the $O(h^\infty)$ accurate DD alters the $n = 2$ FD/FE tridiagonal stencil to the DD form

$$\text{DD: } fQ_X - \nu Q_{XX} + h^2(fQ_{XXX}/6 - \nu Q_{XXXX}/12) = 0 + H.O.T.$$

wherein "*H.O.T.*" denotes higher-order terms not detailed.

NS model ODE provides alternative expressions for each Q derivative in this DD that exceeds second order. Retaining ν a constant during this operation

$$Q_{XXX} = f_X Q_X + fQ_{XX} + H.O.T.$$
$$\nu Q_{XXXX} = \nu(f_{XX}Q_X + 2f_X Q_{XX} + fQ_{XXX}) + H.O.T.$$

then replacing appropriate terms in the prior DD with these identifications and omitting all TE terms less significant than $O(h^4)$ generates

$$\text{DD: } fQ_X - \nu Q_{XX} + h^2 Q_X\left(\frac{ff_X}{12\nu} - \frac{f_{XX}}{12}\right) + h^2 Q_{XX}\left(\frac{f^2}{12\nu} - \frac{f_X}{6}\right) = 0 + O(h^4)$$

the *sought* (!) DD restatement of the NS model ODE FD/FE tridiagonal stencil which is rightfully equated to $0 + O(h^4)$ as the $O(h^2)$ TE term is now embedded in the DD.

Thereby this TS:DD exercise has *rigorously* identified that the cubically *nonlinear continuum calculus* $O(h^2)$ TE functional absent in legacy CFD $n = 2$ tridiagonal stencils is

$$O(h^2)\text{ TE: } h^2\left(Q_X\left(\frac{ff_X}{12\nu} - \frac{f_{XX}}{12}\right) + Q_{XX}\left(\frac{f^2}{12\nu} - \frac{f_X}{6}\right)\right)$$

Its reversed sign insertion into the NS model ODE defines the $O(h^2)$ *truncation error annihilation* (TEA) theory modified $D^m\mathbf{P}$ as the replacement for NS model ODE now also equated to $0 + O(h^4)$

$$D^m\mathbf{P}\text{: } fq_X - \nu q_{XX} - h^2\left[q_X\left(\frac{ff_X}{12\nu} - \frac{f_{XX}}{12}\right) - q_{XX}\left(\frac{f^2}{12\nu} - \frac{f_X}{6}\right)\right] = 0 + O(h^4)$$

Fluid mechanics preferred practice is to render NS/RaNS PDE systems non-dimensional. For isothermal incompressible NS **DP**, this simply replaces kinematic viscosity ν with the inverse of Reynolds number. Thereby, NS TEA theory altered non-D $D^m\mathbf{P}$ in conventional notation

$$D^m\mathbf{P}\text{: } f\frac{dq}{dx} - \frac{1}{\text{Re}}\frac{d^2q}{dx^2} + \frac{h^2}{4}\frac{d}{dx}\left(\frac{dq}{dx}\frac{df}{dx}\right) - \frac{\text{Re}\,h^2}{12}f\frac{d}{dx}\left(f\frac{dq}{dx}\right) = 0 + O(h^4)$$

unequivocally states *any* (!) FD/FE tridiagonal stencil equivalent uniform mesh $D^m\mathbf{P}$ algorithm will generate TS $O(h^4)$ accurate (!) approximations to the NS model **DP** analytical solution.

The TEA term in $D^m\mathbf{P}$ that is independent of Re can be discarded for large Re, the usual practice. Then inserting NS flux vector Jacobian $f \equiv u$ therein identifies the *amazingly robust* (!) functional correlation that exists between TEA altered $O(h^4)$ $\mathrm{D}^m\mathbf{P}$ and temporal TS β term $\mathrm{D}^m\mathbf{P}$

$$\frac{\beta\Delta t}{2}\frac{\partial}{\partial x}\left(u^2\frac{\partial q}{\partial x}\right)\approx\frac{\beta h}{2}\frac{\partial}{\partial x}\left(u\hat{u}\frac{\partial q}{\partial x}\right)\Leftrightarrow\frac{\mathrm{Re}\,h^2}{12}u\frac{d}{dx}\left(u\frac{dq}{dx}\right)$$

Thus quantified is that the *full range* of the Table 4.1 β term correlated legacy CFD stabilizations are in fact *unbeknownst* (!) close functional approximations to rigorous annihilation of the $O(h^2)$ TE. And of truly fundamental theoretical importance, thus identified is that the time TS exercise correlation coefficient β is in fact the Reynolds number dependent *distribution* $\beta \Rightarrow \beta(x) \equiv \mathrm{Re}h(x)/6$ (!) non-D.

Repeating the FD/FE DD:$\mathrm{D}^m\mathbf{P}$ process for compressible NS model ODE simply replaces u with momentum m and $\mathrm{Re}h^2/12$ with $\mathrm{Re}h^2/12\mu^*, \mu^*(x_j)$ the non-D $O(1)$ significance temperature-dependent viscosity distribution. That $O(h^2)$ TEA theory robustness extends to non-uniform discretizations is readily validated. For x axis non-uniform discretization node set $X_{j+1} = \mathrm{pr}\ X_j$, $\mathrm{pr} > 1$ a geometric progression factor, repeating the TS:DD exercise identifies

$$\mathrm{D}^m\mathbf{P}\text{: } fQ_x - \nu Q_{xx} + \frac{h^2(pr^2+pr+1)}{36}\left(Q_x\left(\frac{ff_x}{12\nu}-\frac{f_{xx}}{12}\right)+Q_{xx}\left(\frac{f^2}{12\nu}-\frac{f_x}{6}\right)\right)=0+O(h^3)$$

as NS model TEA altered $\mathrm{D}^m\mathbf{P}$ [9]. That the lead TE term is only one order more significant at $O(h^3)$ quantifies TEA theory $\mathrm{D}^m\mathbf{P}$ alteration retains $O(h^2)$ TE *annihilation* as well as the $O(h^1)$ TE term intrinsic to legacy FD/FE non-uniform mesh tridiagonal stencils.

4.5 Nonlinear fidelity, NS TEA altered $O(h^4)$ $\mathrm{D}^m\mathrm{P}$ GWS^h FE algorithm, Newton Jacobian

For an incompressible fluid the NS state variable $q(x)$ for model $\mathrm{D}^m\mathbf{P}$ ODE is velocity $u(x)$ as is the definition for flux vector $f(q, x)$. The resultant 1-D NS TEA altered $O(h^4)$ $\mathrm{D}^m\mathbf{P}$ ODE

$$\mathrm{D}^m\mathbf{P}\text{: } u\frac{\mathrm{d}u}{\mathrm{d}x}-\frac{1}{\mathrm{Re}}\frac{\mathrm{d}^2u}{\mathrm{d}x^2}+\frac{h^2}{4}\frac{\mathrm{d}}{\mathrm{d}x}\left(\frac{\mathrm{d}u}{\mathrm{d}x}\right)^2-\frac{\mathrm{Re}\,h^2}{12}u\frac{\mathrm{d}}{\mathrm{d}x}\left(u\frac{\mathrm{d}u}{\mathrm{d}x}\right)=0+O(h^4)$$

is immediately recognized as the nonlinear viscous Burgers **DP** absent (of course!) the TEA terms and the equating of $\mathrm{D}^m\mathbf{P}$ to $0 + O(h^4)$. For large Re, thus clarified is that the $O(h^2)$ TE term *unstated* in legacy NS second-order stencil practice is a *cubically nonlinear* functional that is indeed Re-dependent!

For NS **DP** boundary condition (BC) definitions $u(x = 0) = 1$, $u(x = 1) = -1$, the exact solution to Burgers **DP** is a "viscous" shock centered at $x = 0.5$ with thickness linearly dependent on Re. Recalling classic separation of variables techniques, any GWS^h discrete approximation $u^h(x)$ to $u(x)$ is the *union* of the matrix inner product of FE $p=1$ basis $\{N(x)\}$ spanning Ω_e with its node co-located DOF coefficient matrix $\{U\}_e$ over the discretization of NS domain $\Omega^h \equiv \cup_e \Omega_e$

$$\mathrm{GWS}^h\colon\ u^h(x) \equiv \cup_e \{N(x)\}^T \{\mathrm{U}\}_e$$

The *fundamental* operation of an NS GWS^h algorithm code is "DO loop" *assembly* (S_e) of each generated Ω_e matrix into the global algebraic statement. Following numerical linear algebra iterative convergence, the GWS^h solution $u^h(x)$ is thereby a *piecewise* continuous linear distribution.

With this introduction, that the GWS^h FE $p=1$ basis algorithm *precisely* translates the TEA continuum calculus cubic nonlinearity to *algebraic* is easily validated. That incompressible NS velocity distributions are divergence-free admits moving u inside the TEA augmentation divergence operator. The subsequent imposition of the Green-Gauss divergence theorem then generates the GWS^h FE $p=1$ basis Re dependent TEA term element matrix

$$\mathrm{GWS}^h\left(-\frac{\mathrm{Re}\,h^2}{12}\frac{\mathrm{d}}{\mathrm{dx}}\left(uu\frac{\mathrm{d}u}{\mathrm{dx}}\right)\right)_e\colon\ \frac{\mathrm{Re}\,h_e^2}{12}\int_{\Omega_e}\frac{\mathrm{d}\{N\}}{\mathrm{dx}}\{\mathrm{U}\}_e^T\{N\}\{N\}^T\{\mathrm{U}\}_e\frac{\mathrm{d}\{N\}^T}{\mathrm{dx}}\mathrm{dx}\{\mathrm{U}\}_e$$

The $n=1$ FE $p=1$ basis derivative $\mathrm{d}\{N\}^T/\mathrm{dx} = \{1, -1\}/h_e$ is a constant on each Ω_e. Extracting all such *x-independent* data from the GWS^h integrand yields the rearrangement

$$\mathrm{GWS}^h\colon\ \frac{\mathrm{Re}\,h_e^2}{12}\{\mathrm{U}\}_e^T\int_{\Omega_e}\{N\}\{N\}^T\mathrm{dx}\{\mathrm{U}\}_e\frac{1}{h_e^2}\begin{Bmatrix}1\\-1\end{Bmatrix}\{1, -1\}\{\mathrm{U}\}_e$$

for which analytical integration generates the element nonlinear scalar coefficient

$$\{\mathrm{U}\}_e^T\int_{\Omega_e}\{N\}\{N\}^T\mathrm{dx}\{\mathrm{U}\}_e = \frac{h_e}{6}\{\mathrm{U}\}_e^T\begin{bmatrix}2 & 1\\1 & 2\end{bmatrix}\{\mathrm{U}\}_e = (2h_e/6)\left(\mathrm{UL}^2 + \mathrm{ULUR} + \mathrm{UR}^2\right)_e$$

wherein the notation "UL/UR" signifies the "left/right" node speed DOF for any Ω_e.

This nonlinear *non-positive definite* scalar is the Ω_e-dependent multiplier on the FE $p=1$ basis derivative outer-inner matrix product $\{1, -1\}\{1, -1\}^T\{\mathrm{U}\}_e$, which when completed identifies the TEA term element matrix final form

$$\mathrm{GWS}^h\left(O(h^2)\mathrm{TE}\right)_e\colon\ \frac{\mathrm{Re}}{12}\frac{h_e}{3}\left(\mathrm{UL}^2 + \mathrm{ULUR} + \mathrm{UR}^2\right)_e\begin{bmatrix}1 & -1\\-1 & 1\end{bmatrix}\begin{Bmatrix}\mathrm{UL}\\\mathrm{UR}\end{Bmatrix}_e$$

Its *assembly* $S_e\left(\mathrm{GWS}^h\left(O(h^2)\mathrm{TE}\right)_e\right)_{X_j}$ over the Ω_e pair sharing generic node X_j generates the *exact* nonuniform mesh *cubically* nonlinear tridiagonal stencil

$$\frac{\text{Re}}{36}\left(h_\text{L}\left(\text{U}_{j-1}^2+\text{U}_{j-1}\text{U}_j+\text{U}_j^2\right)\text{U}_{j-1} - \left(\begin{array}{c} h_\text{L}\left(\text{U}_{j-1}^2+\text{U}_{j-1}\text{U}_j+\text{U}_j^2\right) \\ +\,h_\text{R}\left(\text{U}_j^2+\text{U}_j\text{U}_{j+1}+\text{U}_{j+1}^2\right)\end{array}\right)\text{U}_j \right.$$
$$\left. +\,h_\text{R}\left(\text{U}_j^2+\text{U}_j\text{U}_{j+1}+\text{U}_{j+1}^2\right)\text{U}_{j+1}\right)$$

wherein h_L, h_R denote the measures of the left, right Ω_e sharing node X_j. Comparison of this GWS^h FE $p = 1$ basis calculus generated cubic nonlinearity *exact* translation to algebraic with conventional FD algebraic stencil practice should be quite informative.

Solution of this NS TEA altered D^m**P** GWS^h element matrix assembled into the global matrix requires a numerical linear algebra decision with *optimal* choice the *quadratic* convergent Newton iteration algorithm. The Burgers TEA altered D^m**P** GWS^h FE $p = 1$ basis Re dependent TEA term element matrix

$$\begin{aligned}
\text{GWS}^h : \frac{\text{Re}\,h_e^2}{12}&\int_{\Omega_e}\frac{\text{d}\{N\}}{\text{dx}}\{\text{U}\}_e^T\{N\}\{N\}^T\{\text{U}\}_e\frac{\text{d}\{N\}^T}{\text{dx}}\text{dx}\{\text{U}\}_e \\
=\frac{\text{Re}\,h_e^2}{12}&\int_{\Omega_e}\{\text{U}\}_e^T\{N\}\{N\}^T\{\text{U}\}_e\text{dx}\frac{\text{d}\{N\}}{\text{dx}}\frac{\text{d}\{N\}^T}{\text{dx}}\{\text{U}\}_e \\
=\frac{\text{Re}\,h_e}{72}&\{\text{U}\}_e^T\begin{bmatrix}2 & 1\\ 1 & 2\end{bmatrix}\{\text{U}\}_e\begin{bmatrix}1 & -1\\ -1 & 1\end{bmatrix}\{\text{U}\}_e
\end{aligned}$$

enables *exact* identification of Newton algorithm Jacobian contributions via matrix differential calculus operations on the trailing DOF column matrix $\{\text{U}\}_e$ in any GWS^h element matrix statement.

The visually obvious Burgers TEA term contribution to the Newton Jacobian is

$$[\text{JAC\_UU}]_e = \frac{\text{Re}\,h_e}{72}\{\text{U}\}_e^T\begin{bmatrix}2 & 1\\ 1 & 2\end{bmatrix}\{\text{U}\}_e\begin{bmatrix}1 & -1\\ -1 & 1\end{bmatrix}$$

The accounting for GWS^h element matrix nonlinear interior $\{\text{U}\}_e$ matrix embeddings is enabled by rearrangement of the scalar matrix product in the Burgers TEA term element matrix to

$$\text{GWS}^h : \frac{\text{Re}\,h_e}{72}\begin{bmatrix}1 & -1\\ -1 & 1\end{bmatrix}\{\text{U}\}_e\{\text{U}\}_e^T\begin{bmatrix}2 & 1\\ 1 & 2\end{bmatrix}\{\text{U}\}_e$$

Differentiation on the newly exposed trailing DOF $\{\text{U}\}_e$ matrices (*plural*) then completes identification of the *exact* TEA term GWS^h element matrix contribution to the Newton Jacobian

$$[\text{JAC\_UU}]_e = \frac{\text{Re}\, h_e}{72}\left(\{U\}_e^T\begin{bmatrix}2 & 1\\ 1 & 2\end{bmatrix}\{U\}_e\begin{bmatrix}1 & -1\\ -1 & 1\end{bmatrix} + 2\begin{bmatrix}1 & -1\\ -1 & 1\end{bmatrix}\{U\}_e\{U\}_e^T\begin{bmatrix}2 & 1\\ 1 & 2\end{bmatrix}\right)$$

To summarize, the GWS^h FE $p = 1$ basis algorithm enables mathematically *precise* translation of Burgers TEA theory altered $D^m\mathbf{P}$ cubic nonlinearity to algebraic, also *exact* identification of *all* element matrix nonlinear contributions to quadratic convergent Newton iteration algorithm Jacobian is quantitatively *validated*. These calculus expressed GWS^h *intrinsic* unstagnations to current CFD practice will truly contribute to the satisfaction of the VVUQ key requirement of quantified *improved outcomes*.

4.6 Validation, Burgers TEA $O(h^4)$ D^mP GWS^h FE DOF monotonicity $\forall Re \le 10^6$

That Burgers TEA altered $O(h^4)$ $D^m\mathbf{P}$ GWS^h FE $p = 1$ basis algorithm generates {U} DOF shock distributions that are *monotone* $\forall \text{Re} \le 10^6$ on *any* Ω^h is readily validated. Uniform x axis discretization into meshes with the odd number of elements M = 21,121,221, each of measure $h = 1/\text{M}$, generates viscous shock resolutions devoid of a mid-point DOF. The GWS^h FE $p = 1$ basis DOF distribution on the extremely coarse M = 21 mesh (Figure 4.4a) affords clear *quantification* that this *inadequately resolved* (!) shock DOF distribution at Re = 10^3 is *resolutely* monotone.

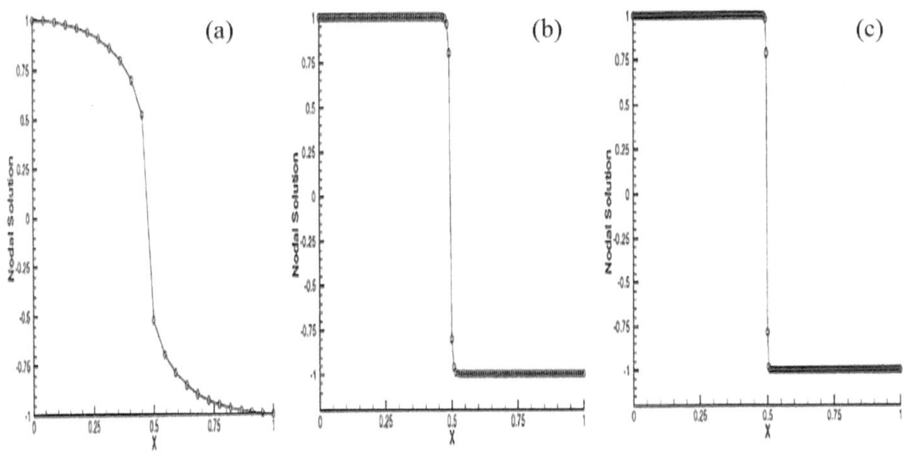

Figure 4.4: Validation, Burgers TEA altered $O(h^4)$ $D^m\mathbf{P}$ GWS^h FE $p = 1$ basis algorithm $u^h(x)$ DOF monotonicity, Re = 10^3, for uniform meshes of measure: a) $h = 1/21$, b) $h = 1/121$, c) $h = 1/221$.

For the $h = 1/121$ mesh, the viscous shock DOF monotone resolution spans three elements Ω_e (Figure 4.4b), while that for the $h = 1/221$ mesh differs *only* in DOF interpolation fidelity of the shock onset and termination (Figure 4.4c). That these two mesh M solutions possess the identical shock interior DOF pair *validates* that the $h = 1/221$ mesh DOF distribution *equates* one to one with the knots of piecewise linear interpolation of the Burgers **DP** *exact* solution at $\mathrm{Re} = 10^3$.

Solution adapted nonuniform mesh refinement sequence $h_{e+1} \equiv \mathrm{pr}h_e$, pr a mirror symmetric progression factor centered at $x = 0.5$, quantifies M = 21 DOF shock resolution accuracy matches that of uniform $h = 1/121$ solution at $\mathrm{Re} = 10^3$ (Figure 4.5a). Shock DOF resolutions for uniform mesh M refinement sequence are validated *monotone* in shock centric DOF distribution zoom at $\mathrm{Re} = 10^4$ (Figure 4.5b). Solution adapted nonuniform mesh M = 464 generates "right angle" shock DOF continuum interpolation at $\mathrm{Re} = 10^6$ (Figure 4.5c). The uniform mesh refinement sequence data for miniscule Re = 10 validates that TEA theory *predicted* $O(h^4)$ convergence rate is secured only when including both TEA terms in $\mathrm{D}^m\mathbf{P}$, Kolesnikov and Baker ([10], Table 2)

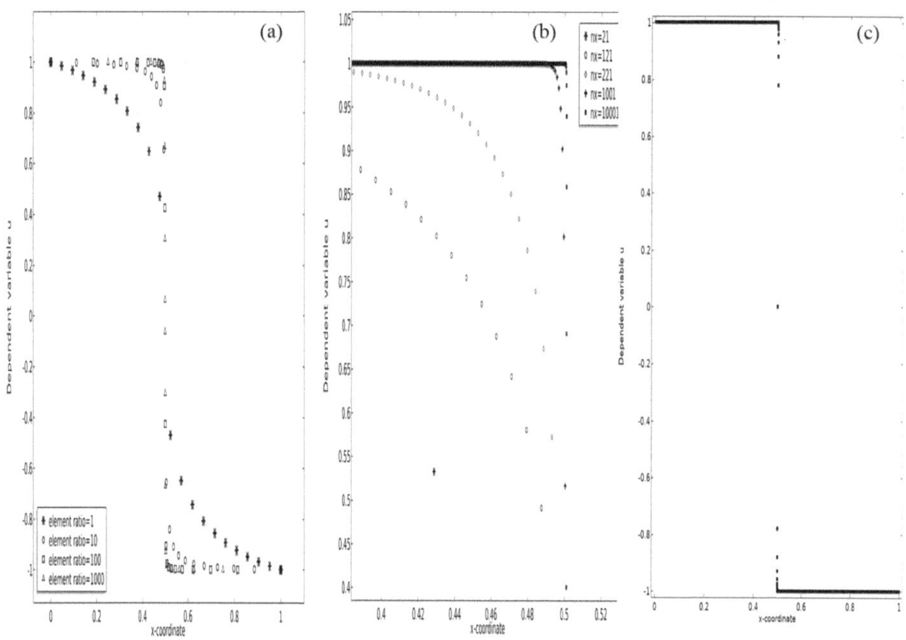

Figure 4.5: Validation, NS $u^h(x)$ state variable DOF monotonicity, Burgers TEA $O(h^4)$ $\mathrm{D}^m\mathbf{P}$ GWS^h FE $p = 1$ basis algorithm, $10^3 \le \mathrm{Re} = 10^6$: a) solution adapted M = 21 nonuniform, $\mathrm{Re} = 10^3$, b) uniform $20 \le \mathrm{M} \le 10^4$, $\mathrm{Re} = 10^4$, c) solution adapted nonuniform mesh M = 464, $\mathrm{Re} = 10^6$.

4.7 Validation, compressible NS TEA $O(h^4)$ PDE *monotone* shock interpolation, $\mathrm{Re} \le 10^5$

Insertion of compressible NS TEA altered $\mathrm{D}^m\mathrm{M}$:$\mathrm{D}^m\mathbf{P}$:$\mathrm{D}^m\mathrm{E}$ non-D $O(h^4)$ ODE system with now quantified $\beta = \mathrm{Re}h/6$ is

$$\mathrm{D}^m\mathrm{M}:\ \frac{\partial\rho}{\partial t}+\frac{\partial}{\partial x}\left[m-\frac{h^2\mathrm{Re}}{12}\widehat{u}\,u\frac{\partial\rho}{\partial x}\right]=0+O(h^4)$$

$$\mathrm{D}^m\mathbf{P}:\ \frac{\partial m}{\partial t}+\frac{\partial}{\partial x}\left[um+\mathrm{Eu}\,p-\frac{\mu^*}{\mathrm{Re}}\frac{\partial u}{\partial x}\right]-\frac{h^2\mathrm{Re}}{12\mu^*}\frac{\partial}{\partial x}\left[u\left(u-\frac{\partial u}{\partial x}\right)\frac{\partial m}{\partial x}\right]=0+O(h^4)$$

$$\mathrm{D}^m\mathrm{E}:\ \frac{\partial E}{\partial t}+\frac{\partial}{\partial x}\left[u(E+p)-\frac{\kappa^*}{\mathrm{Re\,Pr}}\frac{\partial T}{\partial x}\right]-\frac{h^2\mathrm{RePr}}{12\kappa^*}\frac{\partial}{\partial x}\left[u\left(u-\frac{\partial u}{\partial x}\right)\frac{\partial T}{\partial x}\right]=0+O(h^4)$$

The added terms in $\mathrm{D}^m\mathbf{P}$ and $\mathrm{D}^m\mathrm{E}$ result from enforcing compressible DM upon moving u inside the TEA term divergence operator.

Retaining the Riemann shock tube "CFD standard" IC, the compressible NS $O(h^4)$ $\mathrm{D}^m\mathrm{M}$:$\mathrm{D}^m\mathbf{P}$:$\mathrm{D}^m\mathrm{E}$ GWS^h FE $p = 1$ basis algorithm temperature DOF $\mathrm{n}\Delta t = 0.14161\mathrm{s} = \mathrm{T}$ distribution for an M = 1000 uniform mesh at Re = 7943 is strictly *monotone* (Figure 4.6a). Contact surface precursor plateau persistence to $x \approx 0.61$ ($x = 0.625$ is the exact value) is a *truly* significant improvement. The temperature DOF resolution in the contact surface precursor plateau reach is *monotone* and *planar*, the latter indeed *horizontal*. Shock speed DOF *monotone* distribution (Figure 4.6b), confirms TEA theory *annihilates* the Euler $\{\beta_q\}$ ODE algorithm admitted *negative* DOF at the shock foot (Figure 4.3).

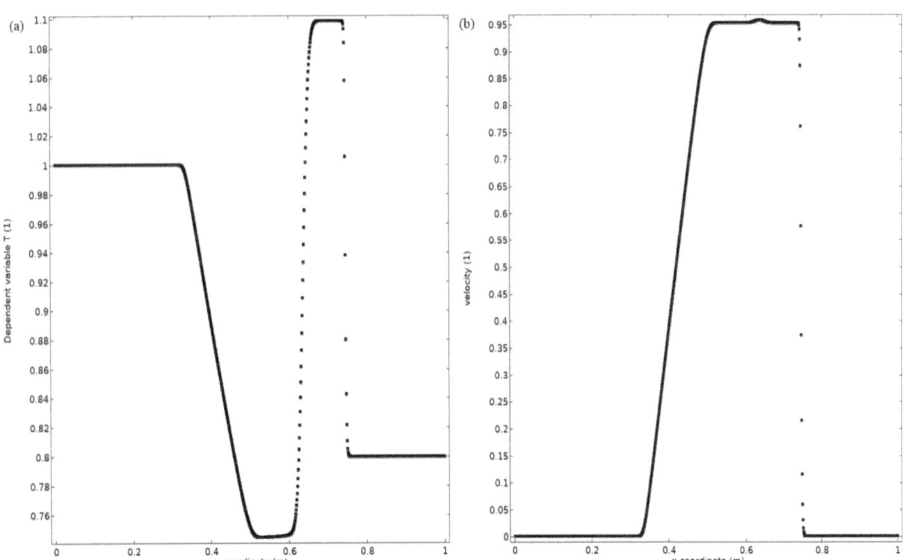

Figure 4.6: Validation, Riemann shock tube NS DOF monotone distributions, TEA altered $O(h^4)\mathrm{D}^m\mathrm{M}$:$\mathrm{D}^m\mathbf{P}$:$\mathrm{D}^m\mathrm{E}$ ODE system GWS^h FE $p = 1$ basis algorithm, uniform M = 1000 mesh, $\mathrm{n}\Delta t = 0.14161\mathrm{s}$, Re = 7943: a) temperature T, b) velocity (speed) u.

NS state variable distributions for uniform mesh refinement sequence M = 10^3 at Re = 15,850 and M = 10^4 at Re = 158,486 are summarized in Figure 4.7. The M = 10^4 mesh contact surface plateau termination at $x = 0.625$ is exact (Figure 4.7d). The right angle breaks in temperature, speed, and density DOF *strictly monotone* distributions (Figure 4.7d,e,f) quantitatively *validate* compressible NS *error freed* CFD mathematics TEA altered $O(h^4)$ D^mM:D^m**P**:D^mE GWS^h FE $p = 1$ basis algorithm for a Re definition approaching reality. Note also that the "bump" in the middle of the high speed plateau at contact surface intercept at Re = 15,860 is eliminated in the Re = 158,486 distribution.

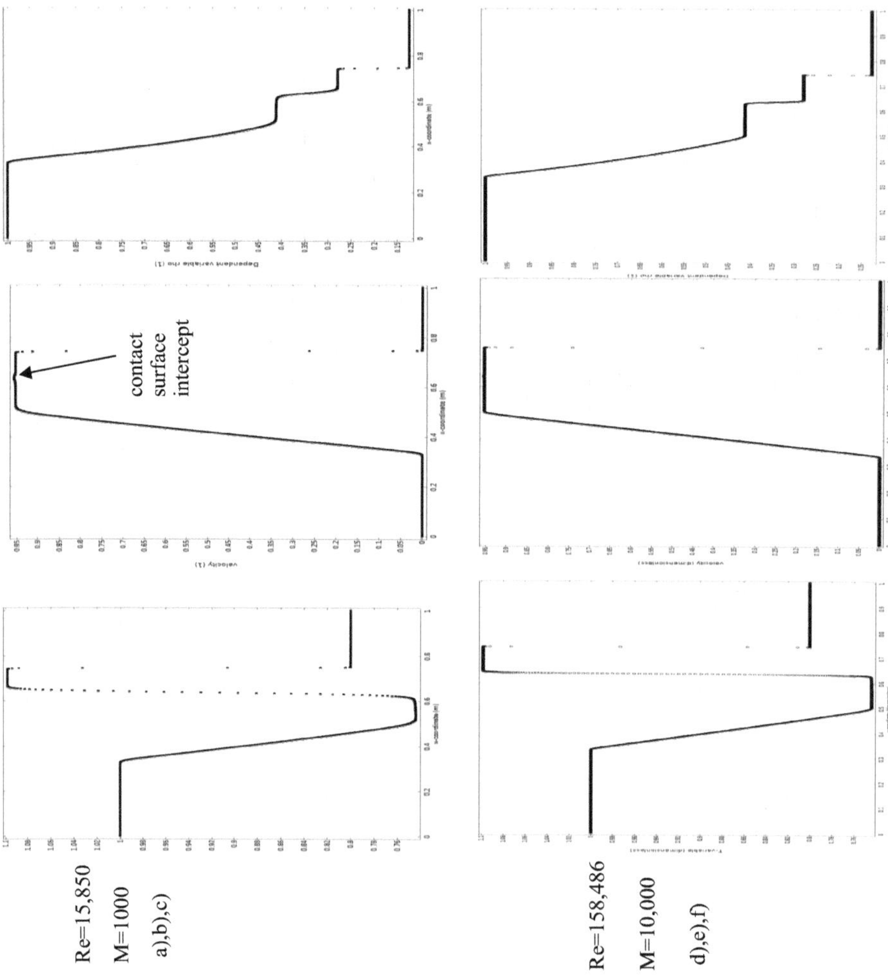

Figure 4.7: Validation, Riemann shock tube compressible monotone DOF distributions, NS TEA altered (h^4) D^mM:D^m**P**:D^mE ODE system GWS^h + θTS FE $p = 1$ basis algorithm, nΔt = 0.14161s, for uniform M = 1000 mesh at Re = 15,850, a) temperature, b) speed, c) density; for uniform M = 10,000 mesh at Re = 158,486, d) temperature, e) speed, f) density.

Regarding *theoretical rigor*, that the DOF number in continuum *interpolation* of the shock is identical for M = 10^3 and M = 10^4 mesh predictions, aside from shock onset and termination refinements, validates that the M = 10^4 compressible NS state variable DOF *monotone* distributions *equate* one-to-one (!) with the knots of piecewise linear interpolation of the NS *exact* state variable nΔt = 0.14161s shock tube distributions at Re = 158,486.

That *no* cost penalty accrues to execution of the NS TEA $O(h^4)$ ODE system GWS^h + θTS algorithm code is inferred by comparison of integration time step size evolution with those of the time TS altered Euler $\{\beta_q\}$ $O(h^2)$ ODE and a commercial code Euler $O(h^2)$ ODE GWS^h stabilized (hence non-Galerkin) algorithm. At Re = 158,486, execution of the NS GWS^h algorithm code admitted a *monotonic* increase in time step Δt to the code specified limit well before termination at nΔt = 0.14161 s. Conversely, the Δt incrementing for both Euler code executions was irregular, not monotone and rarely reached the Δt limit.

Further to *error freed* CFD mathematics validation, the closest approach to the nΔt = 0.14161s *exact* shock coordinate x = 0.750 is accorded to the NS TEA $O(h^4)$ ODE GWS^h solution, the red data in Figure 4.8. In comparison, both Euler $O(h^2)$ ODE algorithms generated shock centric *non-monotone* pressure DOF resolutions due to the negative DOFs at the shock foot encircled in Figure 4.8.

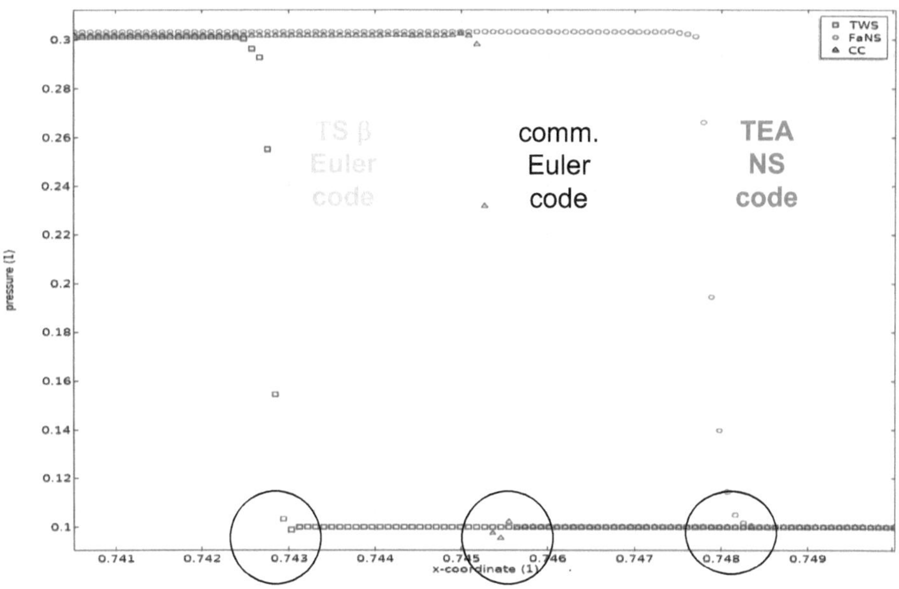

Figure 4.8: Riemann shock tube, CFD algorithm shock centric pressure DOF distributions, uniform M = 10^4 FE p = 1 basis mesh, Re = 158,486, Δt = 0.14161s: squares, Euler $\{\beta_q\}$ $O(h^2)$ ODE; triangles commercial code Euler $O(h^2)$ ODE; circles NS TEA altered $O(h^4)$ ODE.

To conclude, weak formulation compressible NS TEA altered $O(h^4)$D^{m}M:D^m**P**:D^{m}E ODE system *continuous* GWS^h algorithm state variable DOF distributions are validated monotone for a range of FE p = 1 basis domain discretizations. This occurrence *rigor-*

ously quantifies that *continuum* shock interpolation is the mathematically viable, hence ***preferable*** replacement for CFD *phake* physics stabilized shock capture and/or shock fitting mesh movement algorithms! This conclusion will be further quantified in transition to multi-dimensional NS/RaNS $O(h^4)$ PDE system predictions in the ensuing chapters. The continuing need for these legacy stabilization strategies to cure CFD code algebraic instability is now certainly open to serious question.

4.8 $O(h^4)$ *error freed* CFD *mathematics* diagnosed physics of fluids *anomalies*

Two *physics of fluids* anomalies are CFD identified in compressible NS TEA $O(h^4)$ ODE GWS^h predicted Riemann shock tube state variable distributions. For "CFD standard" IC, continuous GWS^h predicted extremum speed plateau transonic Mach number DOF distribution morphs to subsonic near its midpoint (Figure 4.9a). This is due to contact surface temperature excursion induced increase in the local speed of sound, thereon overlaid in light grey, which occurs *without alteration* (!) of fluid speed (Figure 4.9b). While the visual impression of sonic speed alteration of Mach number to subsonic is instantaneous, the DOF number populating this transition quantifies that the distribution is indeed continuous.

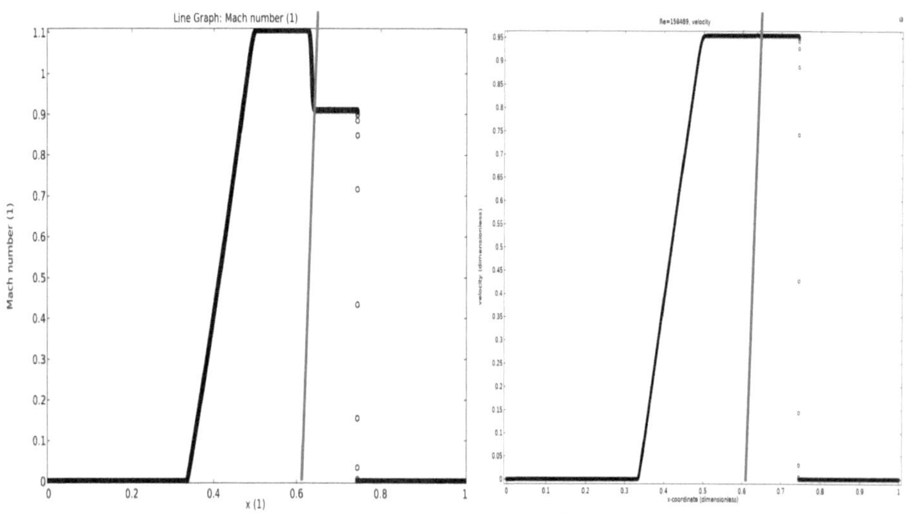

a) Ma^h(*x*,nΔt) DOF, T^h(*x*,nΔt)) superposed

b) u^h(*x*,nΔt) DOF, T^h(*x*,nΔt) superposed

Figure 4.9: Riemann shock tube *physics of fluids* anomalies, compressible NS $O(h^4)$D^{m}M:D^m**P**:D^{m}E GWS^h + θTS FE $p = 1$ basis algorithm, uniform M = 10^4 mesh, Re = 158,486, nΔt = 0.14161s: a) Ma^h(*x*,nΔt) DOF distribution with contact surface T^h(*x*,nΔt) excursion superposed in light grey, b) u^h(*x*,nΔt) DOF distribution, contact surface T^h(*x*,nΔt) superposed.

Thereby, what has been universally cited in the CFD literature as Riemann shock tube extremum speed plateau termination in a shock is instead the *strictly* subsonic $0.9 \geq \mathrm{Ma}^h(x, \mathrm{n}\Delta t) \geq 0$ event. Not withstanding, as the DOF number in resolutely *monotone* interpolation of this subsonic transition confirms it to be *"shock like"* with thickness of $O(\mathrm{Re}^{-1})$. These *error freed* CFD mathematics enabled physics of fluids *anomaly* diagnostics are due strictly to definition/use of the "CFD standard" IC which induces the *remarkable* $T^h(x, \mathrm{n}\Delta t)$ excursion in the contact surface.

References

[1] Baker, A.J., Kim, J.W. (1987). "A Taylor Weak Statement Algorithm for Hyperbolic Conservation Laws," *J Num Mtd Fluids*, 7.

[2] Brooks, A.N., Hughes, T.R.J. (1982). Streamline Upwind Petrov-Galerkin Formulations for Convection Dominated Flows with Particular Emphasis on the Incompressible Navier- Stokes Equations, *Comp Mtd Appl Mech Engr*, 32.

[3] Courant, R., Friedrichs, K., Lewy, H. (1928). Uber die Partiellen Differenzengleichungen der Mathematischen Physik, *Mathematische Annalen*, 100.

[4] Crank, J., Nicolson, P. (1947). A Practical Method for Numerical Evaluation of Solutions of Partial Differential Equations of Heat-Conduction Type, *Cambridge Phil Soc*, 43.

[5] Dendy, J.E. (1974). Two methods of Galerkin type achieving optimum L^2 rates of convergence for first-order hyperbolics, SIAM *J Num Analysis*, 11.

[6] Dendy, J.E., Fairweather, G. (1975). Alternating-direction Galerkin methods for parabolic and hyperbolic problems on rectangular polygons, SIAM *J Num Analysis*, 12.

[7] Engquist, B., Osher, S. (1981). One-Sided Difference Approximations for Nonlinear Conservation Laws, *Math Comput*, 36.

[8] Godunov, S.K. (1959). Finite-Difference Method for Numerical Computation of Discontinuous Solution of the Equations of Fluid Dynamics, *Mat Sb*, 47.

[9] Kolesnikov, A. (2000), "Efficient implementation of high order methods in computational fluid dynamics," PhD dissertation, University of Tennessee

[10] Kolesnikov, A., Baker, A.J. (2001). An efficient high order Taylor weak statement for the Navier-Stokes equations, *J Comp Physics*, 173.

[11] Lax, P., Wendroff, B. (1960). Systems of Conservation Laws, *Comm Pure Applied Math*, 13.

[12] Raymond, W.H., Garder, A. (1976). Selective damping in a Galerkin method for solving wave problems with variable grids, *Monthly Weather Review*, V.104.

[13] Roe, P.L. (1981). Approximate Riemann Solvers, Parameter Vectors and Difference Schemes, *J Comp Physics*, 43.

[14] van Leer, B. (1979). Towards the Ultimate Conservative Difference Scheme, V. A Second Order Sequel to Godunov's Method, *J Comp Physics*, 32.

[15] Wahlbin, L.B. (1974). "A dissipative Galerkin method applied to some quasi-linear hyperbolic equations," RAIRO, 8.

[16] Warming, R.F., Beam, R.M. (1975). "Upwind second order difference schemes and applications in unsteady aerodynamics flows," Proc. AIAA 2nd CFD Conf., 1997.

Chapter 5
Navier-Stokes *error freed* CFD *mathematics*, continuous weak formulation, FE theory, asymptotic convergence, error estimates, validations, monotone *continuum* shock interpolation

Libretto: Rigorous theory predicts that weak formulation *continuous* Galerkin (GWS) CFD algorithms on Navier-Stokes *error freed* PDE mathematics exhibit *optimal performance.*

Asymptotic convergence rates for compressible NS *error freed* PDE mathematics GWS^h FE $p = 1,2,3$ basis algorithms are validated to be $O(h^4, h^4, h^6)$.

Global and local *a posteriori* error estimates for compressible NS $O(h^4, h^4, h^6)$ PDE $GWS^h + \theta TS$ algorithms are derived and validated.

GWS^h FE basis algorithm error quantification, optimal mesh solution validation are precisely responsive to VVUQ topics necessary to guarantee *improved outcomes.*

Error freed CFD mathematics alteration of the 1980s–1990s *finite volume* (FV)-theorized NS codes will generate fourth-order accurate stable *monotone* degree-of-freedon (DOF) distributions on *any* mesh.

To retain pressure an incompressible NS state variable member requires rigorous enforcement of the DM *differential constraint.*

A weak formulation *error freed* DM^h algorithm quantifies secured divergence-freeness order and is validated pertinent to NS velocity predictions in multiply connected $n = 3$ domains.

Keywords: CFD, Navier-Stokes, weak formulation, continuous GWS^h algorithm, finite element, monotone solutions, rigorous theory, *a posteriori* error estimates, optimal mesh identification, VVUQ, DM differential constraint, pressure, multiply-connected domains

5.1 Introduction, unstagnation options for current practice NS (and RaNS) codes

Weak formulation tridiagonal stencil *efficient* Navier-Stokes (NS) truncation error annihilation (TEA) altered $O(h^4)$ PDE system *continuous* GWS^h FE linear ($p = 1$) basis algorithms are precisely responsive to performance criticisms cited for current practice CFD codes, cf., Jameson [8], and Masters [11]:

https://doi.org/10.1515/9783112206430-006

. . . commercial, government CFD codes are based on **80s-90s era** FD algorithms . .

. . . codes are generally limited to ***2nd* order** accuracy . .

. . . perceived **stagnation** of CFD capability *advancement* . . . exists

. . . codes cannot handle turbulence without **modeling** . .

Weak formulation Navier-Stokes PDE unstagnations for the first three issues are chapter derived and validated. The mathematically rigorous ∀Re pertinent analytical alternative to turbulence closure modeling is derived and validated in monograph Part B.

5.2 CFD fundamentals, verification, validation, uncertainty quantification (VVUQ)

Fluid mechanics technical committees of the American Institute of Aeronautics and Astronautics (AIAA) and American Society of Mechanical Engineers (ASME) collaborated in the mid-1990s to define requirements for CFD data to warrant archival publication, AIAA [1]. The resultant "*Verification, Benchmark, Validation* (VBV)" protocol initiated the then *new* (!) requirement to quantify CFD data validity via mesh refinement exercises.

The successor to the VBV identification is the US National Academy of Sciences *Verification, Validation, Uncertainly Quantification* (VVUQ) protocol, NAP [12], which identified topics *required resolved* to ensure complex computational methodologies will indeed generate *improved outcomes*, specifically:

. . . development of goal-oriented *a posteriori* error estimation methods that can be applied to mathematical models more complicated than a linear elliptic PDE

. . . development of algorithms for goal-oriented *error estimates* that scale well on massively parallel architectures, especially given complicated and *adaptive-mesh* grids

. . . methods to estimate *error bounds* particularly when using meshes that cannot resolve important scales, an example is turbulent fluid flow

Weak formulation, vector calculus intrinsic *continuous* Galerkin FE $p = 1,2,3$ basis NS TEA altered $O(h^4)$ PDE algorithms are *optimally* responsive to these VVUQ topics:

. . . mathematically rigorous *nonlinear error estimates*, local *and* global, for genuine NS PDE systems that are *certainly* (!) more complicated than linear elliptic

. . . rigorously derived *asymptotic convergence* theory under uniform and/or *regular* solution adapted nonuniform mesh refinements in *n*-dimensions

. . . *error bounding* in NS PDE system state variable intrinsic norms, error *quantification*

. . . *degradation* of asymptotic convergence theory formalized for *non-smooth* data

. . . rigorous *goal-oriented* criterion for minimal DOF *optimally* accurate mesh solution

. . . precise derivation of jacobian nonlinearities for *quadratic convergent* Newton iteration.

5.3 Derivation, weak formulation continuous Galerkin NS CFD algorithm

Your CFD expert author pair asserts that computer-generated compressible/incompressible Navier-Stokes (NS) state variable distributions can provide *physics of fluids* insight the *fidelity* of which is limited *only* by the error mechanisms (!) intrinsic to CFD algorithm theorization. Fundamentally, as fluid mechanics is a *continuum mechanics* discipline, NS state variable distributions must also be mathematically *continuous* (!) even in the presence of a perceived discontinuity, e.g., a shock. Factually, as Mother Nature has no trouble generating a shock in viscous, heat conducting continuum air (Figure 5.1) then why not CFD?

Figure 5.1: Conical transonic velocity envelope with planar shock termination surrounding a US Navy F/A 18 Hornet in transonic flight, visualized by temperature decrease induced 100% humidity condensation, posted at https://commons.wikimedia.org/wiki/File:FA-18_going_transonic.JPG (Realbigtaco,3.0 CC BY-SA https://creativecommons.org/licenses/CC BY-SA/3.0, via Wikimedia Commons).

As any NS (also RaNS, next chapter) PDE system state variable $\{q(x_j, t)\}$ is a *continuum* distribution, then so *must be* any (!) CFD generated approximation. Reflecting on separation of variables practice for a linear elliptic boundary value (EBV) PDE, any approximation $q^N(x_j, t)$ to a NS state variable member $q(x_j, t)$ can be expressed as the product of N *continuum* spatial functions $\Psi_\alpha(x_j)$, the weak formulation *trial space*, with the same number of time-dependent coefficients $Q_\alpha(t)$

$$q^N(x_j, t) \equiv \sum_{\alpha=1}^{N} \Psi_\alpha(x_j)Q_\alpha(t) = Q_1(t)\Psi_1(x_j) + Q_2(t)\Psi_2(x_j) + \cdots + Q_N(t)\Psi_N(x_j)$$

With *any* NS continuum approximation $q^N(x_j, t)$ thus clearly defined, meeting the VVUQ *improved outcomes* requirement is to somehow *quantify* the generated *error*

$$e^N(x_j, t) \equiv q(x_j, t) - q^N(x_j, t)$$

which obviously (!) is also a *continuum* distribution. Thereby identified is the requirement to rigorously *extremize* $e^N(x_j, t)$, i.e., mathematically render it a *minimum* in some sense.

This requires a means to *measure* error, which a weak formulation (WF) accomplishes by *testing* the approximation error $e^N(x_j, t)$ against a set of *absolutely arbitrary* functions $f_\beta(x_j)$ *for all* ($\forall$) β. Letting $L(q(x_j, t)) = 0$ denote a NS generic scalar member PDE, the candidate WF statement for *testing* the error $e^N(x_j, t)$ is

$$\begin{aligned} \text{WF:} & \int_\Omega f_\beta(x_j)L\left(e^N(x_j, t)\right)d\tau \equiv 0, \; \textit{for} \; \forall f_\beta(x_j) \; \textit{and} \; t \geq 0 \\ & = \int_\Omega f_\beta(x_j)L(q - q^N)d\tau = -\int_\Omega f_\beta(x_j)L\left(q^N(x_j, t)\right)d\tau \equiv 0 \end{aligned}$$

the terminal form of which is due to NS PDE systems being expressed in homogeneous form.

The answer to, "How to evaluate WF $\forall f_\beta(x_j)$?," is, "It's *not* possible (!?)," which identifies the requirement to render WF of practical use. As test functions $f_\beta(x_j)$ are *known*, you choose them (!), interpolation theory exists to approximate $f_\beta(x_j)$ $\forall\beta$ to any *a priori* prescribed accuracy. Selecting the polynomial set $\Phi_\beta(x_j)$, the interpolation for any member $f(x_j)$ of the test function set $f_\beta(x_j)$ is

$$f(x_j) \approx f^{\mathrm{I}}(x_j) \equiv \sum_{\beta=1}^{\mathrm{I}} \Phi_\beta(x_j)F_\beta$$

wherein coefficient set F_β is named the *knots* of $\Phi_\beta(x_j)$ interpolation of $f(x_j)$.

The key attribute of thus altered WF

$$\text{WF:} \int_\Omega f(x_j)L\left(q^N(x_j, t)\right)d\tau \approx \sum_{\beta=1}^{\mathrm{I}} \mathrm{F}_\beta \int_\Omega \Phi_\beta(x_j)L\left(q^N(x_j, t)\right)d\tau \equiv 0$$

is that it admits rigorous *extremization* (!) as the *sole* distinguishing feature of the requirement $\forall f_\beta(x_j)$ now resides *solely* in the knot set F_β. Forming the first derivative of WF with respect to knot set F_β generates the sought *stationary coordinate* of WF in F_β space

$$\text{WS: } \partial \text{WF}/\partial \text{F}_\beta = \partial \left[\sum_{\beta=1}^{\text{I}} \text{F}_\beta \int_\Omega \Phi_\beta(x_j) L(q^N(x_j,t)) d\tau \right] / \partial \text{F}_\beta$$

$$= \int_\Omega \Phi_\beta(x_j) L(q^N(\Psi_\alpha(x_j) Q_\alpha(t))) \equiv \{0\}, \ \textit{for } 1 \le \beta \le \text{I}$$

which is now a *system* of equations, hence equated to a column matrix of zeroes.

This WF generated calculus expressed *weak statement* (WS) rigorously quantifies that but two decisions *remain* (!) to definition of *any* CFD algorithm, i.e.,

the *trial space* set $\Psi_\alpha(x_j)$, the mathematical support for the approximation, and

the interpolation polynomial set $\Phi_\beta(x_j)$, the *test space* in weak form parlance

Choosing function sets $\Psi_\alpha(x_j)$ and $\Phi_\beta(x_j)$ identical causes the WS approximation error $e^N(x_j,t)$ to be the *extremum* as it is rendered *orthogonal*, i.e., "mathematically perpendicular," to the WF trial space set $\Psi_\alpha(x_j)$. (Note: Figure B.1 in Appendix B visualizes this premier mathematical attribute.) This choice defines the *optimal* Galerkin criterion WS herein labeled GWS. Any other definition for $\Phi_\beta(x_j)$ generates a non-Galerkin WS, and insertion of artificial/numerical stabilization operators into an otherwise optimal GWS formulation renders the error no longer orthogonal to $\Psi_\alpha(x_j)$, hence defines a *non-optimal* WS.

5.4 Compressible NS TEA altered *O*(h^4) *error freed* CFD *mathematics* PDE system

The extension of Chapter 4 derived/validated $O(h^2)$ TE annihilation (TEA) theory to *n*-dimensional (*n*-D) compressible/incompressible Navier-Stokes PDE systems simply involves transition to tensor notation and an accounting for NS PDE state variable member coupling. For *n*-D steady NS model **DP** with flux vector $f(q,x) \Rightarrow f_j(q,x_i)$ and viscosity $\mu = \rho\nu$

$$\mathbf{DP}\text{: } f_j(q,x_i)\frac{\partial q}{\partial x_j} - \mu(x_i)\frac{\partial^2 q}{\partial x_j^2} = 0$$

for a uniform measure h Cartesian discretization of the NS PDE domain, *and* maintaining tridiagonal stencil admissibility, repeating the Chapter 4 detailed $n = 1$ TS:DD algebraic process predicts that the non-D $\text{D}^m\mathbf{P}$ for the *n*-D model NS PDE for large Re is, Kolesnikov [9]

$$\text{D}^m\mathbf{P}\text{: } f_j(q,x_i)\frac{\partial q}{\partial x_j} - \frac{1}{\text{Re}}\frac{\partial^2 q}{\partial x_j^2} - \frac{\text{Re}h^2}{12} f_k(q,x_i)\frac{\partial}{\partial x_k}\left[f_j(q,x_i)\frac{\partial q}{\partial x_j}\right] = 0 + O(h^4)$$

Thereby, *any* tridiagonal stencil equivalent CFD algorithm cast on this TEA theory altered NS model *error freed* $\mathrm{D}^m\mathbf{P}$ PDE will generate a Taylor series (TS) $O(h^4)$ accurate prediction on any discretization of n-D domains Ω. Altered to the context of weak formulation theory, the *asymptotic convergence* rate for $\mathrm{D}^m\mathbf{P}$ continuous Galerkin (GWS^h) FE $p = 1$ basis solutions is $O(h^4)$ under uniform and/or *regular* solution adapted nonuniform mesh refinement.

Inserting time-space discretization TS correlation quantified $\beta = h\mathrm{Re}/6$ for time TS β D^mM, the compressible non-D NS model $\mathrm{D}^m\mathbf{P}$ identified n-D TEA $O(h^4)$ PDE for state variable members density, momentum and internal energy $\{q(x_j, t)\} \equiv \{\rho, m_i, E\}^T$ is

$$\mathrm{D}^m\mathrm{M}{:}\ \frac{\partial \rho}{\partial t} + \frac{\partial}{\partial x_j}\left[m_j - \frac{h^2\mathrm{Re}}{12} u_k \widehat{u}_j \frac{\partial \rho}{\partial x_k}\right] = 0 + O(h^4)$$

$$\mathrm{D}^m\mathbf{P}{:}\ \frac{\partial m_i}{\partial t} + \frac{\partial}{\partial x_j}\left[\begin{array}{c} u_j m_i + \mathrm{Eu}\, p\delta_{ij} - \dfrac{\mu^*}{\mathrm{Re}}\left(\dfrac{\partial u_i}{\partial x_j} + \dfrac{\partial u_j}{\partial x_i} - \dfrac{\partial u_k}{\partial x_k}\left(1 + \dfrac{2}{3}\delta_{ij}\right)\right) \\ - \dfrac{h^2\mathrm{Re}}{12\mu^*}\left(u_k\left(u_j - \dfrac{\partial u_l}{\partial x_l}\delta_{j\underline{l}}\right)\dfrac{\partial m_i}{\partial x_k}\right) \end{array}\right] = 0 + O(h^4)$$

$$\mathrm{D}^m\mathrm{E}{:}\ \frac{\partial E}{\partial t} + \frac{\partial}{\partial x_j}\left[\begin{array}{c} u_j(E+p) - \dfrac{\kappa^*}{\mathrm{Re\,Pr}}\dfrac{\partial T}{\partial x_j} - \dfrac{\mu^* u_i}{\mathrm{Re}}\left(\dfrac{\partial u_i}{\partial x_j} + \dfrac{\partial u_j}{\partial x_i} - \dfrac{\partial u_k}{\partial x_k}\left(1 + \dfrac{2}{3}\delta_{ij}\right)\right) \\ - \dfrac{h^2\mathrm{RePr}}{12\kappa^*}\left(u_k\left(u_j - \dfrac{\partial u_l}{\partial x_l}\delta_{j\underline{l}}\right)\dfrac{\partial T}{\partial x_k}\right) \end{array}\right] = 0 + O(h^4)$$

The additional TEA terms in $\mathrm{D}^m\mathbf{P}$ and D^mE result from enforcing DM upon moving velocity vector u_j under the TEA divergence operator. The underbar denoted Kronecker deltas $\delta_{k\underline{l}}$ are tensor index indicators only (no summation). Parameters μ^*, κ^* denote the non-D $O(1)$ significance temperature dependent viscosity, thermal diffusivity distributions.

The mesh measure product h^2 embedded throughout this $O(h^4)$ PDE system is element-dependent data with definitions

... $h_e^2 \equiv 2^{2n/3}\mathrm{det}_e{}^{2/3}$ for FE $p = 1,2,3$ tensor product hexahedron, quadrilaterals

... $h_e^2 \equiv 2^{2n/3}/\mathrm{det}_e{}^{1/3} = 1/(n!)^{2/3}\mathrm{det}_c{}^{1/3}$ for FE $p = 1,2,3$ natural coordinate tetrahedron, triangles

wherein det_e is the determinant of the global-to-element local coordinate system transformation matrix Jacobian.

5.5 Weak formulation continuous Galerkin NS TEA $O(h^4)$ PDE FE basis algorithms

The weak formulation *optimal* continuous GWS algorithm results for function sets $\Psi_\alpha(x_j)$ and $\Phi_\beta(x_j)$ defined identical. The resultant Galerkin error freed CFD mathematics $O(h^4)$ PDE for NS state variable generic scalar member q

$$\mathrm{D}^m Q : L^m(q) = \int_\Omega \Psi_\beta(x_j) \left(\begin{array}{c} \dfrac{\partial q}{\partial t} + \dfrac{\partial u_j q}{\partial x_j} - s \\ - \dfrac{\partial}{\partial x_j} \left[\dfrac{1}{\mathrm{RePa}} \dfrac{\partial q}{\partial x_j} + \dfrac{h^2 \mathrm{RePa}}{12} u_j u_k \dfrac{\partial q}{\partial x_k} \right] \end{array} \right) = 0 + O(h^4) \, \forall \beta$$

embeds the member pertinent non-D parameter Pa. Imposition of the Green-Gauss divergence theorem then *weakens* the differentiability requirement of the trial space $\Psi_\alpha(x_j)$ function set by one order in generating the terminal weak formulation *continuous* GWS weak statement

$$\begin{aligned} \text{GWS:} & \int_\Omega \left[\Psi_\beta \left(\frac{\partial q^N}{\partial t} - s \right) - \frac{\partial \Psi_\beta}{\partial x_j} \left(u_j q^N - \frac{1}{\mathrm{RePa}} \frac{\partial q^N}{\partial x_j} - \frac{h^2 \mathrm{RePa}}{12} u_j u_k \frac{\partial q^N}{\partial x_k} \right) \right] \mathrm{d}\tau \\ & + \int_{\partial\Omega} \Psi_\beta \left(u_j q^N - \frac{1}{\mathrm{RePa}} \frac{\partial q^N}{\partial x_j} - \frac{h^2 \mathrm{RePa}}{12} u_j u_k \frac{\partial q^N}{\partial x_k} \right) \widehat{n}_j \, \mathrm{d}\sigma = 0 + O(h^4), \forall \beta \end{aligned}$$

The transition from continuum weak statement GWS to computable GWS^h accrues to definition of the continuum trial space $\Psi_\beta(x_j)$ as the *union* of finite element (FE) $p = 1,2,3$ completeness degree *trial space bases* $\{N(x_j)\}$, each a set of interpolation polynomials spanning each element Ω_e of domain discretization $\Omega \approx \Omega^h \equiv \cup \Omega_e$. Further, the knots of these FE trial space basis interpolation functions unequivocally define the geometric disposition of the discrete approximation *degrees-of-freedom* (DOF) $\{Q(t)\}_e$ at the nodes of FE p basis-dependent Ω_e and $\partial\Omega_e$.

Inserting this *analytical* transformation of GWS continuum approximation $q^N(x_j, t)$

$$q^N(x_j, t) \equiv \sum_{\alpha=1}^{N} \Psi_\alpha(x_j) Q_\alpha(t) \Rightarrow q^h(x_j, t) \equiv \cup_e \{N(x_j)\}^T \{Q(t)\}_e$$

into GWS generates the terminal spatially discrete *computable* Galerkin weak statement GWS^h

$$\text{GWS}^h\text{: } \mathrm{S}_e \left[\begin{array}{l} \displaystyle\int_{\Omega_e} \left[\{N\} \left(\frac{\partial q^h}{\partial t} - s \right) - \frac{\partial \{N\}}{\partial x_j} \left(u_j q^h - \frac{1}{\mathrm{RePa}} \frac{\partial q^h}{\partial x_j} - \frac{h^2 \mathrm{RePa}}{12} u_j u_k \frac{\partial q^h}{\partial x_k} \right) \right] \mathrm{d}\tau \\ \displaystyle + \int_{\partial\Omega_e \cap \partial\Omega} \{N\} \left(u_j q^h - \frac{1}{\mathrm{RePa}} \frac{\partial q^h}{\partial x_j} - \frac{h^2 \mathrm{RePa}}{12} u_j u_k \frac{\partial q^h}{\partial x_k} \right) \widehat{n}_j \mathrm{d}\sigma \end{array} \right]$$

Of precise significance to *error freed* CFD mathematics, vector differential calculus is maintained throughout the identification of GWS^h to facilitate *exact* integration of *all* nonlinearities. The subsequent matrix *assembly* (S_e) of GWS^h generated element matrices forms the global algebraic matrix statement for linear algebra iterative processing.

5.6 Weak formulation finite volume NS TEA altered $O(h^4)$ PDE algorithm

Current practice CFD codes that function on 1980s–1990s finite volume (FV) algorithms can be cast as a weak formulation by defining the trial space $\Psi_\alpha(x_j)$ as the set of all constants with the test space $\Phi_\beta(x_j)$ the set of all "ones." Inserting this definition into the WF for NS TEA altered $O(h^4)$ PDE for state variable scalar member $q(x_j, t)$ defines the continuous finite volume weak formulation WFFV as

$$\text{WFFV:} \int_\Omega L^m q^N(\Psi_\alpha, \Phi_\beta)\mathrm{d}\tau = \int_\Omega \left(\begin{array}{c} \dfrac{\partial q^N}{\partial t} + \dfrac{\partial u_j q^N}{\partial x_j} - s \\ -\dfrac{\partial}{\partial x_j}\left[\dfrac{1}{\text{RePa}}\dfrac{\partial q^N}{\partial x_j} + \dfrac{h^2\text{RePa}}{12} u_j u_k \dfrac{\partial q^N}{\partial x_k}\right] \end{array} \right) \mathrm{d}\tau = 0$$

Analytical imposition of the divergence theorem then generates the terminal continuous WFFV

$$\text{WFFV:} \int_\Omega \left(\frac{\partial q^N}{\partial t} - s\right)\mathrm{d}\tau + \oint_{\partial\Omega} \left(u_j q^N - \frac{1}{\text{RePa}}\frac{\partial q^N}{\partial x_j} - \frac{h^2\text{RePa}}{12} u_j u_k \frac{\partial q^N}{\partial x_k}\right)\widehat{n}_j \,\mathrm{d}\sigma = 0$$

That this WFFV remains a scalar is then resolved by forming a spatial discretization of NS domain Ω into the union of non-overlapping *finite volumes* $\Omega \approx \Omega^h \equiv \bigcup_{e=1}^{M} \Omega_e$ which generates the terminal *computable* spatially discrete finite volume FV^h algorithm equation *system*

$$\text{FV}^h: \int_\Omega L^m(q^h(x_j, t) \equiv \sum_{e=1}^{M} \int_{\Omega_e} \left(\frac{\partial q^h}{\partial t} - s^h\right)_e \mathrm{d}\tau$$

$$+ \sum_{e=1}^{M} \oint_{\partial\Omega_e} \left(u_j q^h - \frac{1}{\text{RePa}}\frac{\partial q^h}{\partial x_j} - \frac{h^2\text{RePa}}{12} u_j u_k \frac{\partial q^h}{\partial x_k}\right)_e \widehat{n}_j \mathrm{d}\sigma = 0$$

The operational requirement in the terminal discretized FV^h algorithm is to predict the normal *efflux* of the NS state variable, also the normal *efflux* of TEA theory nonlinear alterations, over *the totality* of surfaces bounding *each and every* finite volume cell Ω_e in Ω^h.

The obvious distinction between continuous GWS transition to GWS^h and continuous WFFV transition to FV^h is that the latter results for the trail space constituted of constants rather than the mathematically requisite continuum function set $\Psi_\alpha(x_j)$. This truly challenges rigorous local/global error estimation which accounts for the FV^h CFD community resorting to local TS order of accuracy estimates, as Jameson [8] clearly stated.

An added compute practice complication is that the geometric disposition of $q^h(x_j, t)$ discrete approximation DOF in $\{Q(t)\}_e$ on a finite volume cell has not been identified. The resolution by the FV^h CFD community is that algorithm designers select DOF coordinates for $\{Q(t)\}_e$ to enable difference algebra generated approximate translations of Ω_e volume and $\partial\Omega_e$ surface normal efflux integrations to algebraic. The original FV^h DOF resolution utilized a "staggered mesh" in which algorithm domain Ω_e integral DOF were located at the cell centroid with the surface integral efflux DOF located at the vertex nodes of $\partial\Omega_e$. The successor flux vector/difference split predictor-corrector numerically stabilized FV^h algorithms define the DOF distributions on Ω_e and $\partial\Omega_e$ with principle focus algorithm algebraic stability.

To summarize, these precisely parallel weak formulation continuous CFD algorithm derivations clearly delineate the key mathematical distinctions between GWS^h calculus and FV^h difference algebra translation of NS PDE differential calculus functions to algebraic. The precision of GWS^h intrinsic vector calculus operations is clearly preferable to the "*external* to the code" difference algebra generated discrete cell efflux stabilized shock capturing schemes that ultimately become *the* FV^h code.

5.7 Theory synopsis, weak formulation *error freed* CFD, asymptotic convergence, *optimal* continuous GWS^h FE $p = 1,2,3$ trial space basis algorithms

The mid 50s creation of structural *finite element analysis* (FEA) sought prediction of the displacement distribution for linear elastic deformable continua under various loadings. These FEA algorithms were cast on the continuum mechanics energy conservation principle matrix differential statement, a symmetric quadratic *bilinear form* that contained *only* first-order derivatives, rather than on its calculus of variations predicted *extremum*, a second-order linear EBV PDE system.

When FE trial space basis approximated, the arbitrariness in thus altered energy principle matrix statement resided *solely* in the displacement DOF pre- and post-matrix products of the FEA approximated bilinear form. DOF matrix differential calculus then generated the stationary coordinate of this quadratic form in "DOF space." Solution of the resultant global algebraic matrix statement yielded prediction of the displacement DOF distribution of the structure.

The nonlinear convection term permeating NS (and RaNS) second-order EBV PDE systems precludes direct extension of this energy principle strategy. The alternative is the now *fully detailed* weak formulation continuous Galerkin (GWS^h) NS state variable approximation $\{q^h(x_j)\}$. As with the FEA structures theory, this approximation is required to exist *only* in H^1, the space of all functions with all *first*-order tensor product derivatives square integrable and finite, in clear distinction to the second-order derivative requirement of NS PDE systems.

Thereby, rigorous theorization for the weak formulation continuous Galerkin CFD statement is posed on an abstract bilinear representation of NS compressible DM altered to its full potential *nonlinear* EBV PDE. The mathematical entities involved in this NS CFD theorization are

$$\text{DM:} -\frac{\partial}{\partial x_j}\left(\rho\frac{\partial\varphi}{\partial x_j}\right)=0,\quad \rho=\rho_\infty\left[\frac{T_{0\infty}}{T_\infty}-\frac{\gamma-1}{2\gamma T_\infty}\nabla\varphi\cdot\nabla\varphi\right]^{1/(\gamma-1)}$$

$$\|\varphi,\varphi\|_E\equiv\frac{1}{2}\int_\Omega[\rho\nabla\varphi\cdot\nabla\varphi]\mathrm{d}\tau$$

The mathematically rigorous, hence to an engineer *quite tedious* (!) derivation is fully detailed in Appendix B of this monograph. To summarize, the theory predicts that the *rates* of asymptotic convergence for weak formulation continuous Galerkin FE p-basis implemented algorithm predictions of the NS *nonlinear* potential flow DM PDE in its intrinsic *error* energy norm $\|\varphi,\varphi\|_E$ are equal to 2, 4, and 6 for FE $p=1,2,3$ completeness degree basis GWS^h implementations assuming that the *a priori* given data are *smooth*

$$\text{GWS}^h\text{: }\|e^h\|_E\le C\,h^{2\alpha}\left(\|\text{data}\|_{L2,\Omega},\|\text{data}\|_{L2,\partial\Omega}\right),\ \alpha\equiv\min(p,r-1)$$

Note that the *magnitude* of the DM PDE *error* energy norm is bounded by the L2 norms of these ***data*** distributions on the domain Ω and on its enclosure ∂Ω. The integer r is the measure of data *non-smoothness* which if sufficiently large, i.e., $r-1>p$, will invalidate the error estimate. Examples of the validity of these error estimate features are documented in Appendix B.

Rigorous theory derived convergence exponent $\alpha\equiv\min(p, r-1)$ must then be altered to $\alpha\equiv\min(2,2,3, r-1)$ for *unsteady* NS (and RaNS) TEA altered $O(h^4)$ PDE *system* GWS^h + θTS FE $p=1,2,3$ basis algorithms. This recognizes that the TEA theory altered FE $p=1$ basis tridiagonal stencil has been independently validated to be TS $O(h^4)$ accurate. The energy norm definitions for the compressible and incompressible NS/RaNS **DP** are

$$\text{NS: }\|u_i(t)\|_E\equiv\frac{1}{2\text{Re}}\int_\Omega\frac{\partial u_i(t)}{\partial x_j}\frac{\partial u_i(t)}{\partial x_j}\mathrm{d}\tau,1\le i\le n$$

$$\text{RaNS: }\|u_i(t)\|_E\equiv\frac{1}{2\text{Re}}\int_\Omega\frac{\partial u_i(t)}{\partial x_j}\left[\left(1+\text{Re}^T\right)\frac{\partial u_i(t)}{\partial x_j}\right]\mathrm{d}\tau,1\le i\le n.$$

As solution adapted nonuniform mesh M_n refinement is becoming requisite CFD code practice, the *validated* error *quantification* for the factor of two *uniform* mesh refinement sequence for NS/RaNS TEA $O(h^4)$ PDE GWS^h FE $p=1,2,3$ degree basis algorithms

$$\|e^{h/2}(t)\|_E=\left(\|q^{h/2}(t)\|_E-\|q^h(t)\|_E\right)/\left(2^{f(p)}-1\right),f(p=1,2,3)=4,4,6$$

is required altered to the *regular* solution adapted nonuniform mesh refinement protocol. The *regularity* requirement, Appendix B fully detailed, requires that a nonuniform solution adapted domain $\Omega\cup\partial\Omega$ discretization sequence with measure distributions $h_n(x_i)$ for n, $n+1$ meshes M_n correlates in essence with the number of elements Ω_e populating Ω^h for each mesh M_n. For this premise, *regular* solution adapted nonuniform mesh refinement sequence *error quantification* is code "on-the-fly" execution predictable in NS/RaNS PDE state variable member specific PDE *error* energy norm set $\|\{e^h(t)\}\|_E$

$$\|\{e^h(t)\}\|_{E,\mathrm{M_{n+1}}} = \frac{\|\{q^h(x_j,t)\}\|_{E,\mathrm{M_{n+1}}} - \|\{q^h(x_j,t)\}\|_{E,\mathrm{M_n}}}{(\mathrm{M_{n+1}}/\mathrm{M_n})^{f(p)} - 1}, \; f(p=1,2,3) = 4,4,6$$

for $\mathrm{M}_{n+1}/\mathrm{M}_n$ the ratio of the number of elements populating $\Omega^h \equiv \cup\,\Omega_e$ for each n, $n+1$ in the refinement sequence.

5.8 Validation, shock continuum interpolation, monotonicity, stability, weak formulation compressible NS TEA $O(h^4)$ *error freed* PDE GWS^h FE $p=1$ basis algorithm

Validation of continuous weak formulation compressible NS CFD theory addresses the first AIAA 2024 High Fidelity CFD Workshop manufactured statement of inviscid, non-heat conducting air onset to an isolated surface with the Gaussian bump $y = 0.0625\exp(25x^2)$ centered at $x = 0$ (Figure 5.2a) [5]. Designed specifically for Godunov genre mesh shock fitting Euler PDE algorithm validation, alteration to compressible NS continuum shock interpolation simply involves replacement of the Gaussian bump surface NS no-slip BC with the Euler slip wall homogeneous Neumann BC. The onset Dirichlet BC specifications correlate with $\mathrm{Ma}_\infty \equiv 0.7$ with homogeneous Neumann BCs imposed everywhere else.

The user interface of the COMSOL commercial CFD code (www.comsol.com) enables precise specification of the weak formulation compressible NS TEA $O(h^4)$ PDE continuous GWS^h + θTS FE $p = 1$ basis algorithm. The code also organizes execution of a *regular* solution adapted nonuniform mesh refinement sequence for the Gaussian bump statement. Its geometry with inaugural $\mathrm{M}_1 = 4866$ FE $p = 1$ basis uniform mesh is compared side-by-side with the terminal solution-adapted nonuniform $\mathrm{M}_{13} = 101{,}474$ element mesh in Figure 5.2a. The iterative convergence history of the COMSOL code state variable segregated Jacobian for specification $|\{\delta QI\}^{\mathrm{p}}|_{\max} \le$ E-04 for this refinement sequence is semi-log scaled in Figure 5.2b.

Compressible NS GWS^h steady FE $p = 1$ basis state variable distributions are validated *monotone* to code *realized* iterative convergence $\varepsilon \le$ 2E-03 for each $\mathrm{M_n}$ in sequence $\mathrm{M}_1 \rightarrow \mathrm{M}_{13}$. The Mach number distribution perspective projection superposed over mesh M_{13} (Figure 5.3a) *validates* that shock *continuum* interpolation by the continuous GWS^h algorithm is *monotone* totally *absent* artificial/numerical stabilization. Mesh M_{13} temperature and pressure distribution perspective projections (Figure 5.3b and c) validate NS state variable scalar member resolutions are also monotone to $\varepsilon \le$ 2E-03 significance. Shock centric mesh M_{13} zoom is Figure 5.3d.

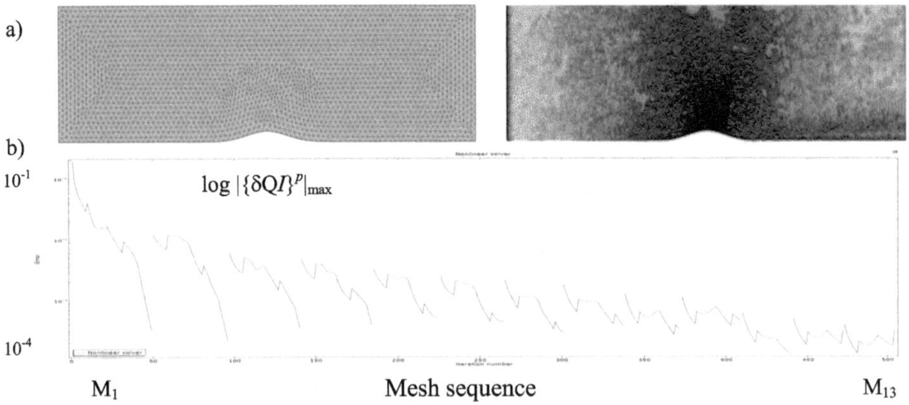

Figure 5.2: AIAA 2024 CFD Workshop Gaussian bump manufactured statement, *regular* solution adapted nonuniform mesh refinement sequence M_n, $1 \le n \le 13$, compressible NS TEA $O(h^4)$ PDE GWS^h + θTS FE $p = 1$ basis algorithm/code, $Ma_\infty = 0.7$, Re = 3400: (a) mesh M_1, M_{13} comparison, and (b) semi-log scaled iteration $|\{\delta QI\}^p|_{max}$ history for meshes $M_1 \to M_{13}$.

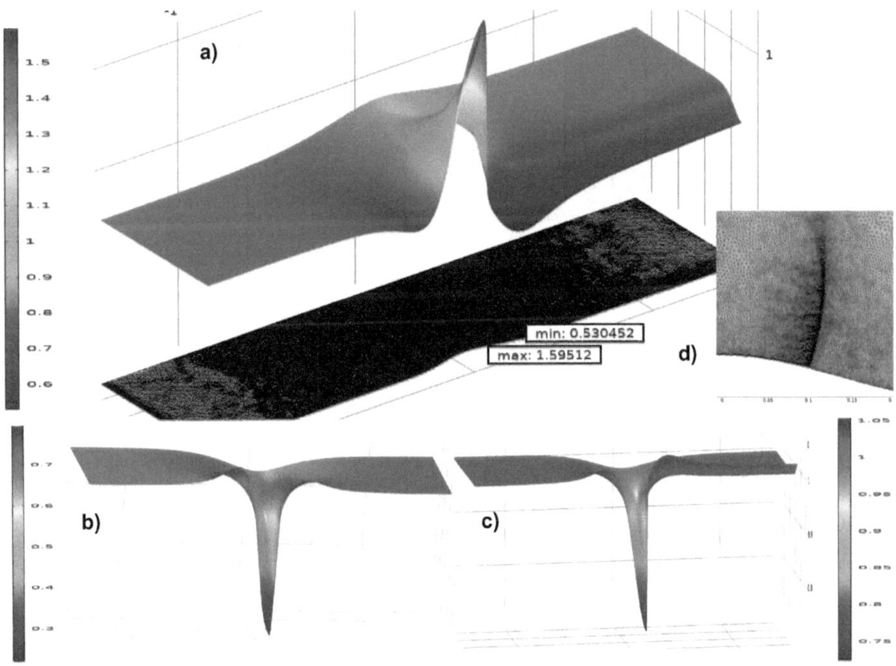

Figure 5.3: Validation, compressible NS *monotone* distribution perspective projections, Gaussian bump, NS TEA $O(h^4)$ PDE GWS^h *regular* nonuniform refinement sequence, mesh M_{13}, $Ma_\infty = 0.7$, Re = 3400: (a) {Ma}, (b) pressure, (c) temperature, and (d) shock centric mesh M_{13} zoom.

The mandated Gaussian bump error indicator is convergence of the stagnation enthalpy *deficit* norm $||H_{0\infty} - H_0(M_n)||_{L2}$. The mesh M_n sequence convergence history is graphed in Figure 5.4a). The perspective of the terminal mesh M_{13} distribution, Figure 5.4b), quantifies it a uniform significance plateau with a miniscule thickness extremum "spike" at the shock. Mesh M_{13} $H_0{}^h$ deficit *error* norm distribution within this "spike" is clearly detailed as the $O(Re^{-1})$ 0.0003 non-D relatively thick monotone shock (Figure 5.4c). As the COMSOL code user interface did not admit insertion of TEA cubic nonlinearities into the linear algebra Jacobian, Re = 3400 was the maximum definition that preserved code stability for the entire $1 \le n \le 13$ M_n sequence.

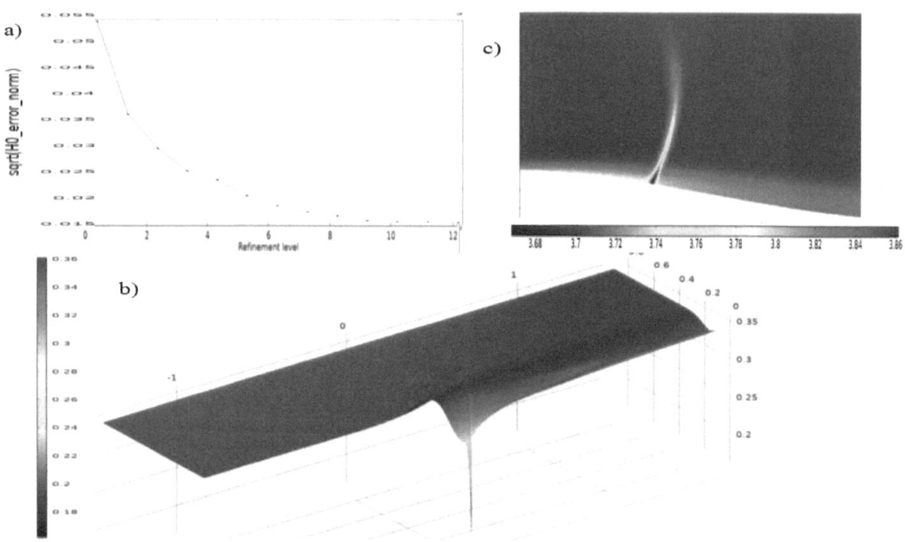

Figure 5.4: Euler stagnation enthalpy *deficit error* L2 norm $||H_{0\infty} - H_0(M_n)||_{L2}$, transonic bump, $Ma_\infty = 0.7$, Re = 3400, compressible NS TEA $O(h^4)$ PDE GWS^h: (a) *regular* nonuniform mesh refinement convergence history, (b) perspective of mesh M_{13} $H_0{}^h$ norm extremum in uniform significance background, and (c) mesh M_{13} $H_0{}^h$ norm distribution shock centric zoom.

5.9 Validation, weak formulation theory, asymptotic convergence, compressible NS TEA-altered $O(h^4)$ *error freed* PDE GWS FE p = 1,2,3 basis algorithms

This *regular* solution adapted nonuniform mesh refinement sequence was re-executed for NS TEA $O(h^4, \Delta t^3)$ PDE GWS^h algorithm FE p = 1,2,3 basis implementations. Each sequence was terminated at mesh M_{17} which were constituted of 101,474, 145,999, 180,397 triangular Ω_e, respectively. Mesh M_{17} Mach number $Ma^h(\mathbf{x}, T)$ distribution global perspective projections are strictly *monotone* for FE p = 1,2,3 bases implementations and vi-

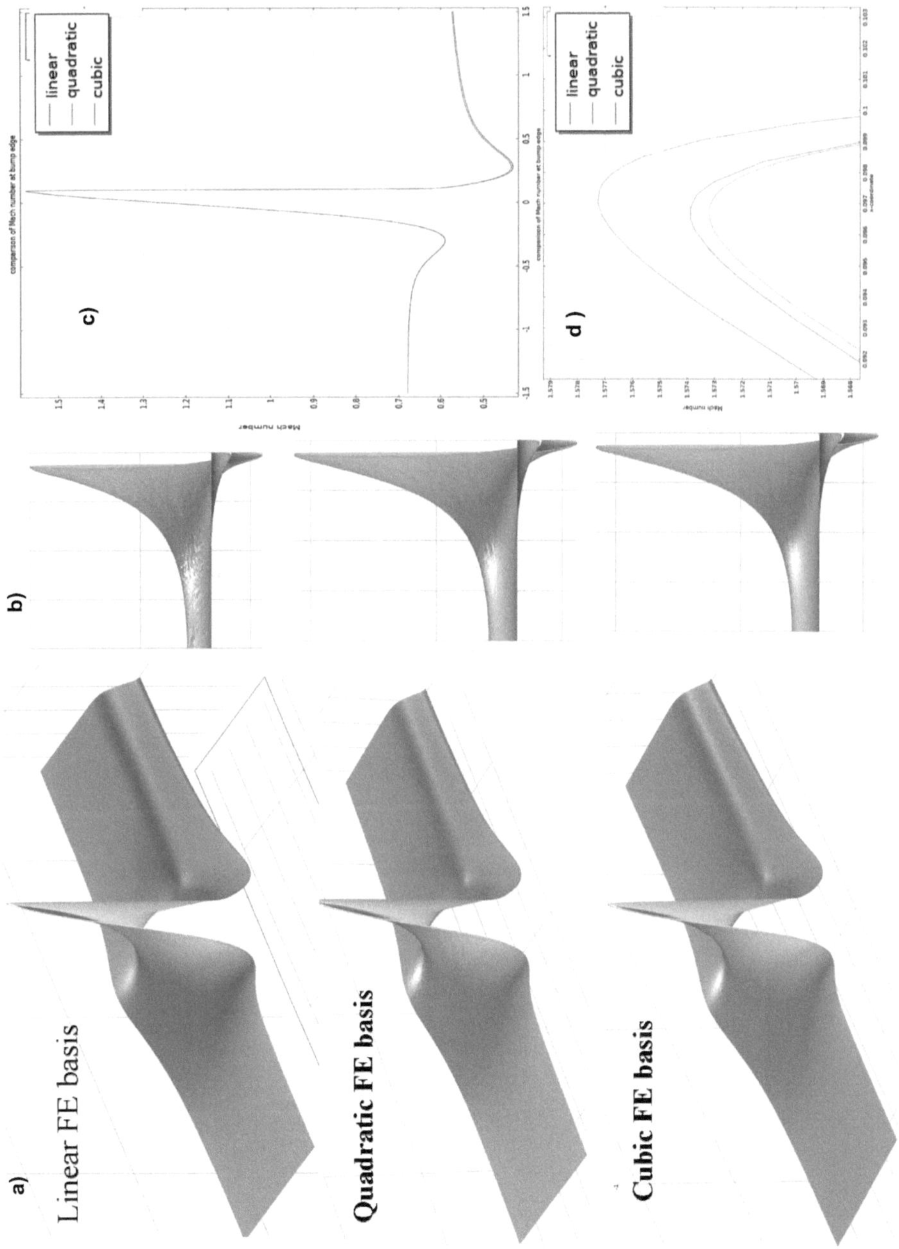

Figure 5.5: Validation, $Ma^h(\mathbf{x}, T)$ monotone distributions, compressible NS TEA $O(h^4)$ PDE GWS^h FE $p = 1,2,3$ basis algorithm, transonic Gaussian bump, mesh M_{17}, $Ma_\infty = 0.7$, Re = 3400: a) global perspectives, b) shock local edge-on perspectives, c) bump surface node $\{Ma(\mathbf{x}_S, T)\}$ DOF interpolation, d) shock centric $\{Ma(\mathbf{x}, T)_{max}$ DOF interpolation super zoom.

sually indistinguishable (Figure 5.5a). Shock edge-on $Ma^h(\mathbf{x}, T)$ distribution perspectives (Figure 5.5b) *validate* GWS^h shock *continuum interpolation* resolution is an FE p basis indistinguishable vertical surface.

Piecewise linear interpolation of FE $p = 1,2,3$ basis $\{Ma(\mathbf{x},T)\}$ DOF distributions over the complete span of the Gaussian bump surface are *validated* monotone, Figure 5.5c). These data *exactly* overly except for FE p-basis dependence at the shock with the DOF *extrema* $Ma_{max} = 1.587966, 1.58780, 1.59034$. The significant digit of distinction in these Ma_{max} correlate precisely with code *realized* iterative convergence $\varepsilon \leq 2E\text{-}03$. Super zoom comparisons of FE p-dependent $\{Ma(\mathbf{x},T)\}$ DOF interpolations in the reach of $\{Ma(x_j,T)\}_{max}$ (Figure 5.5d) are validated monotone with each possessing a *smoothly* turning tangent locus.

Compressible NS $O(h^4)$ D^mE state variable $E^h(T)$ *error* energy norm evaluation at each GWS^h FE $p = 1,2,3$ basis mesh M_n converged solution in this *regular* solution adapted nonuniform refinement sequence is dual log axis graphed in Figure 5.6. As the NS TEA $O(h^4)$ D^mE FE $p = 1$ basis solution is independently *validated* Taylor series (TS) $O(h^4)$ accurate, straight line interpolation of the FE $p = 1,2$ bases mid-range mesh M_n data are validated to possess *identical* $O(h^4)$ asymptotic convergence rates in D^mE *error*

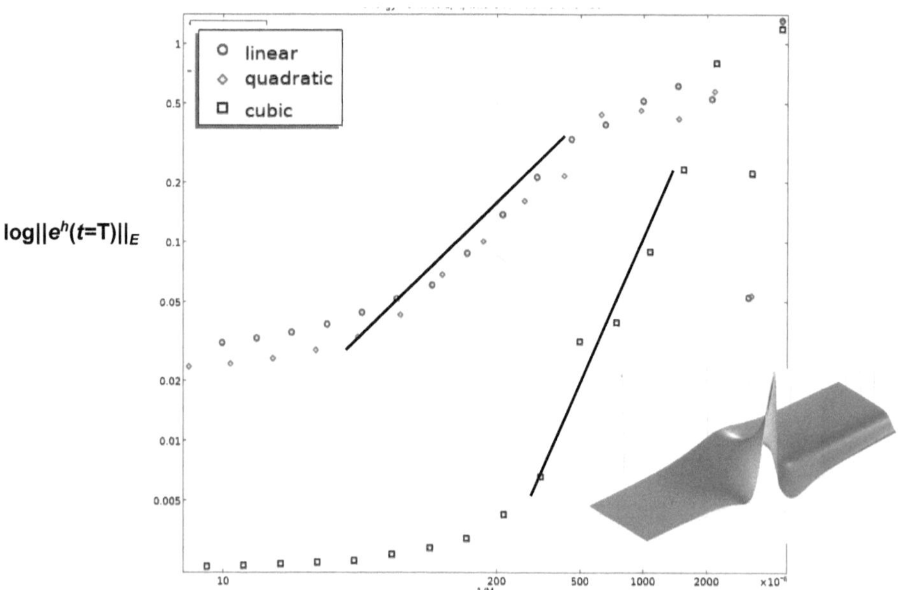

Figure 5.6: Validation, NS weak formulation continuous Galerkin theory, asymptotic convergence, error quantification in D^mE *error* energy norm $\|e^h(E^h(t=T)\|_{E,M_{n+1}}$, compressible NS TEA $O(h^4)$ PDE GWS^h FE $p = 1,2,3$ basis algorithms, *regular* solution adapted nonuniform mesh M_n refinement sequence $0 \leq n \leq 17 = N$, transonic Gaussian bump, $Ma_\infty = 0.7$, $Re = 3400$.

energy norm $\|e^h(E^h(t=\mathrm{T})\|_{E,\mathrm{M}_{n+1}}$ as theory predicted. The asymptotic convergence rate for the GWS^h FE $p = 3$ basis mid-range $\mathrm{M_n}$ error energy norm $\|e^h(E^h(t=\mathrm{T})\|_{E,\mathrm{M}_{n+1}}$ is $O(h^6)$.

5.10 Validation, weak formulation theory, error quantification, compressible NS TEA altered $O(h^4)$ *error freed* PDE GWS^h FE $p = 1,2,3$ basis algorithms

The data distributions in Figure 5.6 clearly quantify that the finer mesh $\mathrm{M_n}$ code executions return very little value for the expenditure of effort. Factually, the mid-range mesh $\mathrm{M_n}$ data in Figure 5.6 not only *validate* weak formulation GWS^h asymptotic convergence theory, but also enable compressible NS TEA $O(h^4)$ PDE GWS^h FE $p = 1,2,3$ basis *error quantification* via the *regular* solution adapted nonuniform mesh refinement sequence.

Weak formulation theory global error quantifications extracted from these FE p-basis dependent data include:

. . . NS D^{m}E *error* energy norm $\|e^h(E^h(t=\mathrm{T})\|_{E,\mathrm{M}_{n+1}}$ $O(h^4, h^4, h^6)$ convergence rates degrade to linear for meshes M_{n+1} past midrange n independent of basis degree

. . . mid-range mesh $5 \le n \le 8$ M_n FE $p = 3$ *error* norm $\|e^h(E^h(t=\mathrm{T})\|_{E,\mathrm{M}_{n+1}}$ is *two orders* smaller than for FE $p = 1,2$ basis predictions

. . . terminal mesh M_{17} FE $p = 3$ basis energy norm error $\|e^h(E^h(t=\mathrm{T})\|_{E,\mathrm{M}_{n+1}}$ is *one order* less significant than those for the FE $p = 1,2$ basis solutions.

To conclude, *regular* solution adapted nonuniform mesh refinement data quantitatively *validate* that weak formulation compressible NS TEA $O(h^4)$ PDE GWS^h FE $p = 1,2,3$ basis algorithms written on *error freed* CFD mathematics generate *continuum monotone interpolation* of a $\mathrm{Ma_S} = 1.59 \pm$ transonic shock, as does Mother Nature (Figure 5.1). This absolutely fundamental VVUQ topics compliant NS CFD algorithm/code unstagnation is the mathematically rigorous and thereby totally ***distinct alternative*** to NS CFD shock “capture” via *phake physics* stabilizations *and/or* Euler CFD shock-mesh realignment with Rankine-Hugoneou shock jump imposition. The continuing *need for* these Godunov genre stabilized shock capture and/or discontinuous Galerkin (DG) shock fitting CFD algorithms is thereby open to serious question.

5.11 Validation, weak formulation theory, optimal mesh identification, compressible NS TEA $O(h^4)$ *error freed* PDE GWS^h FE $p = 1,2,3$ basis algorithms

Execution of *regular* solution adapted *nonuniform* mesh refinement sequences also enables mathematically *rigorous* identification of the minimal DOF *optimal* mesh $\mathrm{M_n}$ supported GWS^h NS solution. The original FEA linear structures theory predicts this to occur when the magnitude of mesh $\mathrm{M_n}$ *element error* energy norm $\|e_e^h(t)\|_{E,\Omega_e}$ distributions transition to *uniform* over $\Omega^h \equiv \cup\Omega_e$, Babuska and Rheinboldt [3].

For $\mathrm{Ma}_\infty = 0.7$ and Re = 3400, the Ω_e distributions of GWS^h FE $p = 1,2,3$ basis D^{m}E *error* energy norm $\|e^h(E^h(t=\mathrm{T})\|_{E,\mathrm{M_{n+1}}}$ are column compared in Figure 5.7 for the regular mesh $\mathrm{M_n}$ refinement solution adapted converged solutions for n = 0,5,8,11,17. Each mesh $\mathrm{M_{n=0}}$ graphic as well details the DOF nodal disposition for n = 2 FE $p = 1,2,3$ natural coordinates bases.

The data presentation in Figure 5.7 quantifies that compressible NS GWS^h D^{m}E state variable member *error* energy norm $\|e_e^h(t=T)\|_{E,M_n,\Omega_e}$ distributions on $\Omega^h \equiv \cup\Omega_e$ that contain a *continuum* interpolated shock will *never* (!) possess uniform significance. Instead, maturation of the shock resolution interpolation under *regular* nonuniform solution adapted mesh refinement sequence $\mathrm{M_n}$ morphs to a sharply defined *extremum* significance (red) curved "line" that dominates all the finer mesh $\mathrm{M_n}$ data in Figure 5.7.

With this established understanding, the significance of continuous GWS^h FE $p = 1,2,3$ basis D^{m}E *error* energy norm $\|e^h(E^h(t=T)\|_{E,\mathrm{M}_{n+1}}$ element distributions that become nominally uniform on $\Omega^h \equiv \cup\Omega_e$ (Figure 5.7) are those for mesh $\mathrm{M_n}$ at $n = 11,8,5$ respectively. The element error norm $\|e^h(E^h(t=T)\|_{E,\mathrm{M}_{n+1}}$ distributions for terminal mesh $\mathrm{M_{17}}$ are FE p-basis visually indistinguishable, as are the p-basis steady $\mathrm{Ma}^h(x_j,\mathrm{T})$ distributions in Figure 5.5.

As the global error norm $\|e^h(t)\|_E$ is but the summation of $\|e_e^h(t)\|_{E,\Omega_e}$ over the mesh, the global approximation verity $\|q(t)\|_E - \|q^h(t)\|_E \equiv \|e^h(t)\|_E$ quantifies that the accuracy of this minimal DOF *optimal* mesh M solution is the *extremum* among all prior mesh M_n predictions. Thereafter, the continuum verity $q(x_j,t) - q^h(x_j,t) \equiv e^h(x_j,t)$ quantifies that this minimal DOF *optimal* mesh $\mathrm{M_n}$ prediction $q^h(x_j,t)$ lies as close as compute-efficient *practical* to the NS exact solution $q(x_j,t)$.

Then, for the significance of global error norm $\|e^h(t)\|_E$ sufficiently small, these optimal mesh M_n NS $q^h(x_j,t)$ GWS^h generated DOF distributions will correspond *one to one* with the knots of FE $p = 1,2,3$ order-consistent piecewise continuous interpolation of the NS exact solution. This was first asserted for the $n = 1$ Riemann shock tube compressible NS uniform $\mathrm{M} = 10^4$ mesh GWS^h solution at Re = 2E05, Chapter 4.7.

Weak formulation theory conclusions drawn from Figures 5.5–5.7 data quantify:

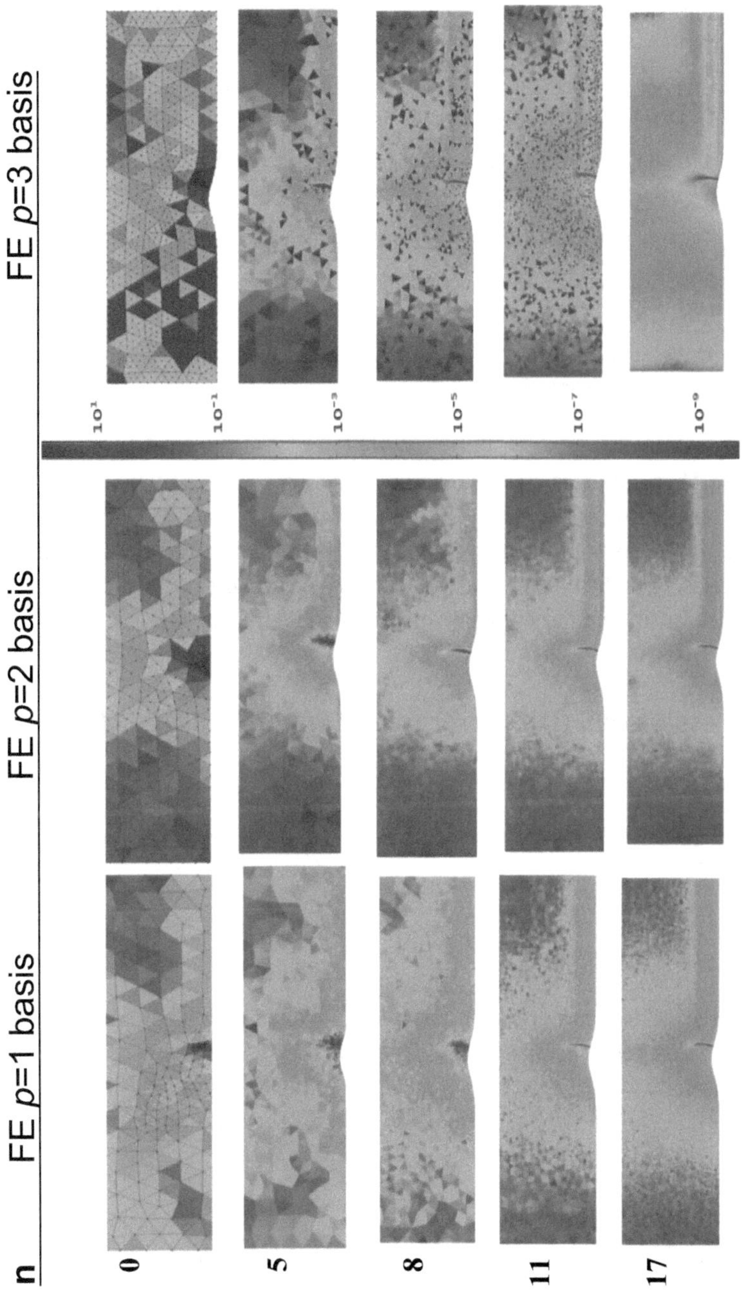

Figure 5.7: Validation, minimal DOF *optimal* mesh M_n solution theory, compressible NS TEA $O(h^4, \Delta t^3)$ PDE GWS^h FE $p = 1,2,3$ basis D^mE state variable Ω_e *error* energy norm $\|e^h(E^h(t = T)\|_{E, M_{n+1}}$ distributions, n = 0,5,8,11,17 meshes M_n, transonic Gaussian bump, $Ma_\infty = 0.7$, Re = 3,400.

FE $p = 3$ basis predictions exhibit overall *superior* accuracy and asymptotic convergence rate for any reasonable mesh M_n.

That FE $p = 1,2$ basis predictions exhibit identical convergence rates confirms that *no need* exists for a FE $p = 2$ basis NS GWS^h implementation.

Error norm $\|e^h(E^h(t = T)\|_{E, M_{n+1}}$ significance equi-distribution distant from a shock is *intrinsic* to the *regular* solution adapted nonuniform mesh refinement protocol.

Mesh M_5 FE $p = 3$ basis error norm significance equals that for FE $p = 1,2$ bases on M_{11}

Mesh M_8 FE $p = 3$ basis error norm is one order smaller than those for FE $p = 1,2$ bases on the terminal mesh M_{17}.

NS GWS^h FE $p = 1,2,3$ generated $\|e^h(E^h(t = T)\|_{E, M_{n+1}}$ and Ma$^h(x_j$,T) distributions are for all intents indistinguishable on terminal mesh M_{17}.

To conclude, compressible NS TEA $O(h^4)$ PDE GWS^h FE $p = 1,2,3$ basis theorized minimal DOF *optimal* mesh M_n solution is code *on the fly* execution identified when the mesh M_{n+1} global *error* energy norm $\|\{e^h(t = T)\}\|_{E, M_{n+1}}$ *departs* from validated weak formulation theory $O(h^4, h^4, h^6)$ asymptotic convergence rates. This code execution identification correlates precisely with the element *error* energy norm $\|e_e^h(t = T)\|_{E, M_n, \Omega_e}$ distribution significance transitioning to nominally uniform on $\Omega^h \equiv \cup \Omega_e$.

5.12 Incompressible NS (INS) *error freed* TEA altered $O(h^4)$ PDE system

The $O(h^2)$ TE annihilation (TEA) theory derivation for n-D Navier-Stokes D^m**P** for NS n-D model D**P** remains

$$\mathrm{D}^m\mathbf{P}\text{: } f_j(q, x_i)\frac{\partial q}{\partial x_j} - \frac{1}{\mathrm{Re}}\frac{\partial^2 q}{\partial x_j^2} - \frac{\mathrm{Re}h^2}{12} f_k(q, x_i)\frac{\partial}{\partial x_k}\left[f_j(q, x_i)\frac{\partial q}{\partial x_j}\right] = 0 + O(h^4)$$

assuming Re is large. That incompressible fluid velocity distributions are divergence-free admits moving flux vector $f_k(q, x_i) = u_k(x_i, t)$ inside the TEA term divergence operator. Resultant unsteady TEA altered non-D incompressible NS $O(h^4)$ D^m**P** for velocity vector $q(x_j, t) \Rightarrow \{u_i(x_j, t\}$ and for the Boussinesq buoyancy assumption, Gebhart et al. [6], is

$$\mathrm{D}^m\mathbf{P}\text{: } \frac{\partial u_i}{\partial t} + \frac{\partial}{\partial x_j}\begin{bmatrix} u_j u_i + P^*|_{n+1}^{p}\delta_{ij} \\ -\frac{1}{\mathrm{Re}}\left(\frac{\partial u_i}{\partial x_j} + \frac{\partial u_j}{\partial x_i}\right) - \frac{h^2\mathrm{Re}}{12} u_j u_k \frac{\partial u_i}{\partial x_k}\end{bmatrix} - \frac{\mathrm{Gr}}{\mathrm{Re}^2}\Theta\widehat{g}_i = 0 + O(h^4)$$

wherein P^* is the pressure kinetic action variable with Θ the potential temperature and Gr the non-D Grashof number.

An incompressible NS PDE system state variable usually includes a scalar member characterized by a linear Poisson PDE, e.g., pressure, streamfunction, velocity-vorticity formulation. For the *n*-D scalar model incompressible NS Poisson PDE

$$\text{DS:} -\nabla^2 q - s(q_{NS}) = 0$$

the source term NS state variable dependence is denoted q_{NS}. Analytical $O(h^2)$ TEA theory readily identifies its $O(h^4)$ alteration to be [9]

$$\text{D}^m\text{S:} -\nabla^2 q - s(q_{NS}) - \frac{h^2}{12}\nabla^2 s(q_{NS}) = 0 + O(h^4)$$

As with any TEA altered NS PDE system, under *regular* solution adapted nonuniform mesh refinement the asymptotic convergence rate for this D^mS PDE continuous GWS^h FE $p = 1$ basis algorithm is $O(h^4)$ on *any* mesh.

5.13 Validation, weak formulation theory, INS TEA altered $O(h^4)$ PDE GWS^h FE $p = 1$ basis algorithm, driven cavity monotone DOF

The incompressible NS vector field theory identified state variable alteration to vorticity and streamfunction $\{q(x_j, t)\} \equiv \{\omega, \psi\}^T$ rigorously satisfies the DM divergence-free velocity constraint and removes the appearance of P^* in $\text{D}^m\mathbf{P}$. Resultant incompressible isothermal NS TEA $O(h^4)$ $\text{D}^m\mathbf{P}$:D^mPSI $O(h^4)$ PDE system for large Re is

$$\text{D}^m\mathbf{P}: \quad \frac{\partial \omega}{\partial t} + \nabla \times \widehat{\psi \mathbf{k}} \cdot \nabla \omega - \frac{1}{\text{Re}}\nabla^2 \omega - \frac{\text{Re}h^2}{12}\frac{\partial}{\partial x_j}\left[u_j u_k \frac{\partial \omega}{\partial x_k}\right] = 0 + O(h^4)$$

$$\text{D}^m\text{PSI:} -\nabla^2 \psi - \omega - \frac{h^2}{12}\nabla^2 \omega = 0 + O(h^4)$$

The 1970s FD CFD community devised the driven cavity *manufactured statement,* a rectangular Cartesian approximation to a rotating shaft journal bearing, to assess algorithm performance. This statement uniquely supports validation of NS PDE TEA theory $O(h^2)$ TE *annihilation.* The BC definition for the driven cavity statement excludes through flow at the intersection of the sliding lid with corners of the cavity vertical walls. This omission renders the vorticity *singular* at each of these corner nodes. This was a non-issue for the 70s era NS FD algorithms as the standard $O(h^2)$ diffusion term stencil did not contain the corner node vorticity DOF. The FD driven cavity archive is Ghia et al. [7].

Convergence to steady state of the textbook NS $O(h^2)$ **DP**:DPSI PDE GWS^h FE $p = 1$ basis algorithm with lid-corner BC induced $\omega^h(x_j, \text{n}\Delta t)$ DOF singularities was secured using a fully coupled quadratic convergent Newton iteration Jacobian. For Re = 2000,

NS cavity flowfield $O(h^2)$ TE dispersion error is manifested as 2Δx period oscillations in the $\omega^h(x_j, \mathrm{n}\Delta t)$ DOF distribution adjacent to the sliding. This is clearly displayed in the $\omega^h(x_j, \mathrm{T})$ distribution perspective projection with mesh overlay, Figure 5.8a), lid-corner extrema truncated for visualization. In stark contrast, the Re = 2,000 NS TEA $O(h^4)$ D^m**P**:D^{m}PSI GWS^h $\omega^h(x_j, \mathrm{T})$ DOF distribution is *monotone* to iterative convergence digit $\varepsilon \leq$ E-04 (Figure 5.8b), lid-corner extrema again truncated.

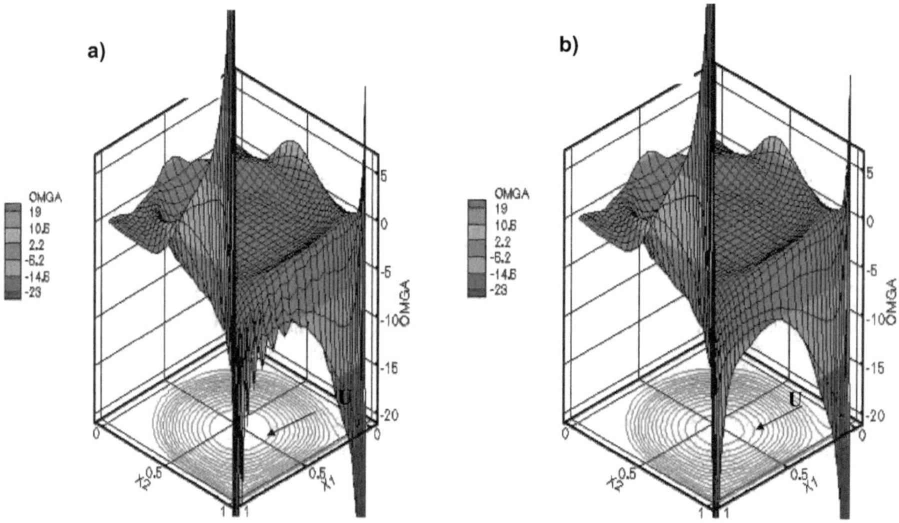

Figure 5.8: FD driven cavity manufactured statement, GWS^h FE $p = 1$ basis algorithm steady $\omega^h(x_j, T)$ DOF distribution perspectives, Re = 2000, modestly solution adapted M = 32^2 mesh: a) for NS $O(h^2)$ D**P**:DPSI PDE system, b) for NS $O(h^4)$ D^m**P**:D^{m}PSI PDE system.

Annihilation of domain discretization generated $O(h^2)$ dispersion error is thus *validated* for an incompressible NS TEA altered PDE system with singular BC specifications. The uniform mesh refinement sequence for the D^m**P**:D^{m}PSI PDE GWS^h FE $p = 1$ basis algorithm at Re = 10 quantified the asymptotic convergence rate in DPSI energy norm $\|\psi^h\|_E$ was $O(h^2)$, [9], as should occur upon omission of the Re-independent TEA augmentation in the definition for D^m**P** (Section 4.5).

For Re = 100, the uniform mesh refinement sequence asymptotic convergence rate in $\|\psi^h\|_E$ approached $O(h^3)$, [10]. That the convergence of TEA theory altered Poisson $O(h^4)$ PDE in the energy norm $\|\psi^h\|_E$ exceeded $O(h^2)$ does *portend* that D^{m}PSI FE $p = 1$ basis $O(h^4)$ convergence will be accorded to incompressible NS PDE GWS^h predictions devoid of singular BC specifications. Asymptotic convergence quantification in $\|\omega^h\|_E$ was intractable as the magnitudes of $\omega^h(x_j, \mathrm{T})$ DOF at each lid-corner node increased without bound during any mesh refinement sequence.

5.14 Validation, weak formulation theory, asymptotic convergence, INS TEA $O(h^4)$ *error freed* D^mE GWS^h FE algorithm, thermal stagnation point

A thermal stagnation point manufactured statement for the convection-diffusion reduction of incompressible NS TEA altered unsteady D^mE for non-D potential temperature Θ is

$$D^mE\colon \frac{\partial\Theta}{\partial t} + u_j\frac{\partial\Theta}{\partial x_j} - \frac{1}{\mathrm{RePr}}\frac{\partial^2\Theta}{\partial x_j^2} - \frac{\mathrm{RePr}h^2}{12}\frac{\partial}{\partial x_j}\left(u_j u_k\frac{\partial\Theta}{\partial x_k}\right) = 0 + O(h^4)$$

assuming that RePr is large. The initial condition (IC) is convection of a fluid of uniform non-D temperature $\Theta(x_j) = 0$ by a non-D velocity vector $u_j \Rightarrow \hat{\mathbf{e}}$ directed parallel to the diagonal of an $n = 2$ unit span square domain Ω. Homogeneous Neumann BCs for Θ are specified everywhere on domain bounding surface $\partial\Omega$ *except* for the *singular* Dirichlet DOF constraint $\Theta(|\mathbf{x}| = 1) \equiv 1$ at the stagnation point corner node located at the confluence of the two outflow boundary segments.

For these *data* the magnitude of DE PDE exact steady exponential solution

$$\Theta(x_j) = \left(1 - exp^{|\mathbf{x}|\mathrm{RePr}}\right) / \left(1 - exp^{\mathrm{RePr}}\right)$$

is nil everywhere except in the *immediate* vicinity of the Dirichlet BC $\Theta(|\mathbf{x}| = 1)$ whereabouts the $\Theta(x_j)$ distribution is sharply dependent on RePr. For RePr $\equiv$ 1000, the directly comparative NS $O(h^2)$DE and TEA altered $O(h^4)D^mE$ GWS^h + ΘTS FE $p = 1$ basis $\Theta^h(x_j, \mathrm{T})$ steady distribution perspectives for uniform meshes $\mathrm{M} = 20^2$ and $\mathrm{M} = 80^2$ are column compared in Figure 5.9.

The NS DE $\Theta^h(x_j, \mathrm{T})$ DOF distributions are rampantly non-monotone due to $O(h^2)$ dispersion error generated DOF with *negative* extrema of −0.5 and −0.4 for this NS state variable *positive definite* member. The exactly comparable D^mE DOF distributions in Figure 5.9 right half are *monotone* to iterative convergence specification $\varepsilon = \mathrm{E}(-04)$. Also clearly visible is that the inadequate resolution of $\Theta^h(x_j,\mathrm{T})$ by the $\mathrm{M} = 20^2$ mesh is eliminated on the $\mathrm{M} = 80^2$ uniform mesh.

The DE PDE *energy norm* definition for the GWS^h FE $p = 1$ basis prediction is

$$\|\Theta\|_E \equiv \frac{1}{2\mathrm{Re\,Pr}}\int_\Omega \frac{\partial\Theta}{\partial x_j}\frac{\partial\Theta}{\partial x_j}d\tau \Rightarrow \sum_{\Omega e}\|\Theta_e^h\|_E = \frac{1}{2\mathrm{Re\,Pr}}\sum_{e=1}^{M}\{\Theta\}_e^T[\mathrm{M2}JJ]_e\{\Theta\}_e$$

For factor of two uniform mesh refinement, NS TEA altered $O(h^4)$ D^mE *asymptotic convergence* rate in DE *error* energy norm $\|\Theta^h\|_E$ is *validated* to be $O(h^4)$ (Figure 5.10a). That the magnitude of the $\mathrm{M} = 160^2$ finest mesh solution $\|\Theta^h\|_E$, the lower right data point in the graphic, is three orders *less* significant than that of the $\mathrm{M} = 80^2$ solution quantifies this $\Theta^h(x_j, \mathrm{T})$ DOF distribution correlates one to one with the knots of the piecewise linear interpolation of the DE exact $\Theta(x_j)$.

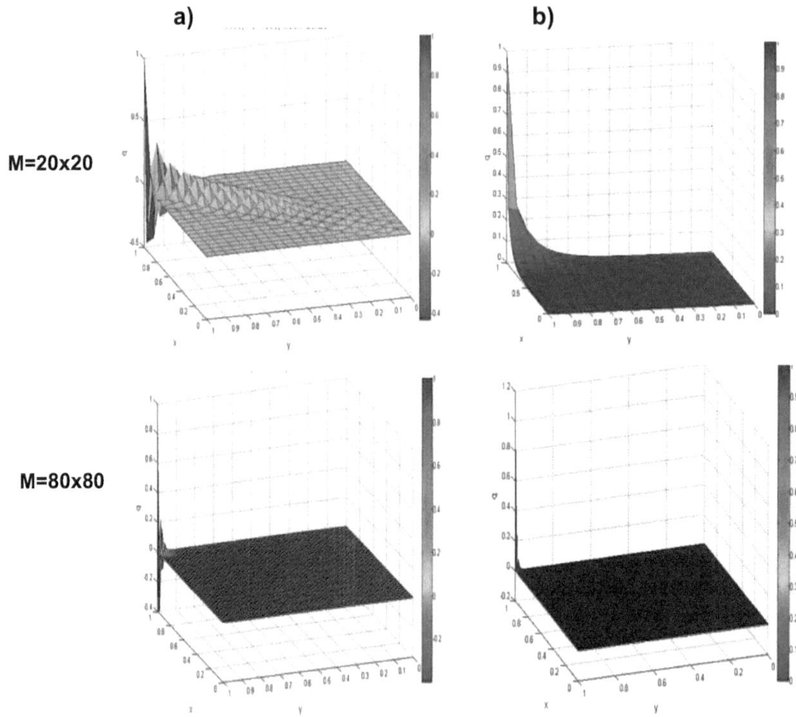

Figure 5.9: Validation, thermal stagnation point, TEA altered NS DE convection-diffusion $n = 2$ manufactured statement, GWS^h + θTS FE $p = 1$ basis steady $\Theta^h(x_j, T)$ perspective projections, uniform M = 20^2, 80^2 meshes, RePr = 1000: a) for $O(h^2)$ DE, b) for $O(h^4)$ D^{m}E.

The *regular* solution-adapted nonuniform mesh refinement sequence utilized $h \Rightarrow \mathrm{pr}h(x_j)$ for pr a progression factor that halved the measure h_e of the outflow corner domain Ω_e that possesses the element *extremum* $\|\Theta_e^h\|_E$ while retaining DOF number constant. The resultant M = 20^2 mesh regular nonuniform solution adapted $\Theta^h(x_j, T)$ distribution is validated resolutely *monotone* via the "L-shaped" DOF distributions existent on each of the domain outflow boundaries $\partial\Omega$ (Figure 5.10b).

5.15 Incompressible NS DM differential constraint, pressure Poisson PDE, DM^h algorithm

For an incompressible NS CFD algorithm to retain pressure a state variable member requires enforcement of the DM divergence-free velocity *differential constraint* $\partial u_j/\partial x_j = 0$. A wide variety of approximate DM^h iterative and non-iterative CFD enforcement strategies have been derived with comparative features/detractions summarized, Williams and Baker, [14]. Of physics of fluids fidelity prediction consequence, that current incompressible NS/RaNS open source/commercial codes do not output the *significant digit* of DM^h algorithm *"divergence-freeness"* constitutes a non-quantified source of error.

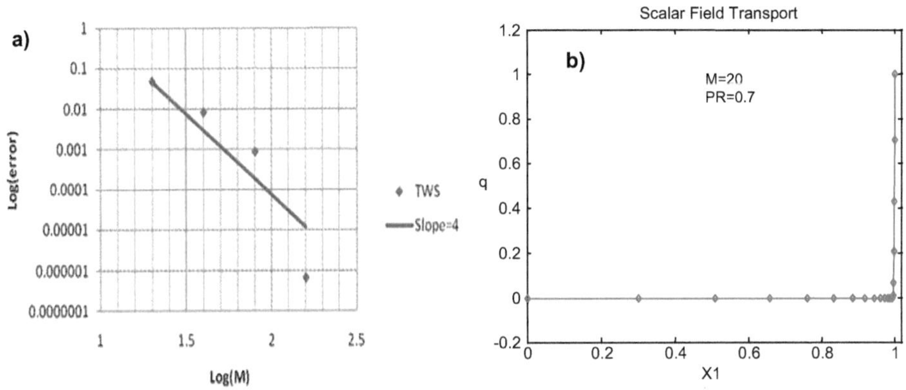

Figure 5.10: Validation, NS TEA $O(h^4)$ D^{m}E PDE GWS^h + θTS FE $p = 1$ basis algorithm, thermal stagnation point manufactured statement, steady $\Theta^h(x_j, \mathrm{T})$ solution, RePr = 1000: a) uniform mesh refinement sequence $O(h^4)$ asymptotic convergence in $\|\Theta^h\|_E$, b) M = 20^2 *regular* solution-adapted nonuniform mesh refinement $\Theta^h(x_j, \mathrm{T})$ distribution on each outflow boundary $\partial\Omega$.

Rigorous mathematical resolution of this *persistent* incompressible NS CFD issue was *finally* enabled by creation of the *nonsymmetric sparse matrix solver,* GMRES the premier issue, [13]. Forming the divergence of incompressible NS velocity potential function $u_j \equiv -\partial\varphi/\partial x_j$ generates the linear DM^h Poisson PDE with the exact *purely* Neumann boundary condition (NBC) valid on the *entirety* (!) of domain Ω closure ∂Ω:

$$\mathrm{DM}^h\!: \ \frac{\partial^2\varphi}{\partial x_k^2} + \frac{\partial u_j^h}{\partial x_j} = 0, \ \ \mathrm{NBC}\left(\frac{\partial\varphi}{\partial x_j} - \left(u_j - u_j^h\right)\right)\widehat{n}_j = 0$$

Therein, $u_j^h(x_i, t_{n+1})$ is *any* discrete approximation of divergence-free velocity $u_j(x_i, t_{n+1})$. That the DM^h PDE+NBC transitions to totally *homogeneous* at coupled D^m**P**:DM^h algorithm time t_{n+1} convergence generates the rank deficient Jacobian matrix that is *solely* tractable using a non-symmetric sparse matrix solver.

The TEA altered incompressible NS D^m**P** is

$$\mathrm{D}^m\mathbf{P}\!: \ \frac{\partial u_i}{\partial t} + \frac{\partial}{\partial x_j}\left[\begin{array}{c} u_j u_i + P^*|_{n+1}^{p}\delta_{ij} \\ -\dfrac{1}{\mathrm{Re}}\left(\dfrac{\partial u_i}{\partial x_j} + \dfrac{\partial u_j}{\partial x_i}\right) - \dfrac{h^2\mathrm{Re}}{12} u_j u_k \dfrac{\partial u_i}{\partial x_k} \end{array}\right] - \frac{\mathrm{Gr}}{\mathrm{Re}^2}\Theta\widehat{g}_i = 0 + O(h^4)$$

The definition for pressure *kinetic impact* variable $P^*(x_i, t)$ that replaces genuine pressure in coupled NS $O(h^4)$D^m**P**:$O(h^2)$DM^h + NBC GWS^h + θTS algorithm during iteration p at time t_{n+1} is

$$P^*(x_i, t_{n+1})|^p \equiv P^h(x_i, t_n) + \frac{1}{\theta \Delta t}\sum_{\alpha=1}^{p} \delta\varphi^h(x_i, t_{n+1})|^\alpha$$

Therein, the *distribution* $\sum_{\alpha=1}^{p} \delta\varphi^h(x_i, t_{n+1})|^\alpha$ is the *accumulation* of the DM^h + NBC iterates generated during the p iterations of coupled $D^m\mathbf{P}$:DM^h + NBC algorithm prior to iteration $p+1$ convergence at time t_{n+1}. Time step implicitness factor definition is limited to $\theta \geq 0.5$ and $P^h(x_i, t_n)$ is the *genuine* kinematic pressure distribution at algorithm time t_n convergence.

No need exists to alter $O(h^2)$ DM^h to TEA $O(h^4)$ D^mM^h as the significance of the *sole* admissible solution is nil. Iterative convergence of this NS $O(h^4)D^m\mathbf{P}$:$O(h^2)DM^h$ + NBC PDE system GWS^h algorithm quantifies that the *significant digit* of incompressible velocity vector divergence freeness is equal to the DM^h *error* energy norm $\|\varphi^h(t_{n+1})\|_E$ which equates to the algorithm convergence specification, typically $\varepsilon \leq$ E-04.

Insertion of thus quantified $O(\varepsilon)$ *divergence-free* velocity $u_j^h(x_i, t)$ into the divergence of NS **DP** enables prediction of the incompressible NS *genuine* kinematic pressure via the DP Poisson PDE with *exact* Neumann BC

$$\text{DP:} \ \frac{\partial^2 P}{\partial x_k^2} + \frac{\partial}{\partial x_j}\left[u_i \frac{\partial u_j}{\partial x_i} + \frac{\text{Gr}}{\text{Re}^2}\bar{\Theta}\hat{g}_j\right] = 0, \qquad \text{NBC:} \left(\frac{\partial P}{\partial x_j} - \frac{1}{\text{Re}}\frac{\partial^2 u_j}{\partial x_k^2}\right)\hat{n}_j = 0$$

valid on the entirety of closure $\partial\Omega$. DP is not eligible for TEA alteration to $O(h^4)$ D^mP as the resultant differentiability requirement well exceeds FE $p=1$ basis admissibility.

NBC requires the *Laplacian* operation on the resolution of the velocity vector normal to outflow $\partial\Omega$ segments (*plural*) of a *multiply connected* domain Ω. GWS^h algorithm Green-Gauss divergence theorem imposition then alters NBC to the wall normal gradient of wall parallel velocity resolution on each outflow $\partial\Omega$ bounding surface, i.e., the attached boundary layer wall *shear stress*. This rigorously derived NBC thereby does make rather *remarkable* (!) fluid mechanics sense.

Prediction of the genuine kinematic pressure $P_h(x_i, t_{n+1})$ distribution using the $O(\varepsilon)$ divergence-free $u_j^h(x_i, t_{n+1})$ is a post process operation following $D^m\mathbf{P}$:DM^h time t_{n+1} convergence. Resultant $P^h(x_i, t_{n+1})$ then replaces $P^h(x_i, t_n)$ in $P^*(x_i, t_{n+1})|^p$ definition for the next time step $O(h^4)D^m\mathbf{P}$:$O(h^2)DM^h$ + NBC PDE system iteration cycle.

5.16 Validation, INS pressure, TEA $O(h^4)$ *error freed* D^mP:$O(h^2)DM^h$ + NBC GWS^h FE $p=1$ basis algorithm, $n=3$ step wall diffuser

For time TS identified β specifications consistent with the *now* realized $\beta = \text{Re}h/6$, *validation* of the β term predecessor of the NS $O(h^4)D^m\mathbf{P}$:$O(h^2)DM^h$ + NBC:DP + NBC PDE system $GWS^h + \theta$TS FE $p=1$ basis algorithm was secured for the $n=3$ close coupled step wall diffuser, Figure 5.11a). For expansion ratio $H/h = 1.9423$, step height $S = 4.9$ mm and channel width W = 180 mm, the extensive experimental data base of [2] enabled thorough quantitative validation of this GWS^h + θTS algorithm, Williams et al. [15].

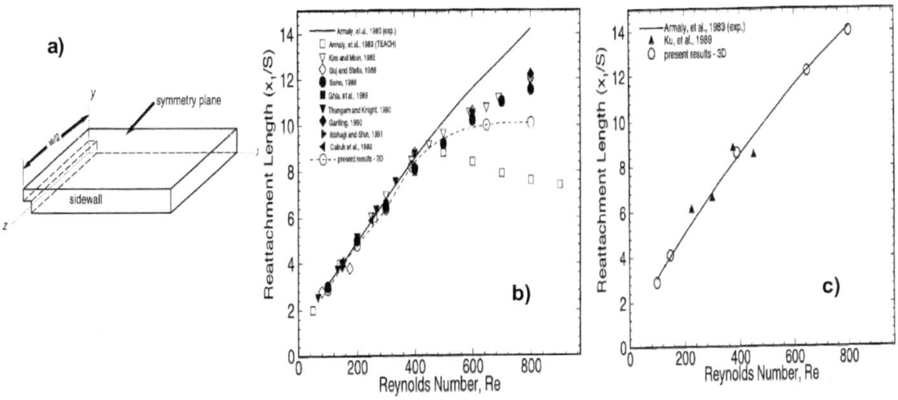

Figure 5.11: Validation, incompressible NS close coupled step wall diffuser: a) $n = 3$ geometry of Armaly et al. [2], for solid line the experimental data, b) legacy CFD codes scalar measure prediction summary, c) scalar measure validation for time TS β term predecessor of NS TEA $O(h^4)$ D^m**P**: $O(h^2)$ DM^h + NBC:DP + NBC PDE, adapted from Williams and Baker [15].

The *scalar* measure the 1980s–1990s CFD community used to document algorithm prediction accuracy for the close coupled step wall diffuser velocity field is the reattachment intercept of the step induced primary recirculation bubble on the diffuser downstream floor symmetry plane. The published record for incompressible NS CFD algorithm/code predictions (Figure 5.11b) quantifies that validations were secured *only* for Reynolds number specifications Re ≤ 400. Conversely, scalar measure validation was secured ∀Re ≤ 800 for the β term predecessor of NS TEA $O(h^4)D^m$**P**:DM^h + NBC:DP + NBC PDE algorithm (Figure 5.11c).

The extensive velocity resolution data base of [2] enabled truly *comprehensive* validation of the β term predecessor to TEA $O(h^4)D^m$**P**:DM^h + NBC:DP + NBC PDE system GWS^h + θTS algorithm/code. Specifically, these predictions quantified that Re ≅ 450 was the *critical* Reynolds number at which this step wall diffuser velocity vector distribution transitioned from essentially two-dimensional to fully three-dimensional.

Thus quantitatively identified is the reason for *uniform* departure of legacy CFD predictions from the step wall scalar validation for Re > 400 (Figure 5.11b). Once past this incompressible NS pressure resolution issue, the β term predecessor NS $O(h^4)$ PDE system GWS^h + θTS algorithm/code solutions quantitatively detailed the existence of multiple diffuser sidewall adjacent vortex structures as a function of 450 ≤ Re ≤ 800. The archive for these $n = 3$ validations is [15].

5.17 Validation, pressure, multiply connected domains, INS $O(h^4)$ *error freed* D^mP: $O(h^2)DM^h$ + NBC:DP + NBC GWS^h + θTS FE $p = 1$ basis algorithm

The rank deficient Jacobian generated at convergence of the incompressible NS TEA $O(h^4)D^mP{:}O(h^2)DM^h$ + NBC:DP + NBC GWS^h + θTS algorithm is tractable only using a nonsymmetric sparse matrix solver. Prior to creation of this type of solver, it was necessary to impose at least one Dirichlet BC (DBC) for prediction of $P^*(x_i, \mathrm{T})$, usually at a $\partial\Omega$ outflow corner node.

For a *multiply connected* channel with off center splitter plate (Figure 5.12a), prior NS approximate DM^h practice pseudo-pressure $P^*(x_i, T)$ prediction requisite DBC is imposed at the lower channel outflow corner node. Only the precisely comparative genuine $P^h(x_i, T)$ distribution (Figure 5.12b), generated by the NS $O(h^4)D^mP{:}O(h^2)DM^h$ + NBC:DP + NBC PDE system $P^h(x_i)$ distribution exhibits the requisite unique channel gradients and exit pressure levels. Total *compromise* of the lower channel $P^*(x_i, T)$ distribution by enforcement of a single DBC DOF is clearly visible hence validated.

At *each* occurrence of domain boundary geometry induced acceleration, deceleration or redirection, the incompressible NS DM velocity divergence-free constraint *instantaneously* generates a fully three-dimensional velocity vector field. The Cartesian resolution of incompressible NS DM

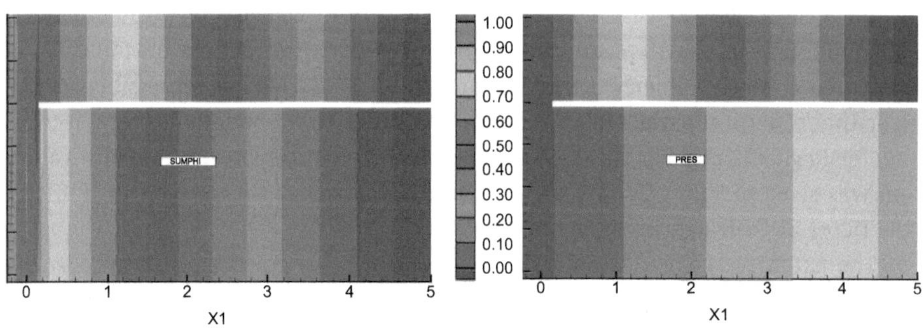

Figure 5.12: Validation, multiply connected domain incompressible NS pressure distributions in unitized scale, GWS^h FE $p = 1$ basis algorithms, Re = 1,000: a) for $O(h^4)D^m\mathbf{P}$ $O(h^2)DM^h$ + NBC + 1DBC compromised $P^*(x_i)$ distribution, b) for $O(h^4)D^mP{:}O(h^2)DM^{h+}$NBC:DP + NBC.

$$\mathrm{DM:}\ \frac{\partial u_j}{\partial x_j} \equiv \frac{\partial u}{\partial x} + \frac{\partial v}{\partial y} + \frac{\partial w}{\partial z} = 0$$

clearly quantifies that a unidirectional velocity vector of magnitude u will *immediately* initiate a velocity vector field in the plane transverse to x at any such encounter. This mathematics of physics of fluids assertion was first validated by turbulent micro jet instantaneous dispersal prediction comparison with smoke mass transport experimental data [4].

Incompressible NS $O(h^4)\mathrm{D}^m\mathbf{P}$:DM^h + NBC:DP + NBC PDE GWS^h + θTS FE $p = 1$ basis algorithm is precisely pertinent to predictions in *multiply connected* fluid distribution systems that possess *a priori* unknown flow partitioning, hence distinct unknown pressure level at each outflow plane. The circular cross-section $n = 3$ fluid distribution manifold graphed at the center bottom of Figure 5.13 is discretized into ~0.8 million FE $p = 1$ tetrahedron. The inflow plane velocity Dirichlet BC is the fully developed laminar profile rotated, i.e., a paraboloid.

Graphs of GWS^h algorithm generated steady velocity vector distributions on transverse planes A-A through F-F surround the geometry (Figure 5.13). Each top half graph presents the section mirror symmetric axial velocity (speed) distribution color flood. Each lower half graph is the mirror symmetric vector distribution on the transverse plane. Clearly displayed is incompressible DM *immediate* initiation of fully three-dimensional velocity vector distributions at each bifurcation in the manifold.

The function of a bronchial tree is mass transport in a seriously multiply connected domain. The interior of the first six *non-coplanar* bifurcations of a human $n = 3$ bronchial tree casting downstream of the trachea (Figure 5.14a) is discretized into ~ 1.6 million FE $p = 1$ basis tetrahedron. Trachea inflow plane velocity Dirichlet BC remains the paraboloid. NS $O(h^4)\mathrm{D}^m\mathrm{P}$:$O(h^2)$DM^h + NBC:DP + NBC PDE GWS^h + θTS FE $p = 1$ basis predicted unsteady velocity vector $u_j(x_i, \mathrm{n}\Delta t)$ evolution at Re = 500 is stopped at solution time $n\Delta t \equiv \mathrm{T}$ when the fluid speed became steady at the outflow plane of the tree sixth generation.

GWS^h algorithm steady $u^h(x_i, \mathrm{T})$ speed distribution visualized on a graphics cutting plane containing the trachea centroidal axis is color flood graphed in Figure 5.14b, non-D color bar at the right. Key observables therein are pre- and post-bifurcation increases (red) in $u^h(x_i, T)$ when traversing from larger to smaller diameter tree branches distant from the trachea. Of course $u^h(x_i, T)$ is distributed throughout the tree branches graphed in shadow (Figure 5.14b), which the graphics cutting plane did not intersect.

5.18 Conclusion, weak formulation Navier-Stokes continuous GWS^h + θTS FE $p = 1,2,3$ basis algorithms cast on *error freed* CFD mathematics

That NS (also RaNS) mathematically rigorous theory derivation addresses TEA altered $O(h^4)$ PDE system GWS^h + θTS FE $p = 1,2,3$ basis algorithms responds precisely to the VVUQ citation on "*reliability of complex computational methodologies*" regarding being *more complicated* (!) than linear elliptic Furthermore, derived and quantitatively *validated* weak formulation NS PDE continuous Galerkin *nonlinear* theory local/

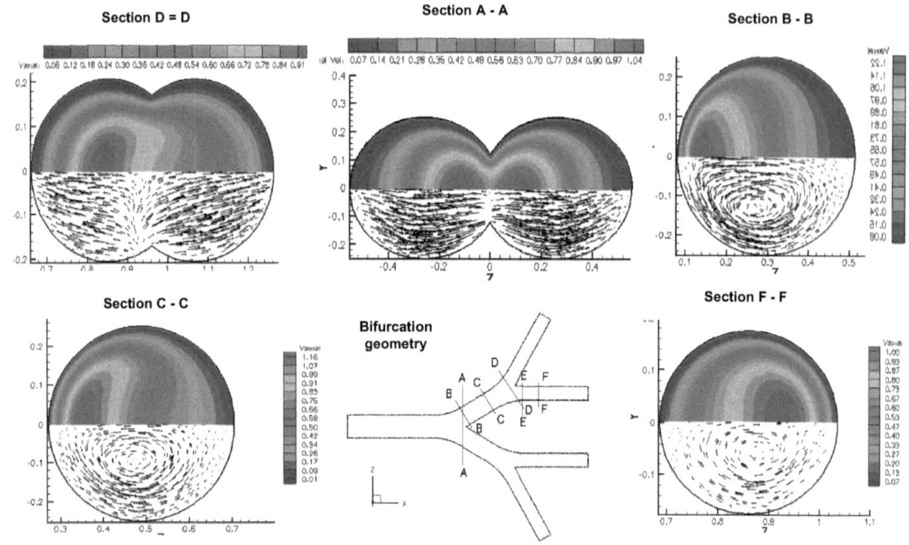

Figure 5.13: Validation, incompressible NS velocity vector $n = 3$ distributions, multiply connected domain, NS $O(h^4)\mathrm{D}^m\mathrm{P}{:}O(h^2)\mathrm{DM}^h$ + NBC:DP + NBC PDE GWS^h + θTS FE $p = 1$ basis algorithm, ~0.8 million tetrahedron, Re = 500: bifurcation geometry at center bottom, velocity speed and vector mirror symmetric distributions on transverse planes A-A to F-F, top half axial velocity magnitude flood, lower half plane vector distribution, adapted from Wong et al. [16].

global *a posteriori error* estimates are also *precisely* responsive to VVUQ topics cited required addressed to ensure *improved outcomes*!

Critical to the retaining of pressure as an incompressible NS state variable member is TEA altered $O(h^4)\mathrm{D}^m\mathbf{P}{:}\mathrm{DM}^h + O(\varepsilon)$ + NBC:DP + NBC PDE system for which the *significant digit* of velocity *"divergence-freeness"* equates to code execution linear algebra convergence requirement. Thus clearly identified is the ***unforgiving verity*** of incompressible NS (and RaNS) CFD code practice:

. . . you get out what you put in *on the boundaries*!

Finally, the *necessary* requirement for *error* prediction/quantification accuracy in NS PDE state variable intrinsic energy norm $\|e^h(t)\|_E$ is that CFD code generated $\{q^h(x_j, t)\}$ DOF distributions be strictly *monotone*. If not, state variable member DOF node-to-node oscillations that are algorithm dispersive error generated will compromise prediction accuracy by the injection of *phake energy* into NS PDE system symmetric quadratic *error* energy norm set $\|\{e^h(t)\}\|_E$.

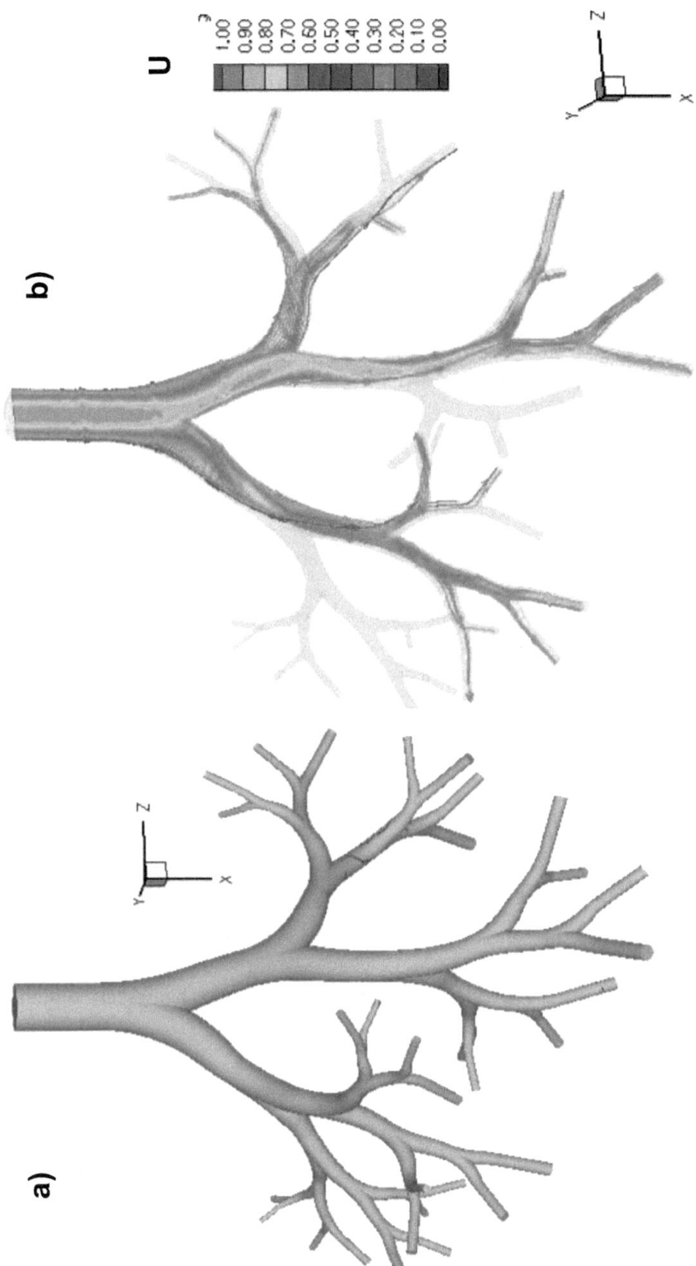

Figure 5.14: Validation, velocity vector speed distribution in non-coplanar six generation human bronchial tree, NS $O(h^4)\mathrm{D}^m\mathrm{P}{:}O(h^2)\mathrm{DM}^h$ + NBC:DP + NBC PDE system GWS^h + θTS algorithm, 1.6 million FE $p = 1$ tetrahedron mesh, Re = 500: (a) tree perspective, trachea inflow plane at top, (b) $u^h(x_i, \mathrm{T})$ speed distribution on the graphics cutting plane containing trachea centroidal axis, adapted from Wong et al. [16].

References

[1] AIAA (1998). "AIAA Guide for the Verification and Validation of Computational Fluid Dynamics Simulations," *American Institute Aeronautics and Astronautics*, Tech Rpt AIAA:G-077-1998.

[2] Armaly, B.F., Durst, F., Pereira, J.C.F., Schonung, B. (1983). "Experimental and theoretical investigation of backward-facing step flow," *J Fluid Mech*. 127

[3] Babuska, I., Rheinboldt, W.C. (1976). "*A-posteriori* error estimates for the finite element method," *J Num Mtd Engr*, 12

[4] Baker, A.J., Orzechowski, J.A., Stungis, G.E. (1985). "Prediction of secondary vortex flowfields generated in an interacting multiple free-jet configuration, *AIAA Journal*, 23.7

[5] Corrigan, A., Zahr, M.J. (2001). "Transonic Flow Through a Channel with a Smooth Gaussian Bump." AIAA 1st High Fidelity CFD Workshop

[6] Gebhart, B., Jaluria, Y., Mahajan, R.L., Sammakia, B. (1988). *Buoyancy-Induced Flow and Transport*, Hemisphere, NY

[7] Ghia, K.N., Ghia, U., Shin, C.T. (1982). "High-Re solutions for incompressible flow using the Navier-Stokes equations and a multigrid method," *J Comp Physics*, 48

[8] Jameson, A. (2012) *Reflections on Four Decades of CFD – A Personal Perspective*, http://dept.ku.edu/~cfdku/JRV/Jameson.pdf

[9] Kolesnikov, A. (2000), "Efficient implementation of high order methods in computational fluid dynamics," PhD dissertation, University of Tennessee

[10] Kolesnikov, A., Baker, A.J. (2001), "An efficient high order Taylor weak statement for the Navier-Stokes equations," J. Comp Physics 173

[11] Masters, J. (2014). "A year of enhancements, plus a groundbreaking vision document," *Aerospace America*, 52(11)

[12] NAP (2012). "Assessing the Reliability of Complex Models: Mathematical and Statistical Foundations of Verification, Validation, and Uncertainty Quantification," National Research Council, National Academies Press, Washington DC

[13] Saad, Y., Schultz, M.H., (1986) "GMRES: A generalized minimal residual algorithm for solving nonsymmetric linear systems," SIAM *J Sci Stat Comp*, 7

[14] Williams, P.T., Baker, A.J., (1996). "Incompressible computational fluid dynamics and the continuity constraint method for the 3-D Navier-Stokes equations," *J Num Heat Trans, Part B*, 29 (entire issue)

[15] Williams, P.T., Baker, A.J., (1997). "Numerical simulations of laminar flow over a 3D backward-facing step," *J Num Mtd Fluids*, 24

[16] Wong, K.L., Ericson, S., Baker, A.J. (2004). "Flow patterns in a multi-generation bifurcation bronchial tree," poster, SIAM Parallel Processing Sci Comput Conf.

Chapter 6
Time-averaged (RaNS) Navier-Stokes *error freed* CFD weak formulation, theory, asymptotic convergence, validations

Libretto: Time-averaged Navier-Stokes (RaNS) PDE system weak formulation continuous Galerkin FE p basis algorithms on *error freed* CFD mathematics exhibit *optimal performance.*

TEA theory altered RaNS PDE systems contain the DOF *rational* product of velocity tensor product divided by turbulent Reynolds number, a fourth-order nonlinearity.

Compressible RaNS TEA $O(h^4)$ PDE system closed with the Spalart-Almarras (SA) eddy viscosity model TEA altered $O(h^4)$ PDE GWS^h + θTS FE $p = 1$ basis algorithm validation addresses the Sajben transonic diffuser for weak and strong *curved shocks.*

RaNS:SA $O(h^4)$ PDE GWS^h *and* a commercial code RaNS:SA $O(h^2)$ PDE numerically stabilized FE $p = 1$ basis algorithm ten stage *regular* mesh refinement sequence data comparisons with experimental data quantify first-order effects validation.

Conversely, GWS^h and non-GWS^h $u_2{}^h(x_j,\mathrm{T})$ and scalar member shock centric state variable distribution comparisons quantify first-/higher-order effects predictions are totally *compromised* by stabilized shock capture

A RaNS:SA $O(h^4)$ PDE GWS^h FE $p = 1$ algorithm thirteen stage regular mesh refinement data generates shock-turbulent boundary layer interaction physics of fluids *fidelity diagnostics.*

Incompressible RaNS:TKE closed TEA $O(h^4)$ PDE GWS^h + θTS FE $p = 1$ basis algorithm monotone DOF *validation* accrues to slug velocity onset to a channel with *singular* no-slip BCs imposed interior to the domain

The spectral frequency distribution of RaNS:TKE monotone state variable *annihilated* error is quantified $O(2h)$ with error *extrema* significance an order larger than that of the state variable.

Keywords: compressible RaNS, TEA, monotone shock interpolation, shock capture corruption, first-/higher-order effects validations/failures, shock-turbulent BL interaction, diagnostics, incompressible RaNS TEA, annihilated error spectral content distribution

6.1 Derivation, compressible RaNS:SA TEA $O(h^4)$ D^{m}M:D^{m}P:D^{m}E: D^{m}SA PDE system

The large Re $O(h^2)$ TE *annihilation* theory alteration to a RaNS model D**P** is, [2]:

https://doi.org/10.1515/9783112206430-007

$$\bar{f}_j(\bar{q}, x_i)\frac{\partial \bar{q}}{\partial x_j} - \frac{1}{\mathrm{Re}}\left(1+\mathrm{Re}^T\right)\frac{\partial^2 \bar{q}}{\partial x_j^2} - \frac{\mathrm{Re}\, h^2}{12\left(1+\mathrm{Re}^T\right)}\bar{f}_k(\bar{q}, x_i)\frac{\partial}{\partial x_k}\left[\bar{f}_j(\bar{q}, x_i)\frac{\partial \bar{q}}{\partial x_j}\right] = 0$$

for Re^T the turbulent Reynolds number. Replacing velocity with momentum and accounting for compressible DM upon moving velocity inside the TEA divergence operator and deleting overbar notation, the compressible RaNS TEA altered $O(h^4)$ D^mM: $\mathrm{D}^m\mathbf{P}$:D^mE PDE system is

$$\mathrm{D}^m\mathrm{M}: \frac{\partial \rho}{\partial t} + \frac{\partial}{\partial x_j}\left[m_j - \frac{h^2 \mathrm{Re}}{12} u_k \hat{u}_j \frac{\partial \rho}{\partial x_k}\right] = 0 + O(h^4)$$

$$\mathrm{D}^m\mathbf{P}: \frac{\partial m_i}{\partial t} + \frac{\partial}{\partial x_j}\left[\begin{array}{l} u_j m_i + \mathrm{Eu} p \delta_{ij} - \dfrac{(\mu^* + \mathrm{RET})}{\mathrm{Re}}\left(\dfrac{\partial u_i}{\partial x_j} + \dfrac{\partial u_j}{\partial x_i} - \dfrac{\partial u_k}{\partial x_k}\left(1 + \dfrac{2}{3}\delta_{ij}\right)\right) \\ \quad - \dfrac{h^2 \mathrm{Re}}{12\{\mu^* + \mathrm{RET}\}_e}\left(u_k\left(u_j - \dfrac{\partial u_l}{\partial x_l}\underline{\delta_{jl}}\right)\dfrac{\partial m_i}{\partial x_k}\right) \end{array}\right] = 0 + O(h^4)$$

$$\mathrm{D}^m\mathrm{E}: \frac{\partial E}{\partial t} + \frac{\partial}{\partial x_j}\left[\begin{array}{l} u_j(E+p) - \dfrac{(\kappa^* + \mathrm{RET})}{\mathrm{Re\,Pr}}\dfrac{\partial T}{\partial x_j} - \dfrac{\mu^* u_i}{\mathrm{Re}}\left(\dfrac{\partial u_i}{\partial x_j} + \dfrac{\partial u_j}{\partial x_i} - \dfrac{\partial u_k}{\partial x_k}\left(1 + \dfrac{2}{3}\delta_{ij}\right)\right) \\ \quad - \dfrac{h^2 \mathrm{Re\,Pr}}{12(\kappa^* + \mathrm{RET})}\left(u_k\left(u_j - \dfrac{\partial u_l}{\partial x_l}\underline{\delta_{jl}}\right)\dfrac{\partial T}{\partial x_k}\right) \end{array}\right] = 0 + O(h^4)$$

for p the thermodynamic pressure, Eu and Pr the Euler and Prandtl numbers, and μ^*, κ^* the temperature-dependent non-D viscosity, thermal diffusivity distributions. Compressible turbulent Reynolds number definition is $\mathrm{Re}^T \equiv \mu^T/\mu_\infty = \rho \nu^T/\mu_\infty$.

Select commercial and government CFD codes offer RaNS PDE system closure with the Spalart-Allmaras (SA) eddy viscosity transport PDE model, [11, 12]. The successor *penultimate* eddy viscosity transport PDE, [1], includes source/sink functional correlations and strain-rate/vorticity tensor product dependencies stated pertinent to BL transition. Eddy viscosity model definition is $\nu^T \equiv \tilde{\nu} f_{v1}$, $f_{v1} \equiv (\tilde{\nu}/\nu)^3\left((\tilde{\nu}/\nu)^3 + c_{v1}^3\right)$, ν the fluid kinematic viscosity and c_{v1} a model coefficient.

The dimensional (D) statement of compressible SA eddy viscosity transport PDE is

$$\mathrm{DSA}: \frac{\partial}{\partial t}\rho\tilde{\nu} + \frac{\partial}{\partial x_j}\left(u_j \rho \tilde{\nu} - \frac{1}{\sigma}\rho(\nu + \tilde{\nu})\frac{\partial \tilde{\nu}}{\partial x_j}\right) - \frac{1}{\sigma}c_{b2}\rho\left(\frac{\partial \tilde{\nu}}{\partial x_j}\right)^2 + \frac{1}{\sigma}\rho(\nu + \tilde{\nu})\frac{\partial \rho}{\partial x_j}\frac{\partial \tilde{\nu}}{\partial x_j}$$
$$- \rho(P - \mathrm{D} + T) = 0$$

where source term $\rho(P - \mathrm{D} + T)$ embeds the correlations relating to transition. Coalescing all source terms into $s(\cdot)$, the TEA altered $O(h^4)$ D^mSA (D)PDE is

$$\mathrm{D}^m\mathrm{SA}: \frac{\partial}{\partial t}\rho\tilde{\nu} + \frac{\partial}{\partial x_j}\left(u_j \rho \tilde{\nu} - \frac{1}{\sigma}\rho(\nu + \tilde{\nu})\frac{\partial \tilde{\nu}}{\partial x_j} - \frac{\sigma h^2}{12} u_j u_k \frac{\partial \rho \tilde{\nu}}{\partial x_k}\right) - s(\sigma, c_{b2}, \rho, \tilde{\nu}, P, \mathrm{D}, T)$$
$$= 0 + O(h^4)$$

That TEA theory augmentation coefficient σ in D^{m}SA is an $O(1)$ constant is totally distinct from Re the numerator coefficient in all other TEA-altered NS/RaNS PDE systems.

Compressible RaNS:SA TEA $O(h^4)$ D^{m}M:D^m**P**:D^{m}E:D^{m}Q PDE system nonlinearity is *fourth order* as turbulent Reynolds number $\mathrm{Re}^T(x_j)$ is a distribution. For TEA term integration altered to the rational tensor product, the RaNS D^m**P** element matrix is

$$\mathrm{GWS}^h\!: -\int_{\Omega_e}\{\mathrm{N}\}\frac{\partial}{\partial x_j}\left(\frac{h^2\mathrm{Re}}{12(\mu^*+\mathrm{Re}^T)}\left(\left(u_j-\frac{\partial u_l}{\partial x_l}\delta_{jl}\right)u_k\frac{\partial m_i}{\partial x_k}\right)\right)^h d\tau$$

$$=\int_{\Omega_e}\left(\frac{h_e^2\mathrm{Re}}{12(\mu^*+\mathrm{Re}^T)}\right)\frac{\partial\{\mathrm{N}\}}{\partial x_j}\left(\{\mathrm{U}J\}_e^T-\{\mathrm{DU}J\}_e^T\right)\{\mathrm{N}\}\{\mathrm{N}\}^T\{UK\}\frac{\partial\{\mathrm{N}\}^T}{\partial x_k}d\tau\{MI\}_e$$

$$\approx\frac{h_e^2\mathrm{Re}}{12}\left(\left\{\frac{\mathrm{U}J^*\mathrm{U}K}{\mu^*+\mathrm{RET}}\right\}_e^T[\mathrm{M30}JK]_e-\left\{\frac{\mathrm{DU}J^*\mathrm{U}K}{\mu^*+\mathrm{RET}}\right\}_e^T[\mathrm{M30}JK]_e\right)\{MI\}_e$$

summation on $1\le J,K\le n$. The distribution for velocity divergence $\{\mathrm{DU}J\}_e$ DOF is weak formulation GWS^h predicted on incompressible DM altered to the *differential statement* (DS)

$$\mathrm{DS}\!:\ \mathcal{D}(u_j)\equiv \mathrm{div}u\ -\ \partial u_j/\partial x_j=0$$

which generates GWS^h element matrix $[\mathrm{M200}]_e\{\mathrm{DU}J\}_e-[\mathrm{M20}J]_e\{\mathrm{U}J\}_e$, for $1\le J\le n$.

Linear algebra *stability-critical* $\mathrm{O}(h^4)$ D^m**P** TEA term momentum vector self-coupling contribution to the Newton algorithm Jacobian element matrix is

$$[\mathrm{JAC\_}\{\mathrm{MI},\mathrm{MI}\}]_e=\frac{\theta\Delta th_e^2\mathrm{Re}}{12}\left\{\frac{\mathrm{U}J^*\mathrm{UK}}{\mu^*+\mathrm{RET}}\right\}_e^T[\mathrm{M30}J\mathrm{K}]_e-\left\{\frac{\mathrm{DU}J^*\mathrm{UK}}{\mu^*+\mathrm{RET}}\right\}_e^T[\mathrm{M30}J\mathrm{K}]_e$$

$$+\frac{\theta\Delta th_e^2\mathrm{Re}}{12}\{\mathrm{MI}\}_e^T[\mathrm{M3K}J0]_e\mathrm{diag}\left(\left[\frac{\mathrm{U}J\delta_{\mathrm{IK}}/\rho}{\mu^*+\mathrm{RET}}\right]_e-\left[\frac{\mathrm{DU}J\delta_{\mathrm{IK}}/\rho}{\mu^*+\mathrm{RET}}\right]_e+\left[\frac{\mathrm{UK}\delta_{\mathrm{I}J}/\rho}{\mu^*+\mathrm{RET}}\right]_e\right)$$

6.2 Validation, weak formulation continuous RaNS:SA TEA $O(h^4)$ *error freed* PDE GWS^h FE $p = 1$ basis algorithm, shock monotone continuum interpolation, stability

The Sajben transonic diffuser [9] is a finite-width channel wind tunnel model generating genuinely two-dimensional velocity distributions on its mid-plane (Figure 6.1a). For sufficiently large Re, interaction of the diffuser contoured and planar surfaces generates transonic shocks possessing significant curvature that presents an *obliqueness* distribution to the far-field onset velocity. Experimental data, Bogar et al. [4], Salmon et al. [10], confirm that oblique shock encounter with the contoured surface turbulent BL causes separation, whereas no turbulent BL separation occurs for shock

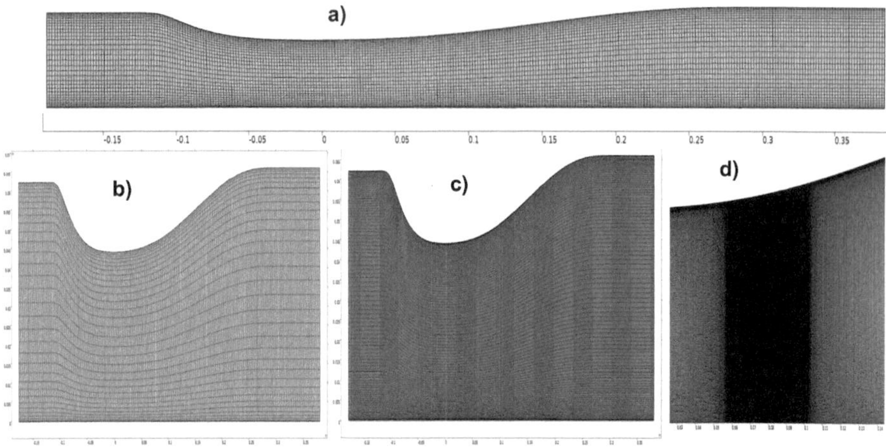

Figure 6.1: Sajben transonic diffuser, regular solution-adapted mesh refinements, compressible RaNS:SA TEA altered $O(h^4)$ D^mM:D^m**P**:D^mE:D^mSA PDE system GWS^h FE $p = 1$ basis algorithm: (a) geometry true scale with inaugural boundary layer solution-adapted mesh M_1, for refined meshes in unitized scale: (b) M_1, (c) M_3, (d) M_{10} shock-centric zoom.

encounter on the planar lower surface. Legacy CFD code predictions are reported, Hsieh et al. [6] and the NPARC Alliance [7].

Figure 6.1a also displays the initial boundary-fitted modestly BL-adapted FE $p = 1$ tensor product $M_1 = 2{,}500$ element mesh. The non-D length scale L is defined the diffuser throat span, $h_{TH} = 1.7322$ non-D, 0.008078 m D. The diffuser inflow and outflow transverse spans are $1.4h_{TH}$ and $1.5h_{TH}$ with axial length equal to 11 h_{TH}. The onset velocity (air) is assumed fully turbulent at $Re = 7 \times 10^5$ based on L. Unitized scale graphics (Figure 6.1b–d) compare mesh M_1 with shock solution-adapted mesh M_3 with 8,000 elements, and the shock-adapted centric zoom of terminal mesh M_{10}.

The true scale color flood of strong shock RaNS:SA TEA $O(h^4)$ PDE GWS^h steady Mach number $Ma^h(x_j,T)$ mesh M_{10} distribution is Figure 6.2a, color bar at the right. The transverse distribution of shock curvature is clearly displayed, as are the post-shock persistence of just supersonic velocity distributions and the large separation bubble along the contoured surface. The right angle breaks in shock-centric $Ma^h(x_j,T)$ color flood perspective zoom with contours (Figure 6.2b), view slightly upstream, quantitatively validates that continuous Galerkin transonic shock continuum interpolation is monotone to code-achieved iterative convergence $\varepsilon \le 2 \times 10^{-4}$.

The abrupt decrease in Mach number distribution to subsonic at the singular instance of velocity normal onset to the shock generates a deep resolutely *monotone* depression with extremum $Ma \approx 0.6$ (Figure 6.2b) perspective projection. The shock-centric distribution of onset velocity in the $\mathbf{u}^h(\mathbf{x},T)$ streamline redirection induced by the shock obliqueness distribution is graphed true scale in Figure 6.2c. The unitized

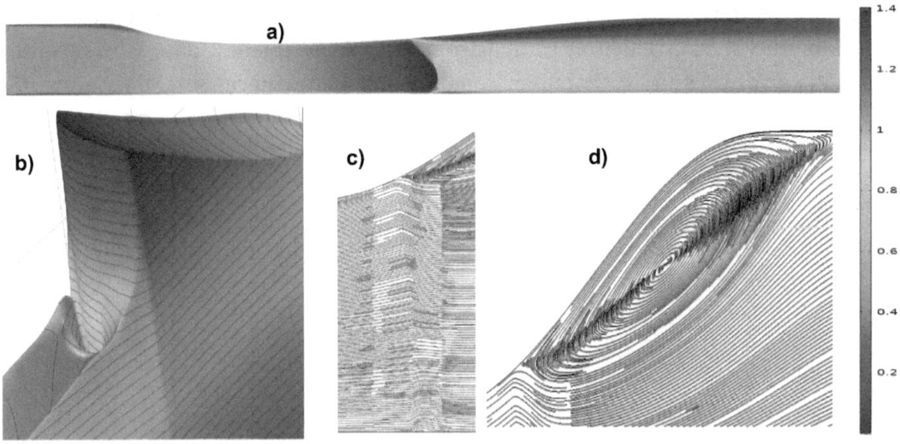

Figure 6.2: Validation, state variable monotonicity, compressible RaNS:SA TEA $O(h^4)$ PDE system GWS^h FE $p = 1$ algorithm, Sajben transonic diffuser strong curved shock, mesh M_{10}, Re $= 7 \times 10^5$: (a) Ma^h(x_j,T) true scale flood, (b) shock-centric Ma^h(x_j,T) distribution in unitized scale perspective with contours, viewpoint slightly upstream, (c) obliqueness distribution induced onset velocity shock local streamline redistribution true scale, (d) contoured surface adjacent separation bubble streamline distribution, unitized scale local zoom.

scale zoom of velocity streamline distribution in the separation bubble adjacent to the contoured surface is Figure 6.2d.

6.3 Validation, first-order effects predictions, same code/same mesh RaNS:SA $O(h^4)$ PDE GWS^h and $O(h^2)$ PDE non-GWS^h numerically stabilized CFD algorithms

A COMSOL commercial code user execution option is a regular solution-adapted nonuniform mesh refinement sequence for the Sajben transonic diffuser using their "company confidential" numerically stabilized (hence non-Galerkin) FE $p = 1$ basis algorithm with SA eddy viscosity transport PDE closure. Specifically pertinent to sought validation of error freed CFD mathematics first- and higher-order effects prediction validations, the COMSOL code user interface enables precise definition of the compressible RaNS:SA TEA $O(h^4)$ PDE GWS^h algorithm except for inclusion of TEA theory fouth-order nonlinearities in the code linear algebra Jacobian. Thus enabled is same code/same mesh *exact* comparison of Galerkin and non-Galerkin CFD algorithm predictions of a transonic shock with obliqueness distribution.

Identical ten stage regular solution adapted nonuniform mesh refinement cycles were executed for the RaNS:SA TEA $O(h^4)$ PDE GWS^h and COMSOL RaNS:SA $O(h^2)$ PDE non-GWS^h stabilized FE $p = 1$ basis algorithms for Sajben diffuser weak and strong shock Mach number onset BC definitions. CFD first-order effects prediction validation invariably involves prediction comparisons with experimental data. The terminal mesh M_{10} axial pressure distribution comparisons with experimental data along the diffuser contoured surface (Figure 6.3a and b) clearly quantify first-order effects prediction validation for both GWS^h and non-GWS^h algorithms.

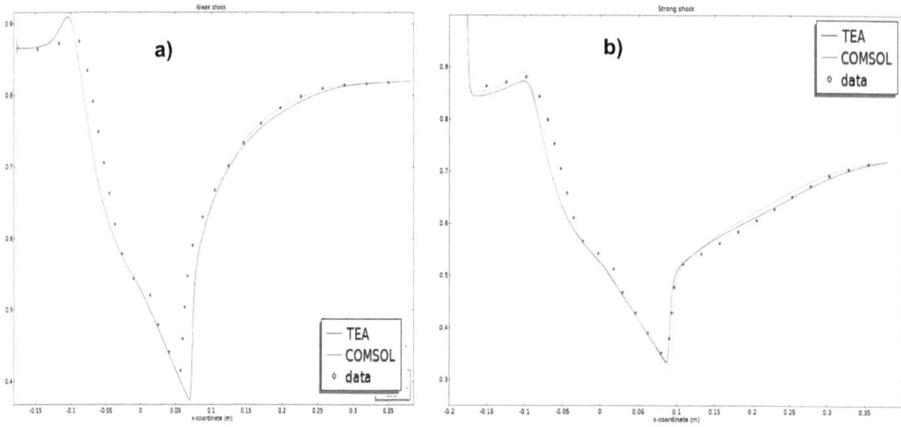

Figure 6.3: Validation, first-order effects CFD predictions, compressible RaNS:SA TEA $O(h^4)$ PDE GWS^h and RaNS:SA $O(h^2)$ PDE non-GWS^h pressure distributions compared with Sajben diffuser contoured surface data, M_{10}, Re = 7 × 10^5: (a) shock $Ma_S = 1.23$ and (b) shock $Ma_S = 1.36$.

Experimental data exist for axial velocity resolution $u_1{}^h(\mathbf{x},T)$ transverse direction distributions at diffuser downstream coordinates $x/h_T = 4.61$ and $x/h_T = 6.64$ for the Sajben diffuser strong shock Mach number onset BC. The RaNS:SA $O(h^4)$ PDE GWS^h generated $u_1{}^h(x/h_T = 4.61, y, t = T)$ distribution exhibits better agreement with these data than does the $O(h^2)$ PDE non-GWS^h prediction (Figure 6.4a). Conversely both GWS^h and non-GWS^h algorithm predicted $u_1{}^h(x/h_T = 6.64, y, t = T)$ distributions bracket the experimental data (Figure 6.4b), and each clearly document persistence of reversed flow to the diffuser outflow plane. This occurrence, first reported in Hsieh et al. [6], is thus independently corroborated by these two totally independent same code/mesh CFD predictions.

Thereby these fundamentally distinct CFD predicted $p^h(\mathbf{x},T)$ and $u_1{}^h(\mathbf{x},T)$ distribution comparisons with experimental data constitute first-order effects validation for both algorithms. Further, these comparisons quantitatively substantiate the assertion made in Chapter 1 that,

> ". . CFD has generated the sum total of all first-order fluid mechanics insight."

To conclude, this same code/mesh CFD exercise also generated the absolutely unique opportunity to compare RaNS:SA TEA $O(h^4)$ PDE GWS^h and $O(h^2)$ PDE non-GWS^h stabilized algorithm predictions of the Mach number and momentum vector resolution distributions on an axial trace through the shock normal onset extremum $Ma_S = 1.36$

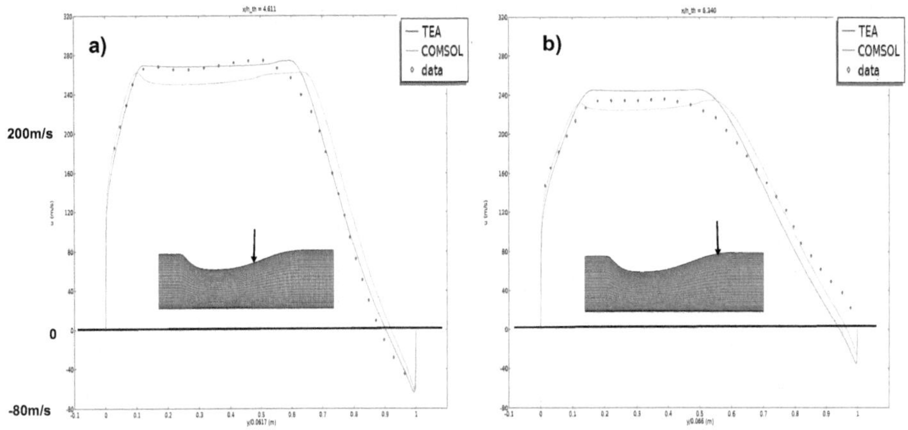

Figure 6.4: Validation, first-order effects CFD prediction, compressible RaNS:SA TEA $O(h^4)$ PDE GWS^h and RaNS:SA $O(h^2)$ PDE non-GWS^h numerically stabilized algorithms predicted velocity axial resolution distribution comparisons with Sajben diffuser strong shock $Ma_S = 1.36$ data, mesh M_{10}, $Re = 7 \times 10^5$: a] $u_1^h(x/h_T = 4.61, y, T)$, b) $u_1^h(x/h_T = 6.64, y, T)$.

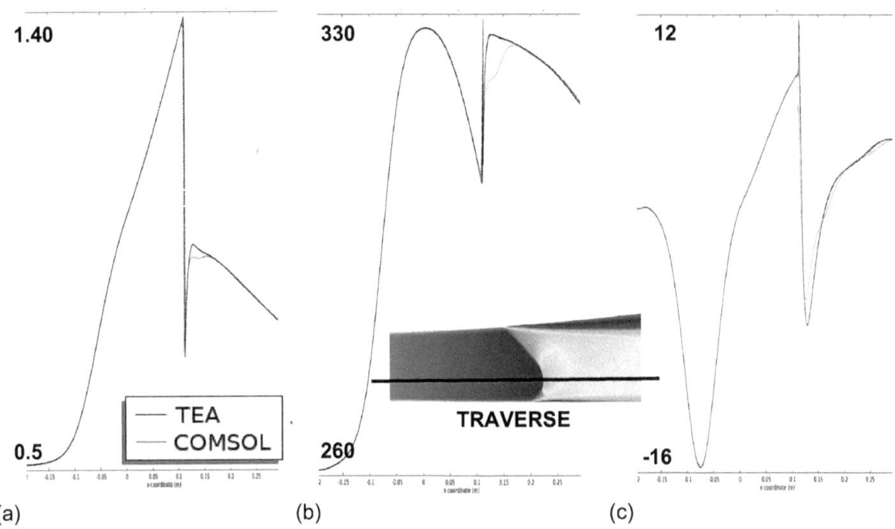

Figure 6.5: Apparent validation, CFD first-order effects prediction, compressible RaNS:SA TEA $O(h^4)$ PDE GWS^h and RaNS:SA $O(h^2)$ PDE non-GWS^h Mach number and momentum vector resolution distributions on the axial traverse through the strong $Ma_S = 1.36$ normal shock, Sajben diffuser, mesh M_{10}, $Re = 7 \times 10^5$, ordinate labels denote the distribution extrema in each graph.

(Figure 6.5). That these dual CFD-generated distributions precisely overlay globally appears to constitute an added first-order effects validation, a conclusion to be proven without merit.

6.4 Validation/failure, GWS^h/non-GWS^h CFD first-, higher-order effects predictions

Close examination of the same code/mesh data in Figure 6.5 confirms existence of clearly visible shock-centric distinctions between the GWS^h and non-GWS^h code distributions. Specifically, the non-GWS^h $\mathrm{Ma}^h(\mathbf{x},\mathrm{T})$ data contour is completely devoid of the post-shock $\mathrm{Ma}^h(\mathbf{x},\mathrm{T})$ extremum present in the error freed CFD GWS^h prediction. This distinction as well exists in the immediate post-shock reach of the $m_1{}^h(\mathbf{x},\mathrm{T})$ resolution distributions, the axial coordinate of which correlates precisely with the post-shock deep subsonic depression with extremum $\mathrm{Ma} \approx 0.6$ for which the GWS^h distribution is validated to be monotone (Figure 6.2b).

Theoretical fluid mechanics predicts that *only* the onset velocity resolution normal to an oblique shock undergoes a decrease in magnitude. RaNS:SA TEA $O(h^4)$ PDE GWS^h predicted shock-centric transverse velocity resolution $u_2{}^h(x_j,\mathrm{T})$ distribution flood with isovels is graphed true scale in Figure 6.6a. The singular instance of onset $u_2{}^h(x_j,T)$ distribution shock normal encounter is clearly displayed and fully detailed by the contours. The GWS^h continuum interpolation of the oblique shock influenced $u_2{}^h(x_j,T)$ distribution on both sides of the shock normal encounter is precisely monotone, and the trajectory of the curved shock possesses a uniformly turning tangent locus throughout.

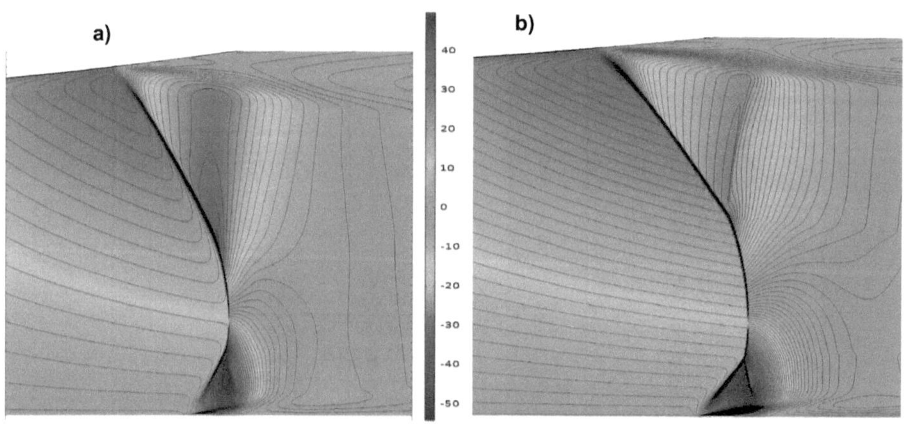

Figure 6.6: Failure/validation, first-/higher-order effects CFD predictions, Sajben diffuser strong curved shock, mesh M_{10}, $\mathrm{Re} = 7 \times 10^5$: (a) RaNS:SA $O(h^4)$ PDE GWS^h $u_2{}^h(x_j,T)$ distribution with isovels; (b) RaNS:SA $O(h^2)$ PDE non-GWS^h $u_2{}^h(x_j,\mathrm{T})$ distribution with isovels.

The same code/mesh M_{10} RaNS:SA $O(h^2)$ PDE numerically stabilized non-GWS^h predicted $u_2{}^h(x_j,T)$ distribution with contours (Figure 6.6b) is totally devoid of the GWS^h generated pre-shock isovel alterations, except directly adjacent to each surface of the diffuser. Additionally, modest tangent slope trajectory discontinuities exist in the non-GWS^h shock distribution at locations just above and below onset velocity normal shock encounter.

Thereby, that these distinctly theorized algorithm generated shock traverse distributions exhibit "essential" global agreement, Figure 6.5, *does not* constitute dual first-order effects validation. This conclusion is demonstrably quantified by direct comparison of GWS^h and non-GWS^h algorithm generated shock centric $u_2{}^h(x_j,T)$ isovel distribution super zoom insertions (Figure 6.7). The $u_2{}^h(x_j,T)$ distribution alteration generated by $O(h^4)$ *error freed* CFD in mathematics PDE GWS^h oblique shock traverse continuum interpolation is smoothly continuous (Figure 6.7a). Conversely, the numerically stabilized shock capture $O(h^2)$ PDE non-GWS^h function has *totally corrupted* the $u_2{}^h(x_j,\mathrm{T})$ oblique shock traverse distribution alteration both pre- and post-shock (Figure 6.7b).

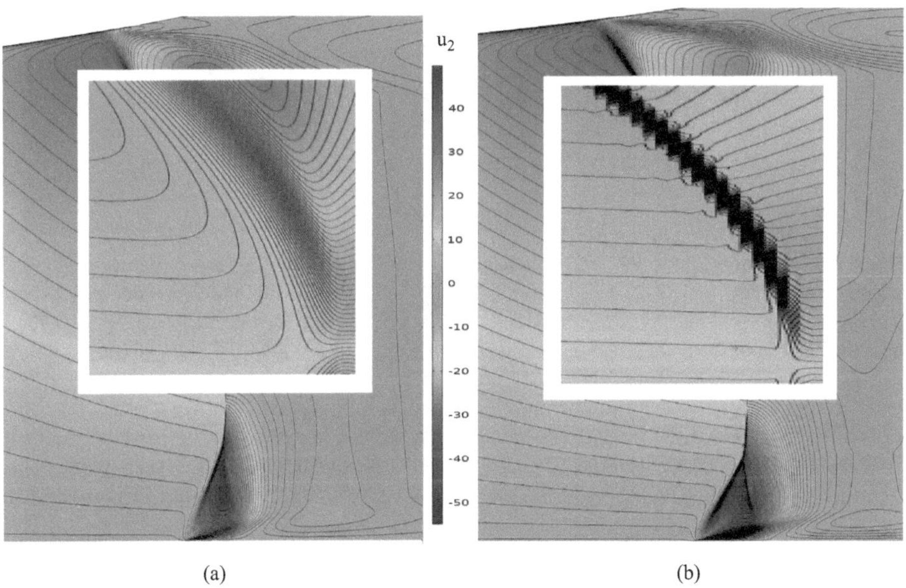

Figure 6.7: Failure/validation, first-/higher-order effects CFD predictions, shock-centric $u_2{}^h(x,y,t = \mathrm{T})$ distribution with contours, Sajben diffuser strong shock, mesh M_{10}, Re $= 7 \times 10^5$: (a) $O(h^4)$ PDE GWS^h shock interpolation insert super zoom, (b) $O(h^2)$ PDE non-GWS^h stabilized shock capture insert super zoom, adapted from Baker and Freels [3].

Of course, non-GWS^h algorithm *phake physics* stabilized shock capture error adversely affects the post-shock fidelity of RaNS:SA PDE state variable scalar member predictions. The $O(h^4)$ PDE GWS^h Ma$^h(x_j,\mathrm{T})$ translucent perspective distribution (Figure 6.8a) quantitatively validates that it is monotone to linear algebra iterative convergence digit. The GWS^h generated pressure $p^h(x_j,\mathrm{T})$ extremum in the post-shock subsonic depression is a sharply peaked monotone distribution (Figure 6.8b). The exactly comparable same code/mesh $O(h^2)$ PDE non-GWS^h stabilized shock-capture

$p^h(x_j T)$ distribution extremum is raggedly non-monotone (Figure 6.8c), hence devoid of the error-freed CFD predicted extremum. These data further quantify non-GWS^h algorithm failure to achieve *either* first- or higher-order effects validation.

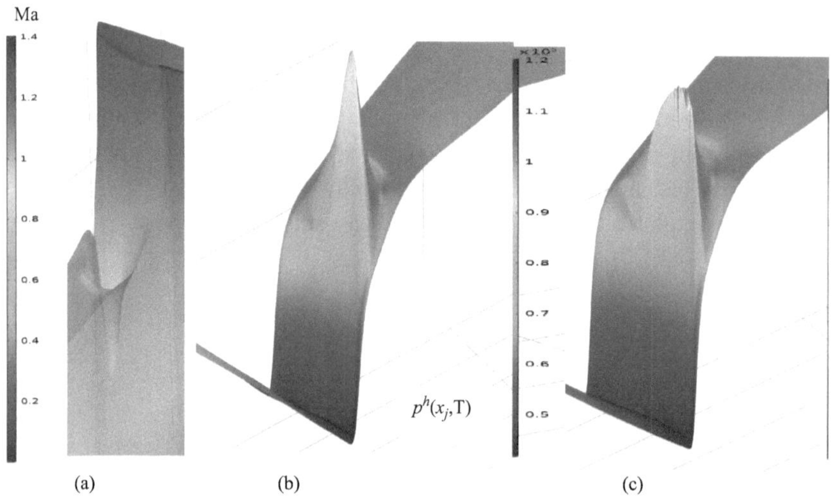

Figure 6.8: Validation/failure, RaNS state variable scalar member monotonicity, Sajben transonic diffuser strong curved shock, mesh M_{10}, Re = 7×10^5: (a) $O(h^4)$ PDE GWS^h Ma$^h(x_j,T)$ subsonic depression translucent perspective zoom, (b) validation, $O(h^4)$ PDE GWS^h pressure $p^h(x_j,T)$ perspective zoom, (c) failure, $O(h^2)$ PDE non-GWS^h stabilized shock-capture $p^h(x_j,T)$ perspective zoom.

The RaNS:SA $O(h^4)$ PDE GWS^h sharply peaked density $\rho^h(x_j,T)$ monotone distribution with extremum (Figure 6.9a) constitutes both first- and higher-order effects prediction validation. The same code/mesh $O(h^2)$ PDE non-GWS^h $\rho^h(x_j,T)$ non-monotone distribution, Figure 6.9b, quantifies total compromise of extremum fidelity by stabilized shock-capture. The $O(h^4)$ PDE GWS^h predicted shock centric $\rho^h(x_j,T)$ shadowed true scale distribution, Figure 6.9c, visually documents the resolution fidelity of subtle higher-order effects enabled using $O(h^4)$ error free CFD mathematics.

To conclude, compressible RaNS:SA TEA $O(h^4)$ PDE system continuous GWS^h FE $p = 1$ basis algorithm written on *error freed* CFD mathematics is *validated* to enable generation of physics of fluids *fidelity monotone* resolution of the sum total ***of all*** first- and higher-order effects resident in the Sajben diffuser strong shock onset BC specification. The same code/mesh $O(h^2)$ PDE system non-GWS^h generated data fully quantify that *phake* physics stabilized shock capture *totally* compromised the fidelity of shock centric first- and higher-order effects predictions. Thus substantiated is the second assertion in Chapter 1 which states,

> ". . error mechanisms intrinsic to legacy CFD practice are the same order as the higher-order effects that are not resolved."

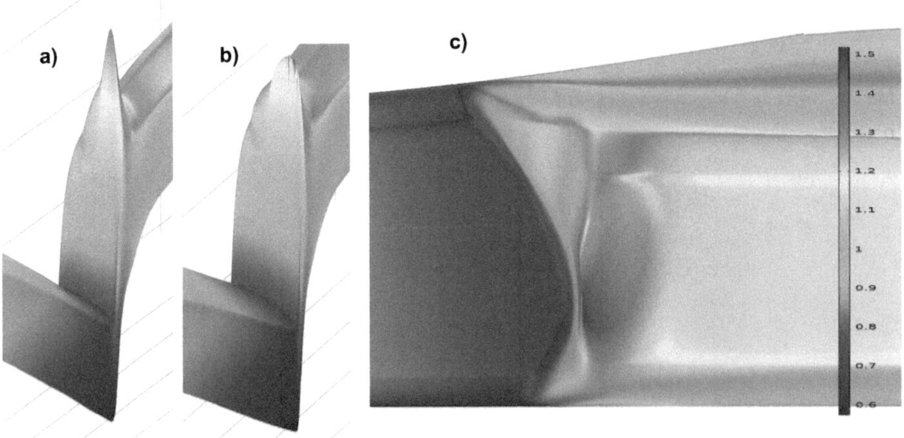

Figure 6.9: Validation/failure, RaNS state variable scalar member monotonicity, Sajben transonic diffuser curved strong shock density distributions, mesh M_{10}, Re = 7×10^5: (a) $O(h^4)$ PDE GWS^h $\rho^h(x_j,T)$ perspective zoom, (b) $O(h^2)$ PDE non-GWS^h numerical shock capture non-monotone $\rho^h(x_j,T)$ perspective zoom, (c) GWS^h post-shock $\rho^h(x_j,\mathrm{T})$ distribution true scale shadowed flood.

6.5 Validation, shock-turbulent BL *interaction* CFD diagnostics, *error freed* PDE prediction of the dominant *physics of fluids* process

An inadequacy cited for the WIND-US code, Georgiadis et al. [5].

> . . . shock/turbulent boundary layer interaction . . .

prompts a repeat of the RaNS:SA TEA $O(h^4)$ PDE GWS^h algorithm regular solution-adapted mesh refinement sequence to the further refined 4×10^6 element FE $p = 1$ basis support for RaNS 2.4×10^7 DOF monotone diagnostics of strong oblique shock-turbulent boundary layer interaction in the Sajben diffuser. The terminal mesh M_{13} $\mathrm{Ma}^h(x_j,\mathrm{T})$ monotone distribution (Figure 6.10a) provides fully detailed resolution of the immediate post-shock distribution of speed that remains just supersonic (light red) distant from farfield velocity shock normal onset. The magnitude of the $\mathrm{Ma}^h(\mathbf{x},T)$ distribution ranging 1.400⇔0.61 (red to light yellow) in the normal shock generated deep subsonic depression is quantified resolutely monotone in translucent perspective projection (Figure 6.10b).

The distribution of shock-centric momentum resolution $m_1{}^h(\mathbf{x},T)$ (Figure 6.10c) quantifies that the extremum 320 kg · m/s D (dark red) persists over the entire transverse span of the *remarkably thin* (!) sliver separating the normal shock from subsonic depression initiation. It is this $m_1{}^h(\mathbf{x},T)$ contour feature that is barely visible in Figure 6.5b. Finally, the cut plane graphic visualization in Figure 6.10d validates the

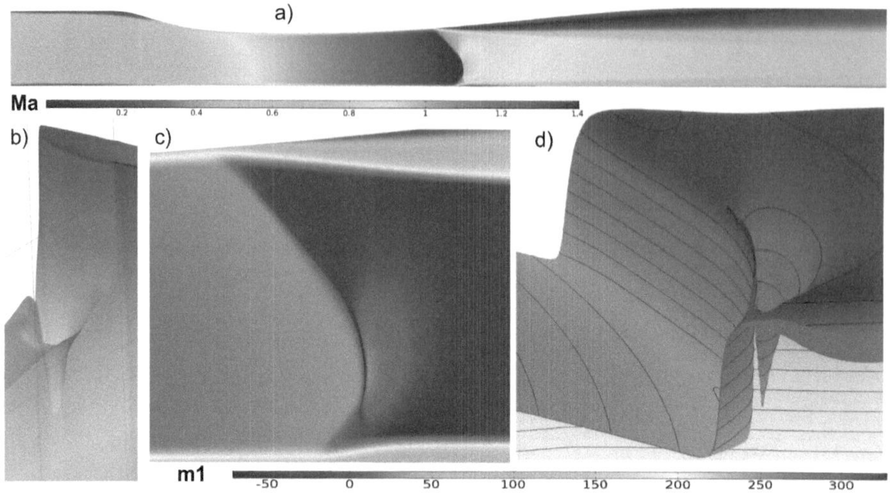

Figure 6.10: Validation, DOF monotonicity, RaNS:SA TEA $O(h^4)$ PDE GWS^h algorithm, Sajben diffuser strong shock, mesh M_{13}, Re = 7×10^5: (a) Ma^h($\boldsymbol{x}$,T) true scale flood, color bar below, (b) subsonic depression Ma^h($\boldsymbol{x}$,T) translucent perspective projection, view upstream, (c) shock-centric $m_1{}^h(\boldsymbol{x},\mathrm{T})$ true scale flood, and (d) cut plane visualized $m_1{}^h(\boldsymbol{x},\mathrm{T})$ extremum monotone perspective.

sharply resolved $m_1{}^h(\mathbf{x},\mathrm{T})$ distribution with Ma = 0.61 extremum is monotone throughout the subsonic depression.

That the $O(h^4)$ PDE system GWS^h generated Ma^h($\mathbf{x}$,T) and $u_2{}^h(\mathbf{x},\mathrm{T})$ distributions adjacent to each Sajben diffuser surface in the near vicinity of oblique shock–turbulent boundary layer interactions are monotone is validated in Figure 6.11a–d. The physics of fluids fidelity resolution of the contoured surface adjacent $u_2{}^h(\mathbf{x},\mathrm{T})$ distribution near the oblique shock–turbulent boundary layer separation point is monotone (Figure 6.11b). The reach of lower surface post-shock adjacent $u_2{}^h(\mathbf{x},\mathrm{T})$ distribution is clearly delineated by contours clustering (Figure 6.11d).

Recall that Reynolds ordering of non-D RaNS PDE systems quantifies the significance of BL velocity transverse resolution $u_2{}^h(\mathbf{x},\mathrm{T})$ is $O(\mathrm{Re}^{-1/2})$. Sajben diffuser experimental data (Figure 6.3) confirm the freestream edge of the pre-shock lower surface BL $u_1{}^h(\mathbf{x},\mathrm{T})$ magnitude is 250 ± m/s D, hence the BL significance of $u_2{}^h(\mathbf{x},\mathrm{T})$ would be 1 ± m/s D. The extremum in the $u_2{}^h(\mathbf{x},\mathrm{T})$ distribution in the lower surface BL at 50 ± m/s D, the deep "red" spot in Figure 6.12a, exceeds BL significance by almost two orders of magnitude. In comparison, at shock encounter initiation of BL separation on the diffuser contoured surface, the $u_2{}^h(\mathbf{x},\mathrm{T})$ extremum at 10± m/s D (yellow-green) exceeds BL significance by an order of magnitude.

Error freed CFD mathematics has thus diagnosed that the *physics of fluids* mechanism *dominant* in shock–turbulent boundary layer interaction is shock impact generation of a $u_2{}^h(\mathbf{x},\mathrm{T})$ plume with magnitude sufficient to force large-scale mass ejection from the turbulent BL into the immediate free stream which may or may not induce

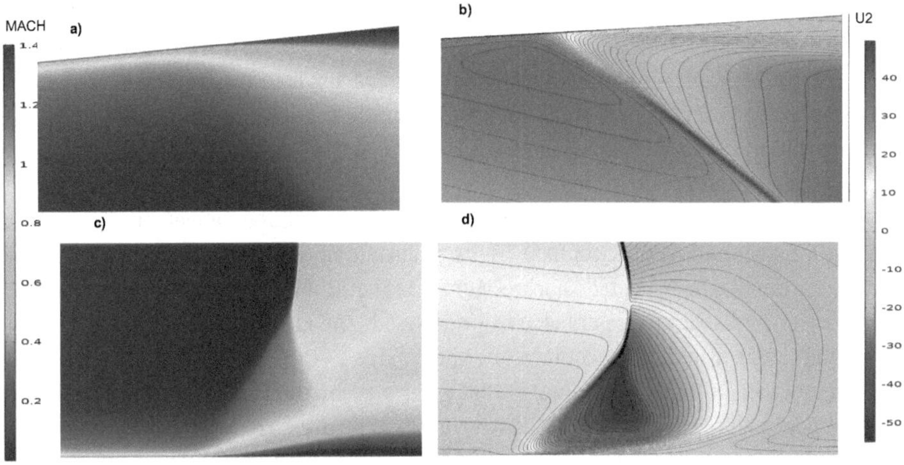

Figure 6.11: RaNS:SA TEA $O(h^4)$ PDE GWS^h shock-centric state variable shock–turbulent boundary layer interaction flood zooms, Sajben diffuser, mesh M_{13}, Re = 7×10^5, upper contoured surface: (a) $Ma^h(\boldsymbol{x},T)$, (b) $u_2{}^h(\boldsymbol{x},T)$ with contours; lower surface: (c) $Ma^h(\boldsymbol{x},T)$, (d) $u_2{}^h(\boldsymbol{x},T)$ with contours.

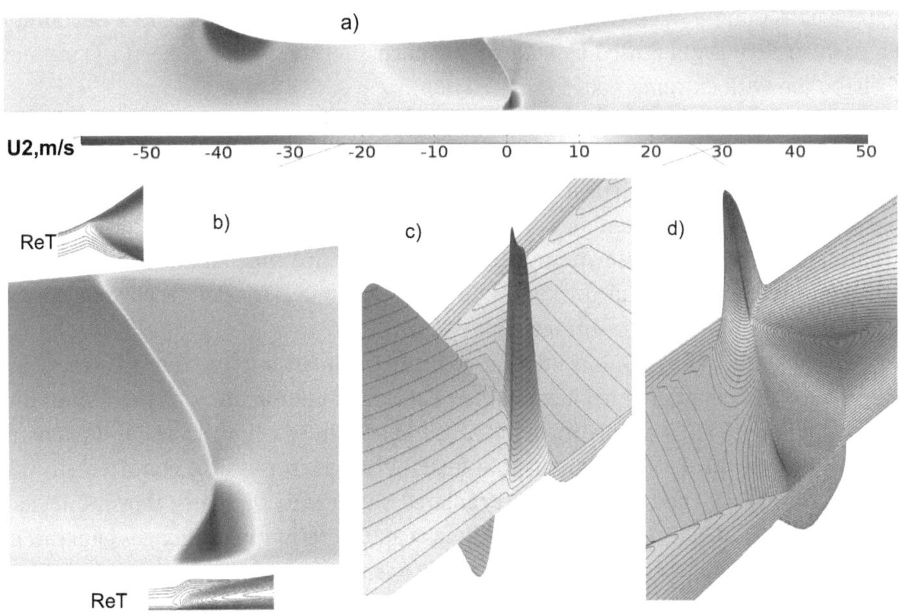

Figure 6.12: Diagnostics, error-freed CFD physics of fluids characterization of shock–turbulent boundary layer interaction, Sajben transonic diffuser, RaNS:SA TEA $O(h^4)$ PDE GWS^h algorithm, mesh M_{13}, Re = 7×10^5: (a) $u_2{}^h(\boldsymbol{x},T)$ distribution true scale flood, color bar below, (b) shock-centric $u_2{}^h(\boldsymbol{x},T)$ flood zoom with adjacent Re^{Th} snippets; diffuser lower surface shock interaction induced BL $u_2{}^h(\boldsymbol{x},T)$ distribution perspective projection zooms, (c) viewed upstream, and (d) view slightly downstream transverse from the contoured surface.

BL separation. The $u_2{}^h(\boldsymbol{x},\mathrm{T})$ distribution enclosing this 50± m/s D extremum with adjacent BL SA closure $\mathrm{Re}^{\mathrm{T}h}$ snippets is detailed in Figure 6.12b. Diffuser lower surface adjacent $u_2{}^h(\mathbf{x},\mathrm{T})$ *monotone* distribution perspective projections, Figure 6.12c and d, quantitatively validate this error freed CFD mathematics diagnostic. The full size color graphic of Figure 6.12d on the monograph cover incisively quantifies shock contact and monotonicity attributes.

Observe that the bounding surface of the $u_2{}^h(\mathbf{x},\mathrm{T})$ distribution envelope encasing the 50± m/s D extremum directly contacts (!) the "subsonic side" of the continuous GWS^h generated continuum *interpolated* shock (Figure 6.12d). And identically, the vertical surface in the $u_2{}^h(\boldsymbol{x},\mathrm{T})$ distribution perspective underneath the clearly visible footprint of the shock curvature also directly contacts the shock. This $u_2{}^h(\boldsymbol{x},\mathrm{T})$ distribution with −32 ± m/s D extremum (light-to-medium blue) is that generated by the shock curvature trajectory in the upper half of the diffuser, as clearly visible in Figure 6.12b.

The Spalart-Allmaras TEA-altered $\mathrm{D}^m\mathrm{SA}$ $O(h^4)$ PDE GWS^h predicted eddy viscosity $\widetilde{\nu}^h(x,\mathrm{T})$ distribution Figure 6.13a quantifies the extremum (red) permeates the centroidal region of the separation bubble. Data alteration to non-D turbulent Reynolds number $\mathrm{Re}^{hT}(\boldsymbol{x},\mathrm{T})$ unitized scale flood, (Figure 6.13b) quantifies that the 6500± (red) extremum is an annular ring with an essentially nil magnitude central core enveloping the extremum in $\widetilde{\nu}^h(x,T)$. The abrupt increase in BL thickness at shock encounter on each surface is quantified by coincident sharp alteration in each $\mathrm{Re}^{hT}(\boldsymbol{x},\mathrm{T})$ surface distribution with extrema 1,000± (Figure 6.13b).

SA closure $O(h^4)$ PDE GWS^h generated $\mathrm{Re}^{hT}(\boldsymbol{x},\mathrm{T})$ distribution is clearly validated *monotone* in the global perspective projection (Figure 6.13c). The shock impact-induced minimal width vertical column in the lower surface adjacent contour is just perceptible with the thickening uniform from the BL freestream to the $\mathrm{Re}^{hT}(\mathbf{x},\mathrm{T})$ extremum (Figure 6.13d). Viewpoint alteration to surface normal confirms that the shock impact induced local thickening monotonically decreases to zero well before encounter with the diffuser surface (Figure 6.13e).

To conclude, the weak formulation continuous Galerkin algorithm cast on $O(h^4)$ *error freed* CFD mathematics has enabled rigorous quantification of the dominant-order *physics of fluids* effects in shock-turbulent BL interaction. That the diagnostic quantifies that the RaNS state variable distribution is in physical contact with the "subsonic side" of the continuum interpolated shock fully validates that the legacy CFD practice of *phake physics* stabilized shock capture seriously challenges generation of both first and higher-order physics of fluids *fidelity* predictions.

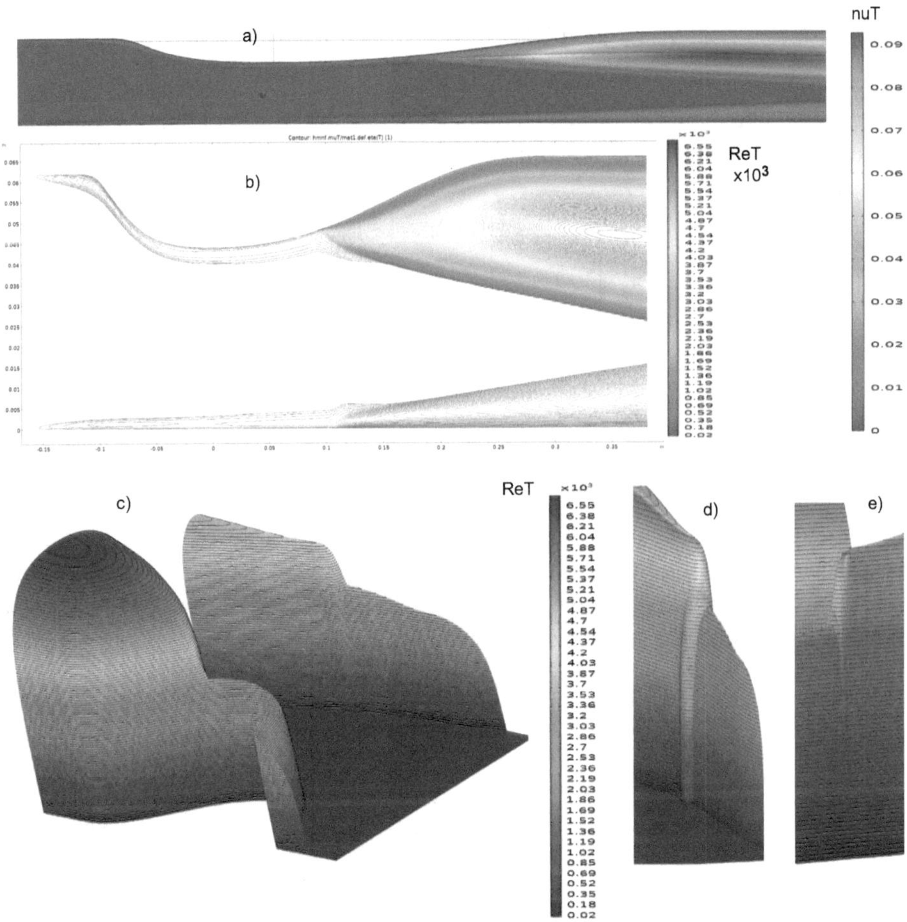

Figure 6.13: SA closure TEA $O(h^4)$ PDE GWS^h generated steady distributions, Sajben transonic diffuser strong shock, mesh M_{13}, Re = 7×10^5: (a) $\tilde{\nu}^h$ $(x, y, t = T)$ true scale flood, (b) $Re^{hT}(\boldsymbol{x},T)$ unitized scale flood, (c) $Re^{hT}(\boldsymbol{x},T)$ unitized scale perspective, (d) lower surface shock-centric $Re^{hT}(\boldsymbol{x},T)$ zoom, viewed from freestream, and (e) $Re^{hT}(\boldsymbol{x},T)$ perspective zoom, viewed from lower surface.

6.6 Derivation, incompressible RaNS TEA $O(h^4)$ D^mP:D^mTKE:$O(h^2)$ DM^h PDE system continuous GWS^h FE basis algorithm, Newton Jacobian

For large Re, *analytically* derived $O(h^2)$ TE *annihilation* theory altered incompressible RaNS model **DP** remains

$$\bar{f}_j(\bar{q},x_i)\frac{\partial \bar{q}}{\partial x_j} - \frac{1}{\mathrm{Re}}\left(1+\mathrm{Re}^T\right)\frac{\partial^2 \bar{q}}{\partial x_j^2} - \frac{\mathrm{Re}h^2}{12\left(1+\mathrm{Re}^T\right)}\bar{f}_k(\bar{q},x_i)\frac{\partial}{\partial x_k}\left[\bar{f}_j(\bar{q},x_i)\frac{\partial \bar{q}}{\partial x_j}\right] = 0$$

Referencing Chapter 5 NS development and deleting overbar notation, the non-D unsteady incompressible RaNS TEA $O(h^4)$ $\mathrm{D}^m\mathbf{P}$ is

$$\mathrm{D^m}\mathbf{P}:\ \frac{\partial u_i}{\partial t} + \frac{\partial}{\partial x_j}\left[\begin{array}{l} u_j u_i + P^*|_{n+1}^{p}\,\delta_{ij} - \left(\dfrac{1+\mathrm{Re}^T}{\mathrm{Re}}\right)\left(\dfrac{\partial u_i}{\partial x_j} + \dfrac{\partial u_j}{\partial x_i}\right) \\ \qquad - \dfrac{h^2\mathrm{Re}}{12\left(1+\mathrm{Re}^T\right)}\left(u_j u_k \dfrac{\partial u_i}{\partial x_k}\right) \end{array}\right]$$
$$- \frac{\mathrm{Gr}}{\mathrm{Re}^2}\,\Theta\widehat{g}_i = 0 + O(h^4)$$

That turbulent Reynolds number $\mathrm{Re}^T(x_j)$ distribution resides in the $\mathrm{D}^m\mathbf{P}$ TEA term denominator elevates the $O(h^4)$ $\mathrm{D}^m\mathbf{P}$ PDE nonlinearity to fourth order.

Selected turbulent closure model for incompressible RaNS TEA $O(h^4)$ PDE system is the turbulent kinetic energy-isotropic dissipation function (k-ε) PDE system. For the resultant state variable $\{q(x_j, t)\} = \{\varphi, u_i, \Theta, k, \varepsilon; P\}^T$, the TEA altered $O(h^4)$ PDE system $\mathrm{D}^m\mathrm{Q}$ for RaNS:TKE state variable scalar members is

$$\mathrm{D^mQ}:\ \frac{\partial}{\partial t}\begin{Bmatrix}\Theta \\ k \\ \varepsilon\end{Bmatrix} + \frac{\partial}{\partial x_j}\begin{Bmatrix} u_j\Theta - \dfrac{1}{\mathrm{Re}}\left(\dfrac{1}{\mathrm{Pr}} + \dfrac{\mathrm{Re}^T}{\mathrm{Pr}^T}\right)\dfrac{\partial\Theta}{\partial x_j} - \dfrac{h^2\mathrm{RePr}^T}{12\left(1+\mathrm{Re}^T\right)}u_j u_k\dfrac{\partial\Theta}{\partial x_k} \\ u_j k - \dfrac{1}{\mathrm{RePr}}\left(1+\dfrac{\mathrm{Re}^T}{\mathrm{C}_k}\right)\dfrac{\partial k}{\partial x_j} - \dfrac{h^2\mathrm{RePr}}{12\left(1+\mathrm{Re}^T/\mathrm{C}_k\right)}u_j u_k\dfrac{\partial k}{\partial x_k} \\ u_j\varepsilon - \dfrac{1}{\mathrm{RePr}}\left(1+\dfrac{\mathrm{Re}^T}{\mathrm{C}_\varepsilon}\right)\dfrac{\partial\varepsilon}{\partial x_j} - \dfrac{h^2\mathrm{RePr}}{12\left(1+\mathrm{Re}^T/\mathrm{C}_\varepsilon\right)}u_j u_k\dfrac{\partial\varepsilon}{\partial x_k}\end{Bmatrix}$$
$$-\begin{Bmatrix} s(\Theta) \\ s(k) \\ s(\varepsilon)\end{Bmatrix} = \{0\} + O(h^4)$$

for Pr the Prandtl number and a turbulent Prandtl number model Pr^T. Closure model coefficients C_k, C_ε and entries in source matrix $\{s(\bullet)\}$ are all well known, cf., [14].

Green-Gauss divergence theorem imposition on GWS^h TEA $O(h^4)\mathrm{D}^m\mathbf{P}{:}\mathrm{D}^m\mathrm{Q}$ PDE systems generates the element matrix

$$\int_{\Omega_e}\{\mathrm{N}\}\frac{\partial}{\partial x_j}\left(\frac{h^2\mathrm{Re}\,(u_j u_k)}{12\left(1+\mathrm{Re^T}\right)}\frac{\partial u_i}{\partial x_k}\right)^h d\tau$$
$$= \frac{h_e^2\mathrm{Re}}{12}\int_{\Omega_e}\frac{\partial\{\mathrm{N}\}}{\partial x_j}\left(\frac{\{UJ\}_e^{\mathrm{T}}\{\mathrm{N}\}\{\mathrm{N}\}^{\mathrm{T}}\{UK\}_e}{\{1+\mathrm{RET}\}_e^{\mathrm{T}}\{\mathrm{N}\}}\right)\frac{\partial\{\mathrm{N}\}^{\mathrm{T}}}{\partial x_k}\{\mathrm{UI}\}_e d\tau$$

summation on all lower/upper case tensor index pairs. As with the compressible NS algorithm, the compute *efficient* alternative for rational tensor product *interpolation* is interpolation of the associated DOF *rational product*

$$\frac{\{UJ\}_e^T\{N\}\{N\}^T\{UK\}_e}{\{1+RET\}_e^T\{N\}} \approx \left\{\frac{UJ*UK}{1+RET}\right\}_e^T\{N\}$$

The terminal statement for $D^m\mathbf{P}$ TEA term element matrix upon rational product DOF extraction from the GWS^h integral is

$$GWS^h: -\int_{\Omega_e}\{N\}\frac{\partial}{\partial x_j}\left(\frac{h^2\mathrm{Re}}{12(1+\mathrm{Re}^T)}u_ju_k\frac{\partial u_i}{\partial x_k}\right)^h d\tau$$

$$\approx \frac{h_e^2\mathrm{Re}}{12}\left\{\frac{UJ*UK}{1+RET}\right\}_e^T[M30JK]_e\{UI\}_e, 1 \le I \le n$$

summation on $1 \le J, K \le n$.

State variable member alterations readily identify the TEA $O(h^4)D^mQ$ PDE system element matrices differ only by their scalar multipliers. FE $p = 1$ basis element *hypermatrix* $[M30JK]_e$ was introduced in Chapter 5, and the element measure product h^2 therein remain FE $p = 1,2,3$ basis element-dependent data.

Incompressible RaNS:TKE scalar kinetic pressure utilizes the $O(h^4)D^m\mathbf{P}$:$O(h^2)DM^h$ + NBC PDE system validated in Chapter 5. Insertion of thus quantified $O(\varepsilon)$ divergence-free velocity $u_j^h(x_i, t)$ into the divergence of RaNS $D\mathbf{P}$ alters the *genuine* kinematic pressure NS DP Poisson PDE, with *exact* NBC valid on the entirety of domain closure $\partial\Omega$, to

$$DP: \frac{\partial^2 P}{\partial x_k^2} + \frac{\partial}{\partial x_j}\left[u_i\frac{\partial u_j}{\partial x_i} - \frac{(1+\mathrm{Re}^T)}{\mathrm{Re}}\frac{\partial u_j}{\partial x_j} + \frac{\mathrm{Gr}}{\mathrm{Re}^2}\Theta\hat{g}_j\right] = 0, \ NBC: \left(\frac{\partial P}{\partial x_j} - \frac{1}{\mathrm{Re}}\frac{\partial^2 u_j}{\partial x_k^2}\right)\hat{n}_j = 0$$

Linear algebra stability for RaNS $O(h^4)D^m\mathbf{P}$:D^mQ:$O(h^2)DM^h$ + NBC:DP + NBC PDE system $GWS^h + \theta TS$ algorithm at large Re is ensured by inclusion of the TEA fourth-order non-linear contributions into the Newton algorithm Jacobian. The $D^m\mathbf{P}$ state variable self-coupling element matrix contribution to the Newton Jacobian is matrix differential calculus identified to be

$$[JAC\_\{UI\},\{UI\}]_e = \frac{\Theta\Delta t\, h_e^2\mathrm{Re}}{12}\left(\left\{\frac{UJ*UK}{1+RET}\right\}_e^T[M30JK]_e + \{UI\}_e^T[M3KJ0]_e\mathrm{diag}\left[\frac{UK\,\delta_{\underline{IJ}}}{1+RET}\right]_e\right.$$

$$\left. + \{UI\}_e^T[M3KJ0]_e\mathrm{diag}\left[\frac{UJ\,\delta_{\underline{IK}}}{1+RET}\right]_e\right)$$

Hypermatrix pre- and post-column matrix interchange in GWS^h $D^m\mathbf{P}$ matrix again leads to the second and third terms with tensor index sequence-altered hypermatrix set $[M3KJ0]_e$, $1 \le (J,K) \le n$. Underbar denoted Kronecker deltas remain indicators *only* (no summation). The Newton Jacobian element matrics for $\mathrm{D}^m\mathbf{P}$ state variable member *off-diagonal* coupling are

$$[\mathrm{JAC\_\{U}I\}, \{\mathrm{U}J\}]_e = \frac{\Theta\Delta t\, h_e^2\mathrm{Re}}{12}\left(\{\mathrm{U}I\}_e^T[\mathrm{M3}KJ0]_e\mathrm{diag}\left[\frac{\mathrm{U}K\,\delta_{ij}}{1+\mathrm{RET}}\right]_e\right)$$

$$[\mathrm{JAC\_\{U}I\}, \{\mathrm{U}K\}]_e = \frac{\Theta\Delta t\, h_e^2\mathrm{Re}}{12}\left(\mathrm{U}I\}_e^T[\mathrm{M3}KJ0]_e\mathrm{diag}\left[\frac{\mathrm{U}J\,\delta_{IK}}{1+\mathrm{RET}}\right]_e\right)$$

Alteration of RaNS $\mathrm{D}^m\mathbf{P}$ state variable member couplings to the $\mathrm{D}^m\mathrm{Q}$ PDE system state variable then identifies appropriate TKE element matrix contributions to the Newton Jacobian.

6.7 Validation, RaNS:TKE $O(h^4)$ PDE system GWS^h + θTS FE $p = 1$ basis algorithm

RaNS:TKE $O(h^4)\mathrm{D}^m\mathbf{P}$:$\mathrm{D}^m\mathrm{Q}$:$O(h^2)\mathrm{DM}^h$ + NBC:DP + NBC PDE system GWS^h + θTS FE $p = 1$ basis algorithm validation addresses isothermal incompressible fluid entering a rectangular cross-sectional channel from a reservoir. The resultant inviscid slug profile onset velocity encounters the RaNS no-slip BC at the channel entrance plane which is located *interior* to the RaNS:TKE PDE system domain Ω, Figure 6.14. This *singular* BC specification injects *large* (!) wall normal gradients into RaNS state variable $\{q(x_j, t)\} = \{\varphi, u_j, k, \varepsilon; P\}^T$ distributions at the no-slip BC plane which in turn induces $O(h^2)$ truncation error local distributions possessing *significant* (!) magnitudes.

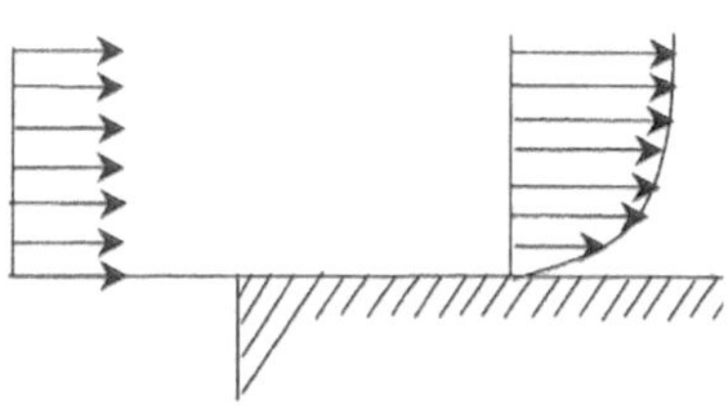

Figure 6.14: Inviscid slug profile farfield velocity onset to a channel.

Slug profile onset velocity BC is $u_1 = 122$ m/s D (400 f/s) which for air generates unit Reynolds number Re/L = 4E + 06 and Re ≈ E + 05 based on channel half span L = 0.025 m. The entrance plane of the channel is located at mesh M axial plane 18 of the RaNS domain Ω^h constituted of 52 mesh planes. Upper/lower horizontal surface no-slip velocity BC imposition at this plane immediately energizes the $\mathrm{D}^m\mathrm{Q}$ TEA $O(h^4)$ PDE GWS^h + θTS algorithm. A regular solution-adapted nonuniform $h(x_j)$ mesh exercise quantified that the plane 18 centric channel walls adapted M = 52 × 32 × 2 (axial, transverse, and lateral) FE $p = 1$ basis $\Omega^h = \cup\,\Omega_e$ mesh supports adequate resolution of the mirror symmetric k,ε model closed turbulent incompressible BL pair at sufficient distance downstream from imposition of the no-slip BC.

Domain onset plane velocity resolution DOF non-D Dirichlet BCs are {U1} = {1.0}, {U2} = {0} = {U3} with {1 + RET} = {9} the non-zero Dirichlet initial condition (IC) required for initiation of the D^mQ $O(h^4)$PDE algorithm. On Ω^h domain axial mesh planes 1 through 18, the homogeneous Neumann BC wrapper for the velocity resolution eliminates lateral direction leakage. Thereafter, homogeneous Neumann {U3} BCs are retained to enforce the unbounded lateral span of channel domain Ω^h. The RaNS $O(h^4)$ PDE system IC is projection of the onset slug profile {U1} and {1 + RET} DOF data onto all domain Ω^h nodes except for the no-slip BC nodes located on each channel surface.

The incompressible RaNS:TKE $O(h^4)D^m\mathbf{P}:D^mQ:O(h^2)DM^h$ + NBC:DP + NBC PDE system $GWS^h + \theta TS$ algorithm/code required ~2,000 time steps $n\Delta t$ to clear the state variable IC error using the time accurate $\theta = 0.5$ $O(\Delta t^3)$ trapezoidal rule. Thereafter evolved a *slowly* time varying, periodic precisely mirror symmetric RaNS state variable DOF distribution for an additional 5,000 time steps $n\Delta t$ for convergence $|\{\delta Q(n\Delta t)\}|_{\max} \leq \varepsilon = E-04$. Total elapsed time to $5{,}000\Delta t \equiv T \approx 4$ s correlates with onset flowfield clearing of the channel length ~ 100 times Unsteady RaNS $O(h^4)D^m\mathbf{P}:D^mQ:O(h^2)DM^h$ + NBC:DP + NBC PDE system $GWS^h + \theta TS$ FE $p = 1$ basis algorithm/ $O(\Delta t^3)$ code is thus *validated* large Re long time stable *totally absent* artificial stabilization.

The mid-period unsteady RaNS:TKE $GWS^h + \theta TS$ generated state variable $\left\{q^h(x_j, T)\right\}$ *monotone* DOF distribution snapshot supports algorithm validation. Velocity axial resolution $u_1^h(x_j, T)$ distribution snapshot unitized scale perspective with mesh overlay (Figure 6.15a) view downstream quantifies that onset slug profile initiation to a mirror symmetric, eventually fully turbulent BL profile distribution is *instantaneous*, recall the Chapter 5 statement on the *immediacy* of incompressible DM.

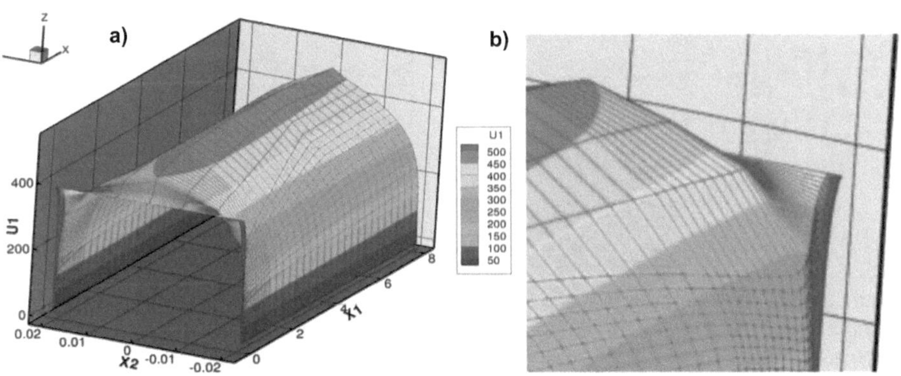

Figure 6.15: Validation, $u_1^h(x_j, T)$ monotone distribution snapshot unitized scale perspective projection with mesh overlay, reservoir outlet rectangular cross-sectional channel, RaNS TEA $O(h^4)D^m\mathbf{P}:D^mQ:O(h^2)$ DM^h + NBC:DP + NBC PDE system $GWS^h + \theta TS$ FE $p = 1$ basis algorithm, Re ≈ E + 05,: a) global view downstream; b) no-slip BC plane 18 centric zoom, view upstream.

That the abrupt alteration of the $u_1^h(x_j, T)$ distribution at mesh plane 18 no-slip BC imposition is *strictly monotone* is validated in the perspective zoom (Figure 6.15b) view upstream. This distribution further quantifies the *robustness* of TEA theory $O(h^2)$ TE annihilation efficacy on the not very well designed (!) Ω^h mesh. Thus validated is the Chapter 1 assertion,

. . DOF distributions are monotone on ***any*** (!) mesh M . . .

for TEA theory altered $O(h^4)$ NS/RaNS PDE system tri-diagonal stencil equivalent algorithms.

The distribution of RaNS:TKE isotropic dissipation function ε is the *acid test* for $O(h^2)$ truncation error annihilation validation as the dissipation *extremum* is located at the wall node DOF which renders its prediction *intensely* influenced by no-slip BC initiation. RaNS: TKE TEA altered $O(h^4)\mathrm{D}^m\mathbf{P}{:}\mathrm{D}^m\mathrm{Q}{:}O(h^2)\mathrm{DM}^h$ + NBC:DP + NBC GWS^h + θTS predicted $\varepsilon^h(x_j, \mathrm{T})$ distribution snapshot is graphed for unitized scale perspective (Figure 6.16a), view downstream, the wall node extrema in which is truncated for visualization. The transition of the nil magnitude onset $\varepsilon^h(x_j, \mathrm{T})$ distribution to the *extremum* indeed occurs at the mesh plane 18 wall nodes. The downstream decay rate of the $\varepsilon^h(x_j, T)$ wall node extremum is *monotone* with transition to nominal x_1-independence occurring at ~ 12% channel travel. The plane 18 centric zoom of $\varepsilon^h(x_j, T)$ near-wall distribution perspective zoom with mesh overlay (Figure 6.16b) quantitatively *validates* that the no-slip BC plane 18 wall centric distribution of $\varepsilon^h(x_j, T)$ is *monotone* throughout.

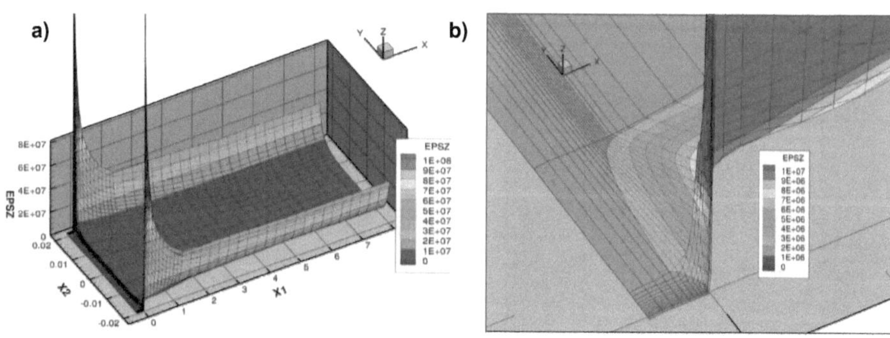

Figure 6.16: Validation, TKE closure model $\varepsilon^h(x_j, \mathrm{T})$ unitized scale perspective projection, RaNS:TKE $O(h^4)$ $\mathrm{D}^m\mathbf{P}{:}\mathrm{D}^m\mathrm{Q}{:}O(h^2)\mathrm{DM}^h$ + NBC:DP + NBC PDE system GWS^h + θTS FE $p = 1$ basis algorithm, reservoir outlet rectangular channel, Re ≈ E + 05: (a) global view downstream, wall extrema truncated; (b) no-slip BC mesh plane 18 centric perspective zoom.

RaNS:TKE GWS^h predicted turbulent kinetic energy $k^h(x_j, \mathrm{T})$ distribution is devoid of a plane 18 wall node DOF extrema, hence also is the $\mathrm{Re}^{T^h}(x_j, \mathrm{T})$ distribution (Figure 6.17a) unitized scale perspective view downstream. The symmetric pair of $\mathrm{Re}^{T^h}(x_j, \mathrm{T})$ distribution DOF *extrema* at 450 (red) at the channel outflow plane nodes confirms a fully turbulent mirror symmetric BL pair. Regarding pressure, downstream of the nil gradient (deep red) distribution existent within the reach of the slug profile inflow (Figure 6.17b) view upstream, the axial gradient of $P^h(x_j, \mathrm{T})$ is a uniform constant with zero lateral gradient throughout These data quantitatively validate the RaNS kinematic pressure NS DP Poisson PDE formulation with *exact* domain NBC.

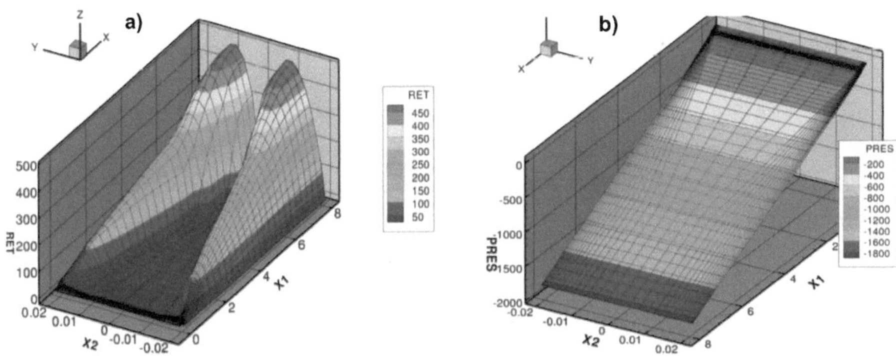

Figure 6.17: Validation, RaNS:TKE TEA $O(h^4)\mathrm{D}^m\mathbf{P}$:$\mathrm{D}^m$Q:$O(h^2)\mathrm{DM}^h$ + NBC:DP + NBC PDE system GWS^h + θTS FE $p = 1$ basis predictions, reservoir outlet rectangular channel, Re ≈ E + 05, unitized scale perspective projections: (a) $\mathrm{Re}^{T^h}(x_j, \mathrm{T})$ distribution, view downstream, (b) kinematic pressure $P^h(x_j, \mathrm{T})$ distribution view upstream.

To conclude, RaNS:TKE TEA $O(h^4)\mathrm{D}^m\mathbf{P}$:$\mathrm{DM}^h$ + NBC:DP + NBC PDE system GWS^h + θTS FE $p = 1$ basis algorithm periodic solution snapshots *validate* existence of resolutely *monotone* incompressible RaNS state variable distributions using a "legacy CFD" $O(h^2)$ PDE *tridiagonal stencil* equivalent *non-stabilized* algorithm on a *non-optimal* (!) mesh M discretization of the onset channel domain Ω^h with domain-interior singular no-slip BC. TEA theory alteration of any legacy CFD RaNS code will reproduce this performance!

6.8 Spectral distribution of TEA annihilated $O(h^2)$ error, RaNS:TKE $O(h^4)$ PDE system

RaNS:TKE state variable member-specific *annihilated* $O(h^2)$ truncation error distributions are the negative of the entries in the $O(h^4)$ PDE system. Alteration to the equivalent differential statement, e.g., for $\mathrm{D}^m\mathbf{P}$

$$\mathrm{DS}\left(e^{h^2}(u_i^h)\right) \equiv e^{h^2}(u_i^h(x_j, t)) - \frac{h^2 Re}{12}\frac{\partial}{\partial x_j}\left\{\left(1 + \mathrm{Re}^T\right)^{-1} u_j^h u_k^h \frac{\partial u_i^h}{\partial x_k}\right\} = 0$$

enables GWS^h algorithm prediction of the distribution of spectral content of the annihilated $O(h^2)$ truncation error. Altered to *rational* tensor product DOF interpolation, Green-Gauss divergence theorem imposition generates the GWS^h element matrix

$$\mathrm{GWS}^h\colon [\mathrm{M200}]_e\{e^{h^2}\_UI\}_e + \frac{h_e^2\mathrm{Re}}{12}\left(\{\mathrm{U}J{*}\mathrm{U}K\,/\,(1+\mathrm{RET})\}_e^T\,[\mathrm{M30}JK]_e\{\mathrm{U}I\}_e\right), 1 \le I \le n$$

summation on $1 \le (J,K) \le n$.

Thus generated $e^{h^2}(u_1^h(x_j, T)$ distribution possesses opposite sign ± 4 (red/blue) *extrema* DOF on the plane 18 no-slip BC wall nodes, just barely visible in perspective super zoom (Figure 6.18a). These DOF exceed $u_1^h(x_j, t)$ non-D significance by an order of magnitude. Plane 18 centric perspective zoom (Figure 6.18b), quantifies the $e^{h^2}(u_1^h(x_j, T)$ distribution is precisely mirror symmetric. That the significance of $e^{h^2}(u_1^h(x_j, T)$ downstream transitions to nil confirms TEA theory *self-assessment* that no continuing need exists for $O(h^2)$ TE annihilation. The super zoom of no-slip BC centric $e^{h^2}(u_1^h(x_j, T)$ distribution (Figure 6.18c), extrema truncated, quantifies the period of the $O(h^2)$ TE distribution is precisely $2h$ which *validates* that the functionality of TEA theory is $O(h)$.

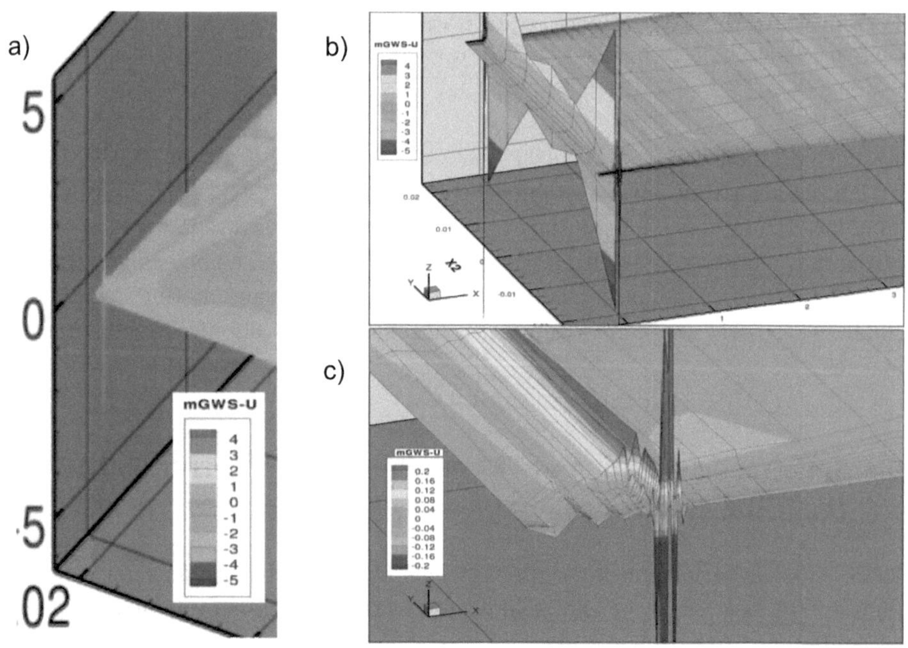

Figure 6.18: Spectral distribution of TEA theory annihilated error $e^{h^2}(u_1(x_j, \mathrm{T})$, slug profile onset to rectangular channel, Re ≈ E + 05: (a) mesh plane 18 centric extrema super zoom, (b) entrance region true scale zoom, and (c) BC mesh plane local super zoom, extrema truncated.

In RaNS:TKE $O(h^4)$ PDE velocity transverse resolution $e^{h^2}(u_2^h(x_j, \mathrm{T})$ mirror symmetric distribution perspective (Figure 6.19a), the DOF *extrema* ± 0.6 (red/blue) well exceed $u_2^h(x_j, T)$ significance The lateral perspective zoom (Figure 6.19b) quantifies error distribution is mirror symmetric and short lived.

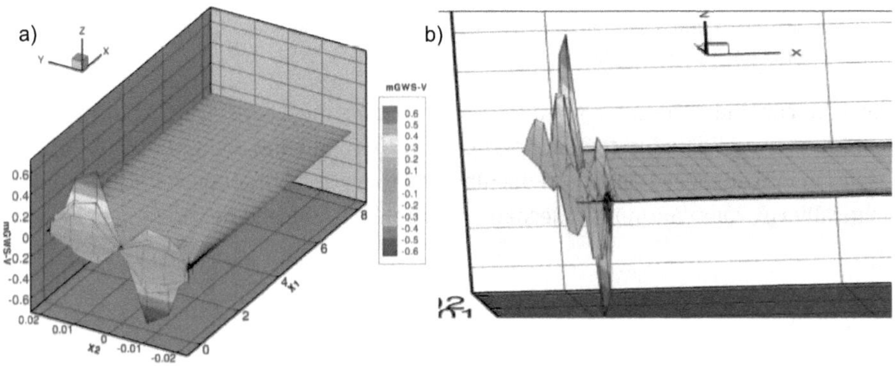

Figure 6.19: Validation, spectral distribution of $u_2^h(x_j, \mathrm{T})$ annihilated truncation error $e^{h^2}\left(u_2^h(x_j, \mathrm{T})\right)$, RaNS: TKE TEA altered $O(h^4)$ $\mathrm{D}^m\mathbf{P}$, entrance channel, Re ≈ E + 05: (a) global perspective unitized scale view downstream, (b) entrance region true scale lateral perspective zoom.

6.9 Conclusion, RaNS *error freed* CFD mathematics $O(h^4)$ PDE systems

To conclude, TEA theory alteration of the textbook appearance of compressible and incompressible NS and RaNS PDE systems is validated to *annihilate* spatial discretization-induced $O(h^2)$ truncation error, hence algorithm algebraic instability. Resultant is *physics of fluids* fidelity *error freed* CFD mathematics generation of stable, at least $O(h^4)$ asymptotic convergent NS/RaNS state variable DOF distributions *monotone* to iteration convergence significant digit on ***any*** mesh! Thereby the continuing *need* (!) for NS/RaNS CFD algorithm *phake physics* stabilization, the *penultimate* focus (!) of seven plus decades of FD/FV CFD algorithm theorization, is thus obviously open to serious question.

And recall from Chapter 1,

> . . . legacy CFD second order numerics . . . has resolved all it can regarding . .
> . . *large scale* first order fluid mechanics knowledge. The *significance* of that remaining,. . . ., i.e., *genuine* physics of fluids insight, is the same order as error mechanisms intrinsic to legacy CFD practice. CFD advancements *unstagnation* require . . . legacy error mechanisms be *annihilated* . . . no alternative exists . . . for extraction of high fidelity *physics of fluids* fidelity higher order insight from CFD predictions.

Clearly, the content of monograph Chapters 5 and 6 collectively *validates* (!) these assertions.

Of truly consequential impact, implementation of these *key unstagnation* identifications into legacy CFD code practice will lead to the refuting of the ***too often expressed*** (!) pessimism about the limitations innate to NS/RaNS CFD, cf., [13].

> There are lots of things you can't compute with sufficient confidence
> with CFD. It's that simple.

Hence, to conclude, the *error freed* CFD *mathematics* theoretical content derived and quantitatively validated in this monograph ***finally*** (!) enables computational fluid dynamics to join advanced textbooks documented theoretical fluid mechanics as a mathematically *rigorous* continuum mechanics discipline in the engineering sciences

References

[1] Allmaras, S., Johnson, F., Spalart, P., (2012). "Modifications and Clarifications for the Implementation of the Spalart-Allmaras Turbulence Model," Proc. ICCFD7-1902.

[2] Baker, A.J., Orzechowski, J.A. (2017). "Analytical $O(h^2)$ Error Annihilation Theory: *FREE* $O(h^4)$ Upgrade for Second Order Numerics Codes," *J Num Heat Trans, Part B*, 71(5).

[3] Baker, A.J., Freels, J.D. (2025). "Physics of fluid fidelity error free computational fluid dynamics mathematics," *Physics of Fluids*, 35,4.

[4] Bogar, T.J., Sajben, M. and Kroutil, J.C. (1983). "Characteristic Frequencies of Transonic Diffuser Flow Oscillations," AIAA J., 21(9).

[5] Georgiadis, N.J., Yoder, D.A., Towne, C.S., Engblom, W.A., Bhagwandin, V.A., Power, G.D., Lankford, D.W., Nelson, C.C. (2009). "Wind-US Code Physical Modeling Improvements to Complement Hypersonic Testing and Evaluation," NASA/TM-2009- 215615, Tech Rpt AIAA:2009-193

[6] Hsieh, T., Wardlaw, A.B., Bogar, T.J., Collins, T., Coakley, T. (1987). "Numerical Investigation of Unsteady Inlet Flowfields," AIAA *J*, 25(1).

[7] NPARC Alliance Validation Archive (2008a). "Sajben Transonic Diffuser: Study #1," http://www.grc.nasa.gov/WWW/wind/valid/ . . .

[8] NPARC Alliance Validation Archive (2008b). "Sajben Transonic Diffuser: Study #2" http://www.grc.nasa.gov/WWW/wind/valid/.

[9] Sajben, M., Kroutil, J.C., Chen, C.P. (1977). "A High Speed Schlieren Investigation of Diffuser Flows with Dynamic Distortion," Tech Rpt. AIAA77:875.

[10] Salmon, J.T., Bogar, T.J., and Sajben, M. (1983). "Laser Doppler Velocimetry in Unsteady Separated Transonic Flow," AIAA J. 21(12).

[11] Spalart, P., Allmaras, S.,(1992). "A One Equation Turbulence Model for Aerodynamics," Tech. Report AIAA92.1370, AIAA Aerospace Sciences conference

[12] Spalart, P., Allmaras, S. (1994). "A One Equation Turbulence Model for Aerodynamics," *Le Recherche Aerospatiale*, 1.

[13] Stumpe, J. (2018). "Why CFD and Wind Tunnels Need Each Other," *Aerospace America*, 56(6).

[14] Wilcox, D.C. (2006). *Turbulence Modeling for* CFD, DCW Industries, La Canada, CA.

Chapter 7
Annihilation of NS/RaNS PDE space-time discretization-induced $O(m^2, m^3)$ phase dispersion and $O(\Delta t^3)$ truncation errors

Libretto: Taylor series (TS) operations enabled NS/RaNS spatial discretization-induced $O(h^2)$ truncation error identification, the annihilation of which eliminates code algebraic instability.

TS operations extension to non-D wave number ($m = \kappa h$) space identifies space-time discretization-generated NS scalar member convective transport dispersion (*aliasing*) error.

Reversed sign insertion of this cubically nonlinear *calculus* functional into NS/RaNS PDE *annihilates* the lead significance $O(m^p)$, $p = 2, 3$ TE terms that generate aliasing error.

Finally, TS manipulations on the NS/RaNS PDE $\text{GWS}^h + \theta = 0.5\text{TS}$ algebraic statement identify the functionals that *annihilate* the trapezoidal rule $O(\Delta t^3)$ TE.

Net is TEA theory-predicted *continuum* calculus alterations to classic appearance NS/RaNS PDE systems that enable tri-diagonal stencil-equivalent prediction of *monotone* DOF distributions TS fourth-order *accurate* in physical space, wave number space and time on *any* mesh.

Keywords: CFD, Navier-Stokes, aliasing error, TEA theory, efficiency, fourth-order stencil

7.1 Characterization, NS scalar convective transport dispersion error mechanisms

A significant liability with current practice CFD codes is the compromised fidelity of mass transport prediction, e.g., the mixing of compressible fluids. This error mechanism arises from the truncation errors generated by space-time discretization of NS/RaNS PDE systems, which compromise the preservation of the Fourier spectral content that is initially present in state variable scalar member distributions.

The propagation of a scalar Fourier resolution, called a wave, by a fluid with velocity $\mathbf{u}$ in the direction of the wavenumber vector $\boldsymbol{\kappa}$ is characterized by the phase velocity $\mathbf{c} \equiv (\mathbf{u} \cdot \boldsymbol{\kappa})\boldsymbol{\kappa}/\kappa^2$ [4]. Wavenumber magnitude $\kappa = |\boldsymbol{\kappa}|$ equates to the number of wave crests existing within the interval 2π in the wave vector direction. In the continuum, the phase velocity magnitude $c = |\mathbf{c}|$ is equal to the fluid speed $u = |\mathbf{u}|$.

https://doi.org/10.1515/9783112206430-008

The means whereby smooth waves propagate scalar variable Fourier content in a dispersive media is characterized by the group velocity $\mathbf{v}_g(\boldsymbol{\kappa}) \equiv \nabla_\kappa \omega$, where ω is the cyclic frequency. In the continuum, the group velocity is independent of the wavenumber $\boldsymbol{\kappa}$, and hence equates to the fluid velocity $\mathbf{u}$. Conversely, the phase and group velocities differ from their continuum values in space–time discretized CFD unsteady scalar convective transport due to the induced dependence of group velocity on wavenumber.

To quantify, the expression for group velocity dependence on wavenumber in 1D

$$v_g(\kappa) \equiv \frac{\partial \omega(\kappa)}{\partial \kappa} = \frac{\partial (c(\kappa)\kappa)}{\partial \kappa} = c(\kappa) + \kappa \frac{\partial (c(\kappa))}{\partial \kappa},$$

states that the phase speed, c, is independent of the wavenumber magnitude, κ, if and only if (*iff*) the group and phase speeds (velocities in n-D) are equal. The occurrence of group velocity not aligned with the wave vector in discretized n-D space will augment this error mechanism.

7.2 Derivation, NS DM scalar transport CFD algorithm amplification factor, $n = 1$

Fourier theoretical analysis addresses the compressible NS time TS-altered DM ODE

$$\mathrm{D}^m\mathrm{M}: \frac{\partial q}{\partial t} + \frac{\partial uq}{\partial x} - \frac{\Delta t}{2}\frac{\partial}{\partial x}\left(\alpha u \frac{\partial q}{\partial t} + \beta u^2 \frac{\partial q}{\partial x}\right) - \frac{\Delta t^2}{6}\frac{\partial}{\partial x}\left(\gamma u^2 \frac{\partial^2 q}{\partial x \partial t}\right) = 0 + O(\Delta t^3)$$

The TS "μ term" is omitted as it is not tridiagonal stencil compatible. For Euler single-stage θ–implicit time integration, superscripted n, n + 1, the NS time TS-altered $\mathrm{D}^m\mathrm{M}$ $\mathrm{GWS}^h + \theta\mathrm{TS}$ FE $p = 1$ basis algorithm assembled global matrix

$$\mathrm{GWS}^h + \theta\mathrm{TS}: \{\mathrm{F}(Q(t)\} = [\mathrm{M}(\alpha,\gamma)]^h \{Q^{n+1} - Q^n\} + \Delta t \left[\theta \frac{\partial \{f^h(\beta)\}}{\partial x}^{n+1} + (1-\theta)\frac{\partial \{f^h(\beta)\}}{\partial x}^{n}\right] = \{0\},$$

identifies the space–time disposition of the coefficient set α, β, γ.

Momentarily neglecting the α, β, γ terms, the $n = 1$ non-D DM $\mathrm{GWS}^h + \theta\mathrm{TS}$ FE $p = 1$ basis element matrix $\{\mathrm{F}(Q(t)\}_e$ assembled (S_e) at the uniform mesh node X_j is the $O(h^2)$ tridiagonal stencil.

$$\begin{aligned} \mathrm{S}_e(\mathrm{GWS}^h + \theta\mathrm{TS})\colon F(Q_j) = \frac{1}{6}\left[\left(Q_{j-1}^{n+1} - Q_{j-1}^{n}\right) + 4\left(Q_j^{n+1} - Q_j^{n}\right) + \left(Q_{j+1}^{n+1} - Q_{j+1}^{n}\right)\right] \\ + \mathrm{C}\left[\theta\left(-Q_{j-1}^{n+1} + Q_{j+1}^{n+1}\right) + (1-\theta)\left(-Q_{j-1}^{n} + Q_{j+1}^{n}\right)\right] = 0 + O(h^2), \end{aligned}$$

wherein the *Courant number* $C \equiv U\Delta t/h$ is the non-D time step. The directly comparable nearest-neighbor finite difference $O(h^2)$ centered space-time Crank-Nicolson (CNFD) algorithm stencil is

$$\mathrm{FD} + \theta\mathrm{TS}:\ \mathrm{F}(Q_j) = Q_j^{n+1} - Q_j^{n} + C\left[\theta\left(-Q_{j-1}^{n+1} + Q_{j+1}^{n+1}\right) + (1-\theta)\left(-Q_{j-1}^{n} + Q_{j+1}^{n}\right)\right] = 0 + O(h^2)$$

for $\theta \equiv 0.5$, Crank and Nicolson [2]. The sole distinction in these stencils is $\mathrm{GWS}^h + \theta\mathrm{TS}$ "mass matrix" nearest neighbor (1,4,1)/6 interpolation of the DOF time derivative.

For any stencil, the generic mesh node DOF Q_j projected onto its time t_n spatially discrete Fourier mode representation

$$Q(j\Delta x, n\Delta t) \equiv Q_j^n = (g^h)^n q^h(j\Delta x) = (g^h)^n e^{i\kappa(j\Delta x)}$$

defines the algorithm amplification factor g^h. Substituting this expression into $\mathrm{F}(Q_j)$ and projecting all $j \pm 1$ and $n + 1$ operands to j and n again generates an algebraic statement $[\bullet](g^h)^n e^{i\kappa(j\Delta x)} = 0$. Recalling the complex variable expressions for sin(•) and cos (•), the vanishing of [•] defines the $n = 1$ DM $\mathrm{GWS}^h + \theta\mathrm{TS}$ FE $p = 1$ basis algorithm amplification factor as

$$\mathrm{GWS}^h + \theta\mathrm{TS}:\ g^h = \frac{1 + (1/2)\cos(\kappa\Delta x) - i(3C(1-\theta)/2)\sin(\kappa\Delta x)}{1 + (1/2)\cos(\kappa\Delta x) + i(3C\theta/2)\sin(\kappa\Delta x)}$$

which is valid $\forall C$ and θ. The directly comparable $\theta \equiv 0.5$ CNFD amplification factor valid $\forall C$ is

$$\mathrm{CNFD}:\ g^h = \frac{1 - i(C/2)\sin(\kappa\Delta x)}{1 + i(C/2)\sin(\kappa\Delta x)}$$

A CFD algorithm is free from artificial diffusion *iff* the magnitude of g^h, i.e., $g^{h*}g^h$, is unity, g^{h*} the amplification factor complex conjugate. For $\mathrm{GWS}^h + \theta\mathrm{TS}$, this results *only* for definition $\theta = 0.5$, which again validates the *optimal* $\mathrm{GWS}^h + \theta\mathrm{TS}$ decision as well as the CNFD default. The admissible alternative selections $\theta = 0$, $0.5 < \theta \le 1.0$ render $g^{h*}g^h < 1$ which quantifies these algorithms possess *artificial numerical diffusion.* Definitions $0 < \theta < 0.5$ render $g^{h*}g^h > 1$ which quantifies these algorithms are algebraically *unstable.*

7.3 Derivation, NS mass transport D^mY $O(m^p)$ TE annihilation, $p = 2, 3$, $n = 1$

Altering unsteady $n = 1$ DM to mass/mole fraction convective transport DY adds a non-D diffusion term with Sc the binary diffusion Schmidt number. The resultant $\theta \equiv 0.5$ time TS (α, β, γ) altered $n = 1$ D^mY ODE:

$$\mathrm{D}^m\mathrm{Y}:\ \frac{\partial q}{\partial t}+\frac{\partial uq}{\partial x}-\frac{1}{\mathrm{ReSc}}\frac{\partial^2 q}{\partial x^2}-\frac{\Delta t}{2}\frac{\partial}{\partial x}\left(\alpha u\frac{\partial q}{\partial t}+\beta u^2\frac{\partial q}{\partial x}\right)-\frac{\Delta t^2}{6}\frac{\partial}{\partial x}\left(\gamma u^2\frac{\partial^2 q}{\partial x\partial t}\right)=0+O(\Delta t^3)$$

when manipulated, as the prior section illustrated, identifies the $\mathrm{D}^m\mathrm{Y}$ ODE $\mathrm{GWS}^h+\theta\mathrm{TS}$ algorithm amplification factor to be

$$g^h=\frac{\begin{array}{c}\left(2+\gamma\mathrm{C}^2-3(1-\theta)\beta\mathrm{C}^2-6(1-\theta)\mathrm{D}\right)\\+\left(1-\gamma\mathrm{C}^2+3(1-\theta)\beta\mathrm{C}^2+6(1-\theta)\mathrm{D}\right)\cos m-\mathrm{i}3\mathrm{C}\left(\frac{1}{2}\alpha+(1-\theta)\right)\sin m\end{array}}{\left(2+\gamma\mathrm{C}^2+3\theta\beta\mathrm{C}^2+6\mathrm{D}\theta\right)+\left(1-\gamma\mathrm{C}^2-3\theta\beta\mathrm{C}^2-6\mathrm{D}\theta\right)\cos m-\mathrm{i}3\mathrm{C}\left(\frac{1}{2}\alpha-\theta\right)\sin m}$$

wherein $\mathrm{D}\equiv\Delta t/\mathrm{ReSc}h^2$ is the diffusion term non-D coefficient. The $\mathrm{D}^m\mathrm{Y}$ ODE $\mathrm{GWS}^h+\theta\mathrm{TS}$ algorithm is "genuine physics" diffusive as the amplification factor magnitude $g^{h*}g^h$ is invariably less than unity.

As derived, the phase speed c is independent of wavenumber κ *only* if the group and phase speeds (velocities in n-D) are equal. Thereby, the theoretical analysis requirement is to *extremize* the wavenumber *independence* of $\mathrm{D}^m\mathrm{Y}$ $\mathrm{GWS}^h+\theta\mathrm{TS}$ *phase speed* error $err(g^h)\equiv g_{exact}-g^h$ This involves TS manipulations in non-D *wavenumber* $m\equiv\kappa h$ space which then requires the resultant complex functions g^h and g_{exact} be multiplied by their complex conjugates followed by clearing of thus generated denominators by expansion of resultant $g^{h*}g^h$ into a TS of sufficient $O(m^{\mathrm{p}})$ in non-D wavenumber.

For the $n=1$ $\mathrm{D}^m\mathrm{Y}$ ODE, the wave number space exercise predicted TS truncated at $O(m^5)$ is, Sahu and Baker [7]

$$\mathrm{GWS}^h+\theta\mathrm{TS}:\ g^h=1-\mathrm{i}\mathrm{C}m+\left[\frac{1}{2}(\alpha-\beta-\mathrm{D})-\theta\right]\mathrm{C}^2m^2$$

$$+\mathrm{i}\left[\left(\frac{1}{4}\alpha(\alpha-\beta)+\frac{\gamma}{6}+(-\alpha+\beta)\theta+\theta^2\right)\mathrm{C}^3+\mathrm{CD}\left(-\frac{\alpha}{2}+2\theta\right)\right]m^3$$

$$+\mathrm{C}^2\left[\begin{array}{c}-\frac{\beta}{24}+\mathrm{C}^2\left(\frac{\alpha^2}{8}(-\alpha+\beta)+\frac{\gamma}{6}\left(-\alpha+\frac{\beta}{2}\right)+\left(\alpha\left(\frac{3}{4}\alpha-\beta\right)+\frac{1}{4}\beta^2+\frac{1}{3}\gamma\right)\theta+\frac{3}{2}(-\alpha+\beta)\theta^2+\theta^3\right)\\+\mathrm{D}\left(\frac{-1}{12\mathrm{C}^2}+\frac{\theta\,\mathrm{D}}{\mathrm{C}^2}+\left(\frac{\alpha^2}{4}+\frac{\gamma}{6}-2\alpha\theta+\beta\theta+3\theta^2\right)\right)\end{array}\right]m^4$$

$$+O(m^5)$$

Similar operations on the $n=1$ DY ODE analytical solution's amplification factor

$$g_{exact}=\frac{q(x,t_{n+1})}{q(x,t_n)}=e^{-\left(\mathrm{i}\mathrm{C}m+\mathrm{D}m^2\right)}$$

generates the $n=1$ DY exact solution amplification factor TS to $O(m^6)$

$$g_{exact} = 1 - iCm - \left(\frac{C^2}{2} + D\right)m^2 + iC\left(\frac{C^2}{6} + D\right)m^3 + \left[C^2\left(\frac{C^2}{24} + \frac{D}{2}\right) + \frac{D^2}{2}\right]m^4$$
$$-iC\left(\frac{C^4}{120} + \frac{DC^2}{6} + \frac{D^2}{2}\right)m^5 + O(m^6)$$

Now ignoring the “D” terms, as “physical” diffusion is *not* the focus, the TS expansion to $O(m^5)$ of the thus-identified $n = 1$ D^mY $GWS^h + \theta TS$ algorithm phase speed error $err\,(g^h) \equiv g_{exact}\, g^h$

$$GWS^h + \theta TS\text{: } err(g^h) = -\left[1 - 2\theta + (\alpha - \beta)\right]\frac{C^2 m^2}{2}$$
$$+ i\left[\frac{1}{6} - \frac{1}{4}\alpha(\alpha-\beta) - \frac{\gamma}{6} - (-\alpha+\beta)\theta - \theta^2\right]C^2 m^3$$
$$+ \left[\frac{C^2}{24} + \frac{\beta}{24} - C^2\left(\begin{matrix}\frac{\alpha^2}{8}(-\alpha+\beta) + \frac{\gamma}{6}\left(-\alpha + \frac{\beta}{2}\right) + \frac{3}{2}(-\alpha+\beta)\theta^2 + \theta^3 \\ + \left(\alpha\left(\frac{3}{4}\alpha - \beta\right) + \frac{1}{4}\beta^2 + \frac{1}{3}\gamma\right)\theta\end{matrix}\right)\right]C^2 m^4 + O(m^5)$$

is amenable to non-D wavenumber $m = \kappa h$ independence *extremization* as a function of parameters α, β, γ, θ, and C. As the $O(m^1)$ TE term is absent in $err(g^h) \equiv g_{exact} - g^h$, *any* tri-diagonal stencil equivalent D^mY algorithm will generate a TS $O(m^2)$ accurate prediction ∀(α, β, γ, θ, C), CNFD the specific example. The TS expansion $O(m^2)$ TE coefficient vanishes ∀(α = β,C) by defining θ≡0.5, which reaffirms the *optimality* of the trapezoidal rule. The TS $O(m^3)$ TE coefficient then vanishes for the definition γ≡−0.5 ∀α = β and θ≡0.5. The TS $O(m^4)$ TE coefficient cannot then be made to vanish for these selections.

7.4 Derivation, NS mass transport D^mY PDE $O(m^p)$ TE annihilation, $p = 2, 3$, n-D

Required is $GWS^h + \theta TS$ algorithm phase speed error $err(g^h) \equiv g_{exact} - g^h$ wavenumber *independence* extremization in n-D. For TS theory altered D^mY PDE absent the diffusion term

$$D^mY\text{: } \frac{\partial q}{\partial t} + u_j\frac{\partial q}{\partial x_j} - \frac{\Delta t}{2}\frac{\partial}{\partial x_j}\left[\alpha u_j\frac{\partial q}{\partial t} + \beta u_j u_k\frac{\partial q}{\partial x_k}\right] - \frac{\Delta t^2}{6}\frac{\partial}{\partial x_j}\left[\gamma u_j u_k\frac{\partial^2 q}{\partial x_k \partial t}\right] = 0$$

generation of the closed form TS wave number space expansion proved tractable only for the $n = 2$ Cartesian reference frame. Resultant DY exact solution

$$q(x,y,t) = q_0(x,y,t_0)e^{-i(\kappa_1(x-ut)+\kappa_2(y-vt))}$$

amplification factor is

$$g_{exact} = \mathrm{e}^{-im\left(C_x \cos\eta + C_y \sin\eta\right)}$$

wherein (C_x, C_y) is the *Courant vector* **C** Cartesian resolution, the magnitude of which is the non-D time step. The resultant DY analytical solution amplification factor TS expansion to $O(m^4)$ is

$$\begin{aligned} g_{exact} &= 1 - \mathrm{i}[C_x\cos(\eta) + C_y\sin(\eta)]m \\ &\quad - \frac{1}{2}[C_x\cos(\eta) + C_y\sin(\eta)]^2 m^2 \\ &\quad + \frac{\mathrm{i}}{6}[C_x\cos(\eta) + C_y\sin(\eta)]^3 m^3 + O(m^4) \end{aligned}$$

D^mY $GWS^h + \theta TS$ FE $p = 1$ tensor product basis algorithm generated $q^h(x, y, t)$ assembled at a generic mesh node (X_j,Y_k) enables the completion of the $n = 2$ wave number TS analysis. For Fourier representations

$$\begin{aligned} Q_{j-1,k-1} &= Q(x - \Delta x, y - \Delta y) \equiv Q_{j,k}\mathrm{e}^{-\mathrm{i}(\kappa_1 \Delta x)}\mathrm{e}^{-\mathrm{i}(\kappa_2 \Delta y)} \\ Q_{j+1,k+1} &= Q(x + \Delta x, y + \Delta y) \equiv Q_{j,k}\mathrm{e}^{\mathrm{i}(\kappa_1 \Delta x)}\mathrm{e}^{\mathrm{i}(\kappa_2 \Delta y)} \end{aligned}$$

For wavenumber resolutions $\kappa_1 = \kappa \cos(\eta)$, $\kappa_2 = \kappa \sin(\eta)$, the D^mY PDE $GWS^h + \theta TS$ algorithm state variable scalar member amplification factor is [7]

$$g^h = \frac{\begin{array}{c}\left(b_{j-1}e^{-i\kappa_1} + b_j + b_{j+1}e^{+i\kappa_1}\right)e^{-i\kappa_2} + \left(b_{j-1}e^{-i\kappa_1} + b_j + b_{j+1}e^{+i\kappa_1}\right) \\ + \left(b_{j-1}e^{-i\kappa_1} + b_j + b_{j+1}e^{+i\kappa_1}\right)e^{+i\kappa_2}\end{array}}{\begin{array}{c}\left(a_{j-1}e^{-i\kappa_1} + a_j + a_{j+1}e^{+i\kappa_1}\right)e^{-i\kappa_2} + \left(a_{j-1}e^{-i\kappa_1} + a_j + a_{j+1}e^{+i\kappa_1}\right) \\ + \left(b_{j-1}e^{-i\kappa_1} + b_j + b_{j+1}e^{+i\kappa_1}\right)e^{+i\kappa_2}\end{array}}$$

wherein the j, $j \pm 1$ indexed coefficient set (a, b) is dependent on α, β, γ, θ, C_x, and C_y.

$GWS^h + \theta TS$ algorithm phase velocity *error* analytical quantification proved tractable *only* for a uniform mesh, hence, $\Delta x \equiv h \equiv \Delta y$ and $\kappa_1 h \equiv m \equiv \kappa_2 h$. The ponderously lengthy TS expansion of g^h to $O(m^4)$ is fully detailed in [7], Appendices A and B. The significant $n = 2$ distinction in phase speed *error* $err(g^h) \equiv g_{exact} - g^h$ resides in the $O(m^2)$ TE term coefficient

$$GWS^h + \theta TS\colon\ err(g^h) = \begin{bmatrix} \frac{C_x^2}{2} + C_x C_y + \frac{C_y^2}{2} - \frac{1}{2}(\alpha + \beta)\left(C_x^2 + C_y^2\right) \\ - \alpha C_x C_y - C_x^2\theta - 2C_x C_y \theta - C_y^2\theta \end{bmatrix} m^2 + O(m^3) + O(m^4)$$

which now vanishes *only* for $\alpha \equiv 0 \equiv \beta$. The identification $\gamma \equiv -0.5$ again eliminates the $O(m^3)$ TE coefficient, and the $O(m^4)$ coefficient cannot then be made to vanish.

To summarize, $n = 2$ non-D wave number space TS manipulations quantify the *optimal* time TS identified coefficient definitions are $\theta \equiv 0.5$, $\alpha \equiv 0 \equiv \beta$, and $\gamma \equiv -0.5$. Thereby, $\text{GWS}^h + \theta\text{TS}$ FE $p = 1$ basis algorithm predictions for TEA-altered non-D D^mY PDE

$$\text{D}^m\text{Y}: \left(1 - \frac{\gamma \text{C}^2 h^2}{6}\frac{\partial}{\partial x_j}\left(u_j u_k \frac{\partial}{\partial x_k}\right)\right)\frac{\partial Y}{\partial t} + u_j \frac{\partial Y}{\partial x_j} - \frac{1}{\text{ReSc}}\frac{\partial^2 Y}{\partial x_j^2} = 0 + O(h^2, m^4, \Delta t^3)$$

will indeed be TS $O(m^4)$ accurate in wave number space.

7.5 Validation, space-time discretization aliasing error annihilation, NS $O(m^p)$ D^mY, $p = 2, 3$

The 1970s meteorological CFD community defined a manufactured statement to assess the impact of phase velocity error on the fidelity of CFD-predicted atmospheric mass transport. Named the "rotating cone," it mimics a nuclear reactor "burp" into an atmospheric wind field that is rotational, that is, a vortex [3]. Absent physical diffusion, the NS DY PDE analytical solution is exact preservation of the Fourier spectral content resolved by DOF interpolation of the $Y(x_j, t = 0)$ IC distribution after any number of circulations around the horizontal plane. The non-D wind field Courant vector distribution is $\mathbf{C}(x,y) = (\Delta t/h) r\omega \hat{\mathbf{e}}_\theta$, where ω is the angular velocity with unit vector $\hat{\mathbf{e}}_\theta$ with r the radial distance from the polar coordinate system origin at the domain Ω center. Thereby, $|\text{C}(r)|$ is the *spatial distribution* of time step $\Delta t(x_j)$

For the NS TEA altered D^mY PDE absent the physical diffusion term

$$\text{D}^m\text{Y}: \left[1 - \frac{\gamma \Delta t^2}{6}\frac{\partial}{\partial x_j}\left(u_j u_k \frac{\partial}{\partial x_k}\right)\right]\frac{\partial Y}{\partial t} + u_j \frac{\partial Y}{\partial x_j} = 0 + O(h^2, m^4)$$

and enforcing the Green-Gauss divergence theorem on the γ term, the $O(h^2, m^4)\text{D}^m\text{Y}$ PDE $\text{GWS}^h + \theta\text{TS}$ FE $p = 1$ basis algorithm element matrix for *optimal* $\theta \equiv 0.5$ and $\gamma \equiv -0.5$ is

$$\text{GWS}^h + \theta = \text{TS}: \{\text{F}(Q(t)\}_e$$

$$= \int_{\Omega e}\left[\{N\}\{N\}^T + \frac{\gamma \Delta t^2}{6}\frac{\partial \{N\}}{\partial x_j}\left(N\}^T \{\text{U}J\}_e \{\text{U}K\}_e^T \{N\}\frac{\partial \{N\}^T}{\partial x_k}\right)\right] \text{d}x\text{d}y(\{Q^{n+1} - Q^n\}_e)$$

$$+ \frac{\Delta t}{2}\int_{\Omega e}\left(\{\text{U}J\}_e^T \{N\}\{N\}\frac{\partial \{N\}^T}{\partial x_j}\right)\text{d}x\text{d}y(\{Q^{n+1} + Q^n\}_e)$$

$$\cong \left([\text{M}200]_e - \frac{\Delta t^2}{3}\{\text{U}J{*}\text{U}K\}_e^T\ [\text{M}30JK]_e\right)(\{Q^{n+1} - Q^n\}_e)$$

$$+ \frac{\Delta t}{2}\{\text{U}J\}_e^T\ [\text{M}300J]_e(\{Q^{n+1} + Q^n\}_e), 1 \le J, K \le n$$

wherein interpolation of the γ term velocity tensor product has been replaced with interpolation of the velocity tensor product DOF. The hypermatrix set $[\mathrm{M30}JK]_e$ remains as previously defined.

The standard non-D domain Ω^h for the rotating cone is the uniformly discretized $\mathrm{M} = 32 \times 32$ unit square spanned by the FE $p = 1$ basis discretization. Mass fraction boundary conditions are homogeneous Dirichlet on $\partial\Omega^h$ segments with inflow and homogeneous Neumann on outflow boundary segments. The release mass fraction $Y(x_j, t = 0]$ initial condition (IC) is the axi-symmetric $n = 2$ cone generated by the rotation of a Gaussian around its extremum.

Mesh resolution of this cone IC is *intentionally* defined minimal with only the vertex node mass fraction DOF at 100%. The IC cone perspective view with the mesh overlay is graphed in Figure 7.1a, with the color bar in percent. For the standard non-D time step $|\mathrm{C}(r_\mathrm{C})| \equiv 0.41$ at the IC $Y(x_j, t = 0)$ distribution centroid, the integration time step size Δt is defined such that, after exactly "n" equal steps, the DOF distribution of the accurate $Y^h(x_j, \mathrm{n}\Delta t \equiv \mathrm{T})$ prediction will coincide *exactly* with the IC $Y(x_j, t = 0)$ DOF disposition after one clockwise circulation.

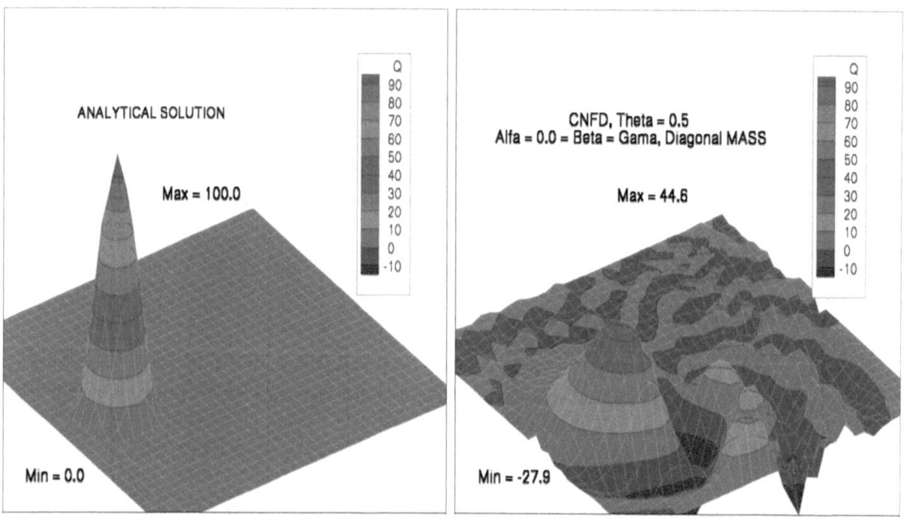

a) DY $O(m^\infty)$ analytical solution

b) DY $O(h^2, m^2, \Delta t^3)$ PDE CNFD $Y^h(x_i, \mathrm{T})$

Figure 7.1: Validation, mass transport prediction fidelity, NS DY PDE rotating cone manufactured statement, $Y^h(x_i, \mathrm{T})$ after one clockwise circulation in the horizontal plane, uniform $\mathrm{M} = 32 \times 32$ mesh, $|\mathrm{C}(r_\mathrm{C})| \equiv 0.41$ non-D at the IC centroid: (a) $O(m^\infty)$ DY analytical solution, (b) $O(h^2, m^2, \Delta t^3)$ DY CNFD solution, −28% (dark blue) extremum aft of $Y^h(x_j, \mathrm{T})$, truncated for visualization.

The TS wave number space operation predicts that *any* NS DY PDE tri-diagonal stencil-equivalent CFD algorithm is $O(m^2)$ accurate, CNFD the example. Following one clockwise circulation, the phase velocity *error* $e^h(m^2)$ intrinsic to the NS $O(h^2, m^2)$

DY PDE CNFD algorithm generates the totally *corrupted* terminal $Y^h(x_j,\mathrm{T})$ DOF distribution (Figure 7.1b). Specifically, the IC $Y(x_j,t=0)$ centroid DOF at 100% is reduced to ~45%. The accumulation of $e^h(m^2)$ *aliasing error* disperses thus lost IC $Y(x_j,t=0)$ Fourier content into the three isolated mass packets "left behind" in the circulation track, each with 30±% (light blue, green) DOF extrema. As mass fraction is a positive definite function, even more disastrous are the "left behind" $Y^h(x_j,\mathrm{T})$ distributions with *negative* DOF, the clearly visible elongated streaks in the horizontal plane with −10±% (medium blue) significance to the −28% extremum (dark blue) directly aft of $Y^h(x_j,\mathrm{T})$.

Comparison of the tri-diagonal equivalent stencils quantifies that the *sole* distinction between CNFD $O(h^2,m^2)$ DY and $\mathrm{GWS}^h+\theta\mathrm{TS}$ $O(h^2,m^3)$ DY PDE algorithms is the nearest-neighbor time derivative DOF mass matrix *coupling* of the latter. The impact of one order increase in phase velocity wave number *independence* on solution fidelity is truly remarkable. Following one circulation, the $O(h^2,m^3)$ DY PDE $\mathrm{GWS}^h+\theta\mathrm{TS}$ FE $p=1$ basis distribution $Y^h(x_j,\mathrm{T})$ (Figure 7.2.a) does not quite return to the IC $Y(x_j,0)$ DOF disposition, hence the $Y^h(x_j,\mathrm{T})$ centroid DOF is reduced to 94%. The accumulation of $e^h(m^3)$ error generates the just-visible distortion of the $Y^h(x_j,\mathrm{T})$ distribution from axisymmetry. The accumulation of space–time discretization $O(h^2,m^3)$ error interaction in the $Y^h(x_j,\mathrm{T})$ DOF distribution appears as the regularly spaced "left-behind" ridges of modest −10±% (dark blue) significance to the −15% extremum directly aft of $Y^h(x_j,\mathrm{T})$.

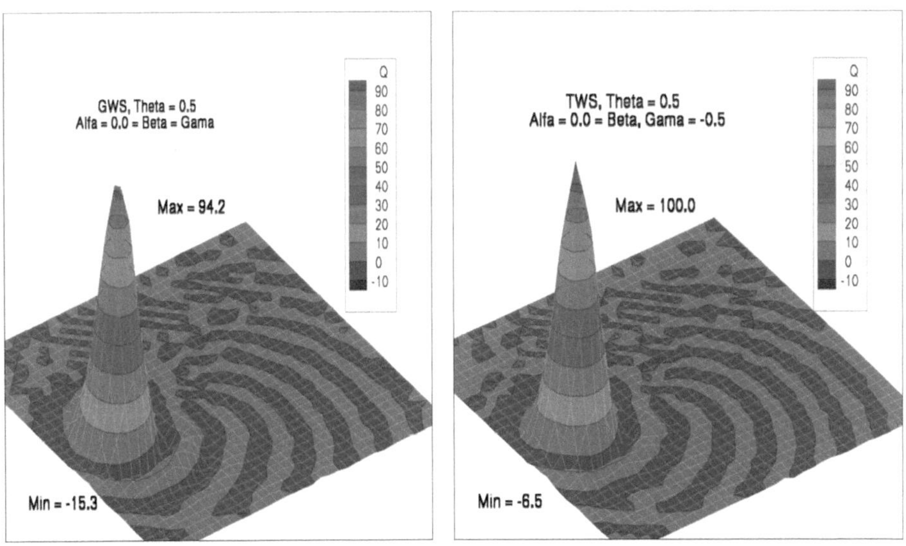

a) $O(h^2,m^3,\Delta t^3)$ DY $Y^h(x_i,\mathrm{T})$ b) $O(h^2,m^4,\Delta t^3)$ $\mathrm{D}^m\mathrm{Y}$ PDE $Y^h(x_i,\mathrm{T})$

Figure 7.2: Validation, mass transport prediction fidelity, NS DY, and $\mathrm{D}^m\mathrm{Y}$ PDE $\mathrm{GWS}^h+\theta\mathrm{TS}$ algorithm, rotating cone manufactured statement, $Y^h(x_i,T)$ after one clockwise circulation in the horizontal plane uniform M = 32 × 32 mesh, $|C(r_C)|\equiv 0.41$ at the IC centroid: (a) $O(h^2,m^3,\Delta t^3)$ DY PDE and (b) $O(h^2,m^4,\Delta t^3)$ $\mathrm{D}^m\mathrm{Y}$ PDE.

The TEA altered $O(h^2,m^4)\mathrm{D}^m\mathrm{Y}$ PDE GWS^h + θTS prediction (Figure 7.2.b) quantifies that the $e^h(m^4)$ phase error is sufficiently reduced such that the *coincidence* of predicted $Y^h(x_j,\mathrm{T})$ DOF and IC $Y(x_j,0)$ DOF distributions is *essentially* exact. The $Y^h(x_j,\mathrm{T})$ distribution centroid node DOF at 100% is precise, as is the $Y^h(x_j,\mathrm{T})$ distribution preservation of the axi-symmetry of the IC $Y(x_j,0)$. The accumulation of space–time discretization $O(h^2,m^4)$ error interaction "leaves behind" modest significance (light blue) $Y^h(x_j,\mathrm{T})$ DOF streaks with −6.5% DOF extremum directly aft of $Y^h(x_j,\mathrm{T})$, a 2.5-factor reduction of the $O(h^2,m^3)$ DY PDE prediction.

The obvious resolution for $O(h^2,m^4,\Delta t^3)$ $\mathrm{D}^m\mathrm{Y}$ PDE GWS^h + θTS space-time error *interaction*-induced negative DOF is DY PDE augmentation with a physical diffusion term to enable alteration to the *optimal* TEA-altered $\mathrm{D}^m\mathrm{Y}$ $O(h^4,m^4)$ PDE. Switching to RaNS for reality, the comparison of textbook RaNS DY with RaNS $O(h^4,m^4)$ $\mathrm{D}^m\mathrm{Y}$ clearly visualizes the impact of *error freed* CFD mathematics

$$\text{textbook RaNS DY:}\ \frac{\partial Y}{\partial t} + u_j\frac{\partial Y}{\partial x_j} - \frac{\partial}{\partial x_j}\left(\frac{1+\mathrm{Re}^T}{\mathrm{ReSc}}\right)\frac{\partial Y}{\partial x_j} = 0$$

$$\text{RaNS TS theorized } \mathrm{D}^m Y\text{:}\ \left[1 - \frac{\gamma\Delta t^2}{6}\frac{\partial}{\partial x_j}\left(u_j u_k\frac{\partial}{\partial x_k}\right)\right]\frac{\partial Y}{\partial t} + u_j\frac{\partial Y}{\partial x_j}$$

$$-\frac{\partial}{\partial x_j}\left[\left(\frac{1+\mathrm{Re}^T}{\mathrm{ReSc}}\right)\frac{\partial Y}{\partial x_j} + \frac{h^2\mathrm{ReSc}}{12\left(1+\mathrm{Re}^T\right)}u_j u_k\frac{\partial Y}{\partial x_k}\right] = 0 + O(h^4, m^4, \Delta t^3)$$

7.6 Derivation, trapezoidal rule $O(\Delta t^3)$ TE-annihilated NS/RaNS GWS^h + θTS algebraic statement

Rigorous annihilation of the trapezoidal rule $O(\Delta t^3)$ TE term in NS/RaNS PDE GWS^h + θTS FE $p = 1$ basis algorithm terminal algebraic statements is the final topic. For the n-D NS model **D**P

$$\mathbf{D}\mathrm{P}\text{:}\ \frac{\partial q}{\partial t} + u_j\frac{\partial q}{\partial x_j} - \frac{1}{\mathrm{Re}}\frac{\partial^2 q}{\partial x_j^2} = 0$$

TS manipulations first detailed in Lax and Wendroff [6], subsequently completed by Kolesnikov [5] following Chaffin [1], identified the trapezoidal rule θ = 0.5 $O(\Delta t^3)$ TE *continuum* functional. For TS time step Δt the linear combination of the partial forward $(1 - \alpha)\Delta t$ step with two backwards steps from $t + \Delta t$ alters the NS model **D**P GWS^h + θTS FE basis algorithm θ-implicit algebraic statement to

$$\text{TS: } q^{n+1} - q^n = \Delta t\left(\theta\frac{\partial q^{n+1}}{\partial t} + (1-\theta)\frac{\partial q^n}{\partial t}\right) - \frac{\Delta t^2}{6}\left[(3\theta-1)\frac{\partial^2 q^{n+1}}{\partial t^2} + (3\theta-2)\frac{\partial^2 q^n}{\partial t^2}\right]$$
$$-\frac{\Delta t^4}{72}\left[-(3\theta-1)^2\frac{\partial^4 q^{n+1}}{\partial t^4} + (3\theta-2)^2\frac{\partial^4 q^n}{\partial t^4}\right] + O(\Delta t^5)$$

The first right-side term is the trapezoidal rule for $\theta = 0.5$. Thereby, the second right-side term identifies the trapezoidal rule $O(\Delta t^3)$ TE term.

Replacing higher-order time derivatives therein with spatial derivatives as documented in Chapter 4, then assembling the resultant $\text{GWS}^h + \theta\text{TS}$ FE $p = 1$ basis element matrix into a tri-diagonal stencil and imposing a symmetrizing operation, forming the limit $\Delta t \Rightarrow \varepsilon > 0$ generates the time TS-altered *continuum* NS model $\text{D}^m\mathbf{P}$ assuming large Re, Kolesnikov ([5], Ch. 4)

$$\text{NS D}^m\mathbf{P}\text{: } \frac{\partial q}{\partial t} + u_j\frac{\partial q}{\partial x_j} - \frac{1}{\text{Re}}\frac{\partial^2 q}{\partial x_j^2} + \frac{h^2}{6\Delta t}\left(\text{Re}\; u_j\frac{\partial q}{\partial x_j} - \frac{\partial^2 q}{\partial x_j^2}\right) = 0 + O(\Delta t^4)$$

Thereby, for *any* tri-diagonal stencil equivalent algorithm on thus time TS-altered NS $\text{D}^m\mathbf{P}$ PDE, the leading TE term for the trapezoidal rule $\theta \equiv 0.5$ implementation is $O(\Delta t^4)$.

The astute observer will recognize that *every* TEA theory altered NS/RaNS PDE system functional possesses a term with the Reynolds number (Re) appearing in the ***numerator***. Thus validated is the monograph's assertion that *error freed* CFD theorizations must address NS/RaNS PDE systems, *not* an Euler reduction!

Coalescing like terms, the $O(\Delta t^4)$ TEA alteration to the preceding RaNS TEA $O(h^4,m^4,\Delta t^3)$ D^mY PDE

$$\text{RaNS D}^m\text{Y: } \left[1 - \frac{\gamma\Delta t^2}{6}\frac{\partial}{\partial x_j}\left(u_j u_k\frac{\partial}{\partial x_k}\right)\right]\frac{\partial Y}{\partial t} + \left(1 + \frac{h^2\text{Re}}{6\Delta t}\right)u_j\frac{\partial Y}{\partial x_j}$$
$$-\frac{\partial}{\partial x_j}\left[\left(\frac{h^2}{6\Delta t} + \frac{1+\text{Re}^T}{\text{ReSc}}\right)\frac{\partial Y}{\partial x_j} + \frac{h^2\text{ReSc}}{12\left(1+\text{Re}^T\right)}u_j u_k\frac{\partial Y}{\partial x_k}\right] = 0 + O(h^4, m^4, \Delta t^4)$$

identifies the *error freed* CFD mathematics RaNS $O(h^4,m^4,\Delta t^4)$ D^mY *optimally efficient*, Newton Jacobian quadratic convergent $\text{GWS}^h + \theta\text{TS}$ FE $p = 1$ basis algorithm *guaranteed* to generate *resolutely monotone* state variable DOF distributions on *any* (!) mesh M that are TS *fourth order* accurate in physical space, wave number space and time.

No doubt exists that the insertion of monograph-identified TEA theory *advancements* into current practice open-source/commercial/government CFD codes will enable the alteration of the 2014 CFD literature observation cited in Chapter 1 to

> . . . perceived **stagnation** of CFD capability *advancement* . . . **no longer** exists.

References

[1] Chaffin, D.J. (1997). "A Taylor weak statement finite element method for computational fluid dynamics," PhD dissertation, University of Tennessee

[2] Crank, J., Nicolson, P. (1947). "A Practical Method for Numerical Evaluation of Solutions of Partial Differential Equations of Heat-Conduction Type," *Cambridge Phil Soc*, 43.

[3] Fabrick, A., Sklarew, R., Wilson, J. (1977). "Point Source Model Evaluation and Development Study," Science Applications Corp. Tech Rpt A5-058-87

[4] Gresho, P.M., Sani, R.L. (2000). *Incompressible Flow and the Finite Element Method*, John Wiley, London.

[5] Kolesnikov, A. (2000), "Efficient implementation of high order methods in computational fluid dynamics," PhD dissertation, University of Tennessee

[6] Lax, P., Wendroff, B. (1960). "Systems of Conservation Laws," *Comm Pure Applied Math*, 13.

[7] Sahu, S., Baker, A.J. (2007). "A Modified Conservation Principles Theory Leading to an Optimal Galerkin CFD Algorithm," *J Num Mtd Fluids*, 55(8).

Chapter 8
Hypersonics, aerothermodynamics, radiation CFD issues

Libretto: Renewed interest in hypersonic aerodynamics, defined for velocity onset Mach number $Ma_\infty \geq 5$, has energized the need to assess aerothermodynamics and radiation CFD capabilities.

Year-end issues of *Aerospace America* (2019:57:11, 2020:58:11, and 2021:58:11) summarize annual advances in hypersonic research activities worldwide.

Certainly, *error freed* CFD *mathematics* will contribute to, "Reducing the need for hypersonic tunnel testing," the title of an op-ed in *Aerospace America* (2022).

This chapter identifies and resolves the aerothermodynamics, compressible multispecies mixing, and radiation issues that uniquely challenge hypersonic CFD prediction *physics of fluids* fidelity.

Rigorously derived and validated is *differential radiosity theory* resolution as the CFD compute *efficient* and *accurate* alternative to classic Stephan-Boltzmann theory.

Data generated from the *University Consortium for Applied Hypersonics* (UCAH) hypersonic shock tube, Texas A&M University, and the recently announced hypersonic wind tunnels at Notre Dame University and University of Tennessee Space Institute will support CFD validation.

Keywords: hypersonics, aerothermodynamics, compressible mixing, radiosity, Newton Jacobian

8.1 Issues uniquely challenging to hypersonic aerodynamics CFD

Worldwide renewed interest in *hypersonics* will require current commercial/government CFD code capability advancements for the prediction of aerothermodynamics-induced compressible multispecies mixing interaction with radiation. Among the published textbooks on high-speed aerodynamics, Shapiro (1954), Dorrance (1962), and Hayes and Probstein (1964), only Anderson (1989) includes a brief appendix exposure to radiation. The repository for reentry gas dynamics CFD is Gnoffo (1998). Hypersonic planetary reentry shape optimization, heat shield design, and thermal protection system CFD topics are detailed in Duffa (2013) and Candler (2019). Literature overviews on current hypersonic research topics are by Button (2020, 2022).

The US Air Force Research Laboratory (AFRL) sponsored BOLT (BOundary Layer Transition) project [7] seeks to document the influence of hypersonic boundary layer (BL) laminar-turbulent transition on missile surface thermal/pressure distributions during atmospheric reentry of a specifically designed and instrumented forebody

https://doi.org/10.1515/9783112206430-009

(Figure 8.1a). While a transonic BL is warm, and a supersonic one is warmer, the surface temperature on which a hypersonic BL is submerged beneath a Mach 8 shock layer can easily reach 4,500± R.

A wind tunnel subscale model coated with heat-sensitive paint quantifies the anticipated temperature distribution (Figure 8.1b) [7]. Although the BOLT reentry missile nose cone geometry *and* the wind tunnel model are each mirror-symmetric, note that the model temperature distribution is not!

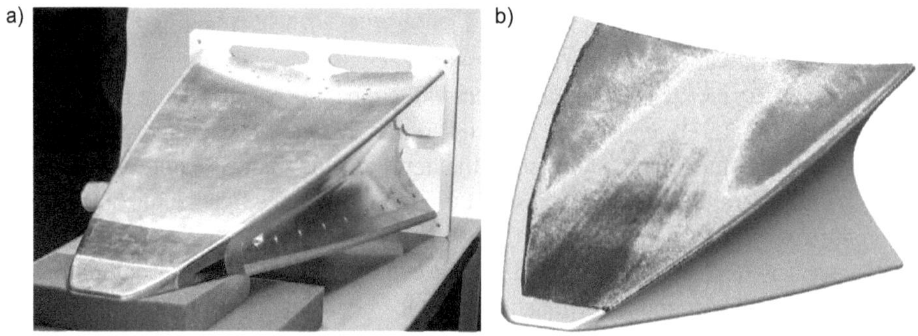

Figure 8.1: AFRL BOLT hypersonic aerothermodynamics project, $Ma_\infty \approx 7$ atmospheric reentry missile: (a) 1 m forebody instrumented for data acquisition, (b) wind tunnel forebody model temperature distribution on attached boundary layer undergoing transition adapted from Button [7], with permission.

The heat flux distribution due to $Ma_\infty = 6$ onset to a wind tunnel model shock layer submerged BL undergoing transition at zero angle of attack is graphed in Figure 8.2 [13]. Clearly documented is that the elevated heat flux distributions (red) visible on the streamwise edges of the model *are* in radiative communication with the surrounding unbounded half space. These data also do not exhibit the mirror symmetry of the

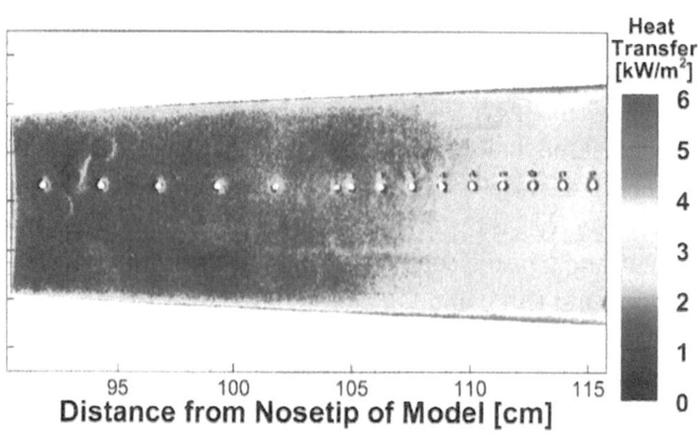

Figure 8.2: $Ma_\infty = 6$ hypersonic BL laminar-turbulent transition heat transfer distribution, sharp cone model at zero angle of attack in quiet wind tunnel adapted from Gonzalez and Medina [13] with permission.

model which further quantifies the complexity of hypersonic shock layer submerged BL separation/transition.

Generation of the *truly high* Mach number velocity vector fields required for hypersonic aerothermodynamics research is currently enabled by the *University Consortium for Applied Hypersonics* (UCAH) shock tube nozzle-diffuser facility at Texas A&M University (Figure 8.3a). Conditioning of shock tube flowfield maturation (Figure 8.3b) is consequentially more involved (!) than was that for the classic Riemann shock tube prediction detailed in Chapter 4.

a)

b)

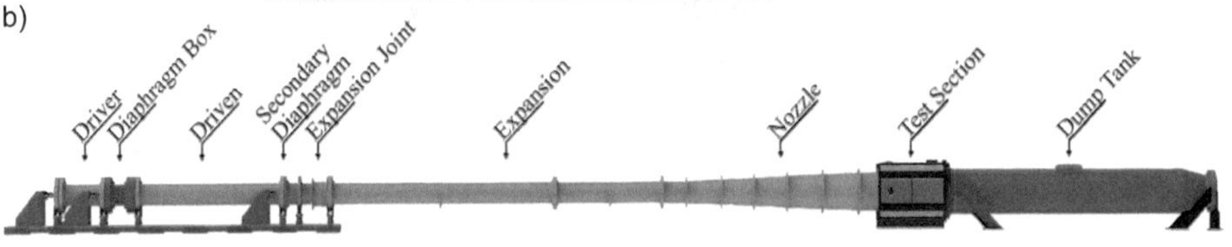

Figure 8.3: Hypersonic shock tube-nozzle-diffuser experimental facility in the University Consortium for Applied Hypersonics (UCAH) at Texas A&M University: (a) facility perspective and (b) fluid dynamics functional design of the UCAH hypersonic shock tube (courtesy of Blair [9]).

Absolutely critical to hypersonic aerothermodynamics flowfield CFD characterization, data generated in the UCAH facility will support quantitative *validation* of proposed CFD *error-freed* mathematics prediction advances, the truly efficient, and cost-effective alternative to instrumentation of hypersonic reentry vehicles (!), (Button [7]). Further to this subject, the US Navy funded Mach 10 hypersonic wind tunnel to be located at the University of Notre Dame in Indiana was announced in November 2024 and shortly thereafter the US Air Force funded hypersonic wind tunnel at the University of Tennessee Space Institute.

Atmospheric gases are essentially transparent to radiation. However, the Mach number distribution in the BL submerged beneath a hypersonic shock layer undergoes separation *and* transition on a surface that is an active participant in radiative conjugate heat transfer with the surrounding half space. As NS/RaNS PDE systems are universally elliptic boundary value, the thermal state of surface to which the submerged BL is attached is the *non-homogeneous* Robin boundary condition (RBC) for the RaNS DE state variable member. Rigorous *error freed* CFD-bounded domain well-posed mathematics theorization is absolutely requisite to ensure *physics of fluids fidelity* CFD prediction of hypersonic *aerothermodynamics* and/or *radiation*-influenced BL transition, separation then turbulent profile reattachment.

8.2 Aerothermodynamics impact on hypersonic BL CFD prediction

An extensive hypersonic aerodynamics data base exists for cone-cylinder wind tunnel missile models. That of Martellucci and Weinberg [15] supported *validation* of a hypersonic, variable γ air aerothermodynamics-approximated $n = 3$ RaNS PDE statement using an implicit space marching (*parabolic*) "PRaNS" $GWS^h + \theta TS$ FE $p = 1$ basis algorithm, approximately factored (AF) Jacobian code characterization of $Ma_\infty = 8$ onset to a sharp cone/cylinder/cone wind tunnel model [3].

The $n = 3$ domain Ω^h for hypersonic PRaNS PDE $GWS^h + \theta TS$ algorithm predictions (Figure 8.4c) is bounded circumferentially by streamwise evolution of the missile body surface and the missile sharp nose cone generated $Ma_\infty = 8$ bow shock. The streamwise *evolution* of BL solution adapted, missile body-bow shock-fitted curvilinear FE $p = 1$ basis 64×18 (wall normal $\times$ azimuthal) mesh transverse plane discretization is graphed in Figure 8.4d. The PRaNS algorithm generated steady, three-dimensional variable γ air aerothermodynamics model state variable member distributions spanned the full reach of the axial evolution of the 3D shock layer from model surface through the submerged fully turbulent BL to the missile nose cone-generated bowshock trajectory.

A time TS β-term-altered compressible Euler PDE $GWS^h + \theta TS$ FE $p = 1$ basis algorithm/code was used to generate the $Ma_\infty = 8$ onset hypersonic axisymmetric inviscid, non-heat conducting perfect gas NS state variable distribution from sharp nose cone-attached bow shock to the PRaNS solution domain initial condition (IC) plane. During PRaNS execution, the Euler code was also used to predict *evolution* of the bow-shock trajectory surface distribution that defined the streamwise evolution of the PRaNS PDE domain farfield boundary and, if required, PRaNS state variable member Dirichlet BC data at the bow-shock surface.

PRaNS PDE turbulence closure model was the *self-starting* supersonic, high temperature inner/outer mixing length theory eddy viscosity model of Baldwin and Lomax [4]. The IC plane for the space-marched PRaNS algorithm coincides with the downstream termination of knurled BL trip of the wind tunnel model (Figure 8.4b).

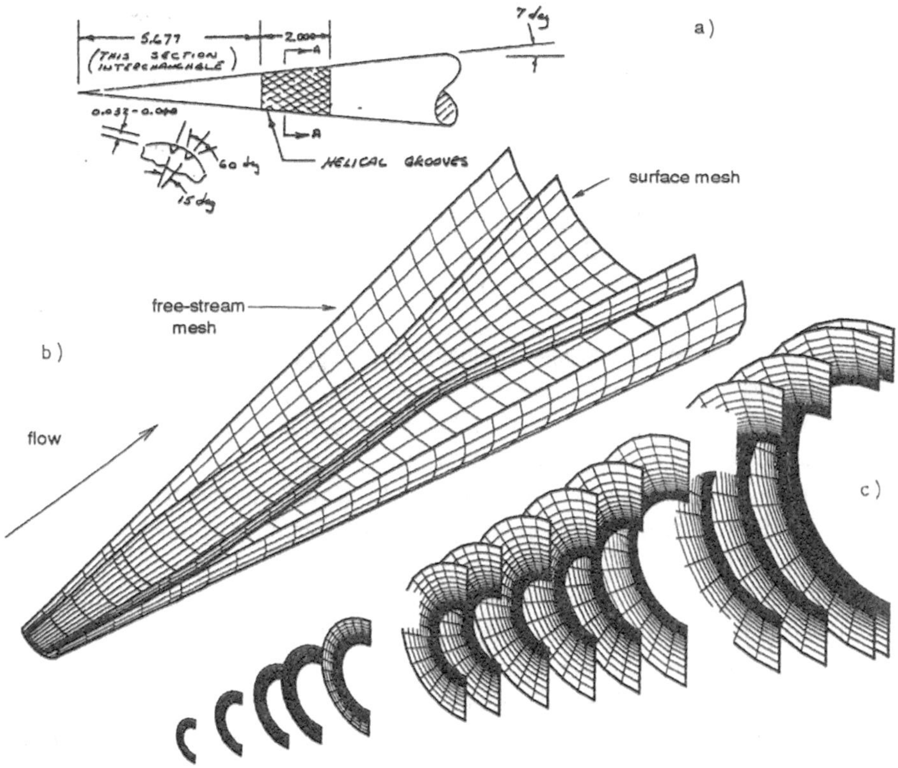

Figure 8.4: $Ma_\infty = 8$ onset to sharp 7°forecone/cylinder/7°aftcone wind tunnel model: (a) wind tunnel model knurled BL trip, (b) PRaNS domain Ω bounded by bow shock-missile surface streamwise evolutions, (c) shock layer streamwise evolution spanning body-fitted solution adapted nonuniform transverse plane FE $p = 1$ basis $M = 64 \times 18$ mesh Ω^h, from Baker et al. [3].

For onset $Ma_\infty = 8$, the Euler code was used to generate the inviscid, non-heat conducting perfect gas state variable distribution on the PRaNS IC plane mesh, the first very tiny (!) mesh graphic in Figure 8.4d. PRaNS algorithm self-generation of the state variable IC *distribution* was initialized by interpolation of the Euler state variable DOF distribution onto the nodes of the PRaNS IC plane mesh. Subsequent PRaNS code execution for a short distance downstream from the IC plane then altered the Euler state variable near surface distribution to the RaNS no-slip BC fully turbulent BL. Thus altered DOF distribution was then projected back to the IC plane mesh as the RaNS viscous, turbulent, heat-conducting perfect gas state variable IC distribution.

That the missile surface of a hypersonic shock layer submerged BL is hot (!) requires CFD accounting for induced air diatomic species dissociation and recombination. The PRaNS algorithm accommodation for thus requisite *aerothermodynamics* impact utilized the high temperature, *equilibrium* variable γ air data base of Tannehill and Mugge [17].

For $\mathrm{Ma}_\infty = 8$ onset, the PRaNS algorithm/code solution-adapted mesh-resolved BL DOF distributions of momentum axial resolution $m_1^h(\eta_2,\eta_1)$, density $\rho^h(\eta_2,\eta_1)$, and velocity axial resolution $u_1(\eta_2,\eta_1)$, compressible NS computed as $u_j \equiv m_j/\rho$, at the missile model mid-cylinder axial coordinate η_1 are *five decades* (!) logarithmic ordinate scaled in Figure 8.5.

The appearance of the $u_1^h(\eta_2,\eta_1)$ BL distribution is certainly "traditional" (Figure 8.5a) in the sense that the magnitude increases monotonically from the wall to the local *boundary layer thickness* $\delta(\eta_2,\eta_1)$, for η_2 classically defined where $u^h(\eta_2,\eta_1)/U(\eta_2,\eta_1)_{\mathrm{shock}} \equiv 0.99$. Thereafter, the $u_1^h(\eta_2,\eta_1)$ distribution magnitude remains uniformly constant throughout the outer reach of the shock layer to bow shock local encounter at Euler predicted $U(\eta_2,\eta_1)$ shock.

That the shock layer BL $u_1^h(\eta_2,\eta_1)$ velocity profile appearance is "nominal" is a surprise, as it is computed algebraically via compressible NS definition $u_j \equiv m_j/\rho$, from the clearly *non-nominal* in appearance shock layer distributions of $m_1^h(\eta_2,\eta_1)$ and $\rho^h(\eta_2,\eta_1)$ (Figure 8.5b and c). The DOF distribution for $m_1^h(\eta_2,\eta_1)$ undergoes a rapid excursion to its extremum located in the mid- reach of the *subsonic* BL whereat $\mathrm{Ma}(\eta_2,\eta_1) \approx 0.6$. The profile then transitions to resolute *η_2-independence* (!) from this extremum to encounter with the bow shock.

This PRaNS predicted hypersonic physics of fluids event is the direct result of algorithm closure with variable γ air model generated air diatomic species dissociation which *profoundly* affects the density $\rho^h(\eta_2,\eta_1)$ distribution throughout the reach of the subsonic BL (Figure 8.5c). The η_2 centric horizontal trace through the $\mathrm{Ma}(\eta_2) \approx 0.6$ DOF (Figure 8.5), quantifies that transition of the $m_1^h(\eta_2,\eta_1)$ distribution to η_2 independence occurs precisely at the inflection point at the $\rho^h(\eta_2,\eta_1)$ distribution extremum.

To correlate these data with conventional transonic compressible NS BL mesh resolution experience, the hypersonic variable γ air aerothermodynamics alteration of the BL mesh resolution η_2 ordinate of PRaNS-predicted BL $\rho^h(\eta_2,\eta_1)$ extremum at Ma $(\eta_2) \approx 0.6$ equates to the BL thickness $\delta(\eta_2)$ divided by *ten thousand*, a four orders of magnitude distinction! In fact, had the $m_1^h(\eta_2)$ DOF data been graphed linear rather than five decades logarithmic, its appearance would match the *right angle* DOF distribution for the incompressible NS DE stagnation point (Figure 5.8b).

Reflecting on the three-plus decades of computational environment maturation since this late 1980s project, the accurate accounting of non-equilibrium high temperature hypersonic aerothermochemistry processes will require prediction of compressible RaNS mixing and reaction processes induced by air diatomic multispecies *non-equilibrium* dissociation and recombination. The optimal candidate might well be the prior chapter identified unsteady *error-freed* CFD *mathematics* TEA-altered $O(h^4,m^4,\Delta t^4)$ $\mathrm{D}^m\mathrm{Y}$ PDE system altered to $\mathrm{D}^m\rho\mathrm{Y}^\alpha$ for prediction of the $1 \le \alpha \le n$ mass/mole fraction species $\rho\, Y^\alpha$

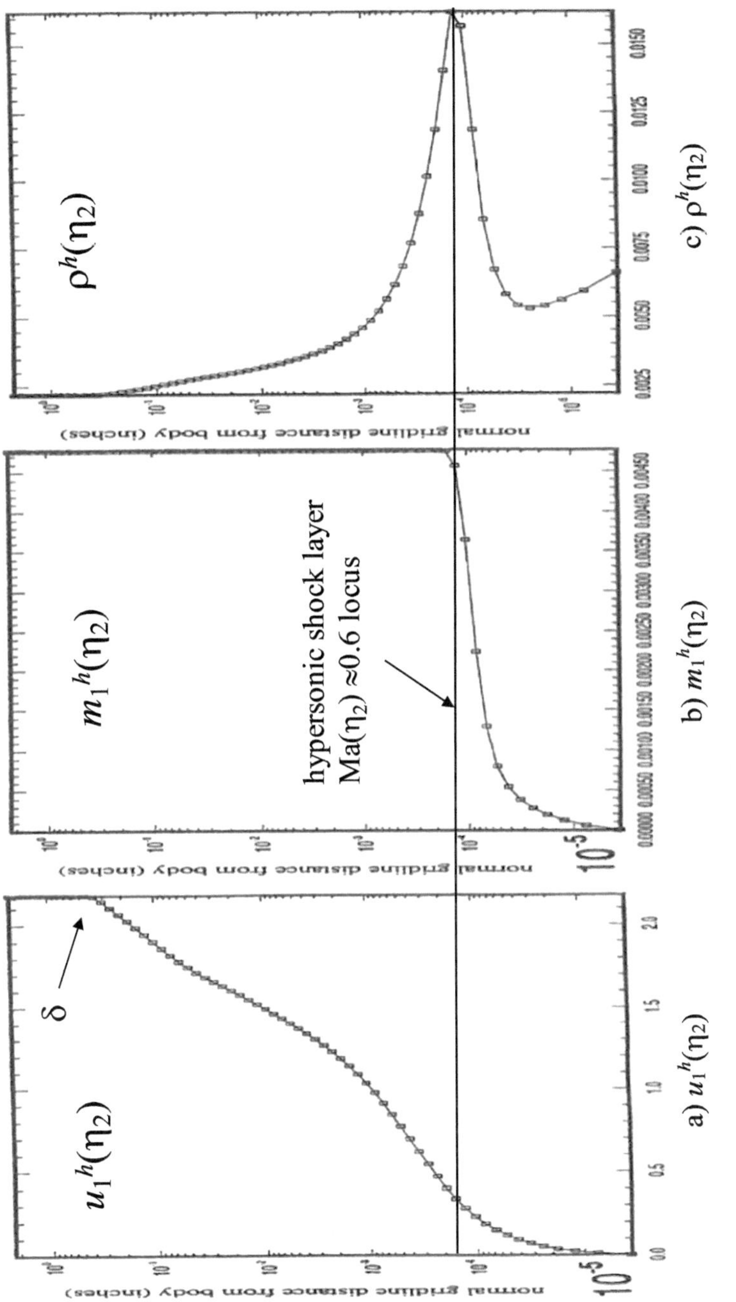

Figure 8.5: $Ma_\infty = 8$ onset PRaNS predicted hypersonic shock layer non-D state variable member DOF distributions five decades semilog scaled, sharp 7°forecone/cylinder/7°aftcone wind tunnel model: (a) velocity axial resolution $u_1{}^h(\eta_2)$, (b) momentum axial resolution $m_1{}^h(\eta_2)$, and (c) density $\rho^h(\eta_2)$, adapted from Baker et al. [3].

$$\text{RaNS D}^m\rho Y^\alpha\text{:}\left[1-\frac{\gamma\Delta t^2}{6}\frac{\partial}{\partial x_j}\left(u_j u_k\frac{\partial}{\partial x_k}\right)\right]\frac{\partial\rho Y^\alpha}{\partial t}+\left(1+\frac{h^2\text{Re}}{6\Delta t}\right)u_j\frac{\partial\rho Y^\alpha}{\partial x_j}$$

$$-\frac{\partial}{\partial x_j}\left[\left(\frac{h^2}{6\Delta t}+\frac{1+\text{Re}^T}{\text{ReSc}}\right)\frac{\partial\rho Y^\alpha}{\partial x_j}+\frac{h^2\text{ReSc}}{12(1+\text{Re}^T)}u_j u_k\frac{\partial\rho Y^\alpha}{\partial x_k}\right]=0+O(h^4,m^4,\Delta t^4)$$

8.3 Radiation, weak formulation Stefan-Boltzmann RBC, Newton Jacobian

A NS/RaNS DE PDE closed with Stefan-Boltzmann (SB) theory for black body radiation generates a fourth-order nonlinear nonhomogeneous Robin BC (RBC). Unfortunately, CFD algorithm implementation of this flux RBC generates a highly nonlinear *dense* global Newton Jacobian *not amenable* to conventional CFD algebraic iteration strategy without enforcement of very significant simplifying assumptions.

Green-Gauss divergence theorem enforcement of weak formulation $\text{GWS}^h+\theta\text{TS}$ FE p basis NS/RaNS DE + RBC algebraic iteration $p+1$ stage generates the $\partial\Omega^h$ element matrix

$$\text{GWS}^h\theta\,\text{TS DE RBC:}\theta\Delta t\left(\frac{\sigma}{\rho c_p}[\text{M200}]_e\left(4\text{det}_{e=i}\,F_{i\to k}\{Q^4\}_{e=i}-4\text{det}_{e=k}F_{k\to i}\{Q^4\}_{e=k}\right)\right)$$

time stations n, $n+1$ θ partitioned. Therein σ is the Stephan-Boltzmann constant, $[\text{M200}]_e$ remains the FE mass matrix on $\partial\Omega_e$, $F_{i\to k}$, $F_{k\to i}$are view factors for surface element i interaction with other surfaces k, with $4\text{det}_{e=i}$, $4\text{det}_{e=k}$ the surface area of each $\partial\Omega_e$.

In reality, radiation occurs among surfaces each possessing a distinct *emissivity* ε, a number less than unity, which alters the process to what is termed *grey body* radiation. Alteration of SB to grey body radiation involves generation of the Gebhart factors $G_{k\to i}$, cf., Miller et al. [16]. In the absence of surface transparency, radiant energy received by surface $\partial\Omega_e$ from surface elements $\partial\Omega_{e,}\forall e=j$ is the summation

$$G_{k\to i}\equiv\varepsilon_i F_{k\to i}+\sum_{j=1}^{n}(1-\varepsilon_j)F_{k\to j}G_{j\to i}$$

Thus generated global algebraic matrix statement for prediction of the $G_{k\to i}\,(\partial\Omega_{e=i})$ distribution is linear but computationally intense as the matrix is dense.

Once computed, the SB grey body *cubically nonlinear* $\text{GWS}^h+\theta\text{TS}$ FE p basis element matrix differential calculus-derived contribution to the Newton Jacobian is

$$\left[\text{JAC}\_\{Q^p{}_{n+1}\}\right]_{e=i}\equiv\frac{\partial}{\partial\{Q\}}\left(\frac{\theta\Delta t\sigma}{\rho c_p}[\text{M200}]_e\left(\varepsilon_i 4\text{det}_{e=i}\{Q^4\}_{e=i}\right.\right.$$

$$\left.\left.-\sum_{k-1}^{n}\varepsilon_k 4\text{det}_{e=k}G_{k\to i}\{Q^4\}_{e=k}\right)\right)$$

$$= \frac{4\theta\Delta t\sigma}{\rho c_p}[\text{M200}]_e\left(\varepsilon_i \text{det}_i \text{diag}\left[Q^3\right]_{e=i} - \sum_{k-1}^{n} \varepsilon_k \text{det}_{e=k} G_{k\to i}\{Q^3\}_k\right)$$

wherein diag denotes a *diagonal* matrix. The matrix assembly of SB Newton Jacobian element contributions over Ω^h generates the *dense* global rank Jacobian $[\text{JAC}\_\{Q^p_{n+1}\}]$ not amenable to conventional CFD numerical linear algebra practice without enforcing major simplifying approximations.

8.4 Radiation, differential radiosity theory weak formulation NBC, Newton Jacobian

The CFD linear algebra *efficient* and approximation-free, hence functionally *accurate* alternative to SB implementation is *differential facet radiosity* theory (Argyris and Szimmat [2]). Radiation $R_k \text{d}A_k$ emanating into a semi-infinite half space from surface *differential facet* $\text{d}A_k$ impinges on differential facet $\text{d}A_i$ (Figure 8.6). The portion of this radiation that is directed towards $\text{d}A_i$ is expressed by the statement of *differential radiosity*

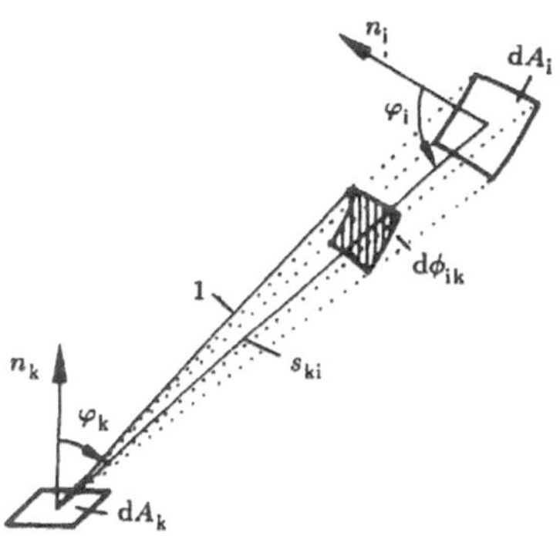

Figure 8.6: Surface differential facets in radiant energy exchange.

$$\text{d}R_{k\to i}\text{d}A_k = R_k \text{d}A_k \cos\phi_k/\pi$$

The emission actually impinging on differential facet $\text{d}A_i$ located at the distance $r_{k\to i}$ from differential facet $\text{d}A_k$ is defined by *Lambert's cosine law* in differential form

$$\text{d}R_{k\to i} = R_k \text{d}A_k \frac{\cos\phi_k \cos\phi_i}{\pi r^2_{k\to i}} \text{d}A_i$$

Radiation emanating from differential facet $\text{d}A_i$ per unit time in the absence of through transmission is the sum of self-emission, according to SB, and that reflected from generic differential surface facet dA_k

$$R_i \mathrm{d}A_i = \varepsilon_i \sigma T_i^4 \mathrm{d}A_i + (1-\varepsilon_i)\mathrm{d}R_{k\to i}$$

Generalizing to n grey body differential surface facets, differential facet $\mathrm{d}A_i$ radiosity R_i is defined by the *Fredholm integral equation* of the second kind

$$R_i = \varepsilon_i \sigma T_i^4 + (1-\varepsilon_i)\left[\sum_{k=1}^{n}\int_{A_k} R_k \frac{\cos\phi_k \cos\phi_i}{\pi r_{k\to i}^2} dA_k\right]$$

for R_i the radiant energy per unit time per unit area.

The energy efflux q_i generated at a surface of a finite facet of area A_i is the difference between emanated radiant energy R_i and the total radiation incident from all exchange finite surface facets in the view of surface A_i

$$q_i = R_i - \sum_{k=1}^{n}\int_{A_k} R_k K_{ik} \mathrm{d}A_k$$

Thereby the net energy efflux from a grey body surface A_i in terms of its *radiosity* R_i is

$$q_i = \frac{\varepsilon_i}{(1-\varepsilon_i)}\left[\sigma T_i^4 - R_i\right]$$

At first view this statement of net radiant energy efflux radiosity appears to not be applicable to black body radiation due to the factor $(1-\varepsilon_i)$ in the denominator, a conclusion to be dispelled.

For an n grey body configuration, and realizing that radiosity is a *distribution* on any finite surface facets $\partial\Omega_e$ with element geometric node colocated DOF $\{\mathrm{R}(t)\}_e$, the weak formulation $\mathrm{GWS}^h + \theta\mathrm{TS}$ FE p basis non-homogeneous Neumann BC (NBC) element matrix

$$\mathrm{GWS}^h + \theta\,\mathrm{TS\,DE\,NBC:}\ \ \theta\,\Delta t\left(\frac{\det_e}{\rho c_p}\frac{\varepsilon_e}{(1-\varepsilon_e)}[\mathrm{M200}]\left(\sigma\{Q^4\}_e - \{R\}_e\right)\right)$$

is a *much less* compute intensive alternative to Stephan-Boltzmann with Gebhart factors. Of fundamental importance, the *sparse matrix attribute* of this radiosity theory NBC $\partial\Omega_e$ matrix persists in its differential calculus-derived contribution to the Newton Jacobian

$$\begin{aligned}[\mathrm{JAC\_}\{Q_{n+1}^p\}]_e &\equiv \frac{\partial}{\partial\{Q\}_e}\left[\theta\Delta t\frac{\det_e}{\rho c_p}\frac{\varepsilon_e}{(1-\varepsilon_\varepsilon)}[\mathrm{M200}]\left(\sigma\{Q^4\}_e - \{R\}_e\right)\right] \\ &= \theta\Delta t \det_e \frac{\sigma\varepsilon_e}{\rho c_p(1-\varepsilon_\varepsilon)}[\mathrm{M200}]diag\left[Q^3\right]_e\end{aligned}$$

which of course persists upon assembly into the quadratic convergent Newton Jacobian matrix.

8.5 Differential radiosity weak formulation, GWS^h algorithm

Radiation exchange depends on the surface facet orientation generated line of sight which is named the *viewfactor* (VF). Lambert's cosine law defines VF as the radiation received at surface differential facet dA_i from differential facet dA_k for $r_{k\rightarrow i}$ the separation distance is

$$\text{VF:}\ F_{k\rightarrow i} \equiv \frac{1}{A_k}\int_{A_k}\int_{A_i}\frac{\cos\phi_k\cos\phi_i}{\pi r^2_{k\rightarrow i}}\,dA_k\,dA_i$$

For planar differential surface facets the VF kernel possesses the symmetry property

$$K_{k\rightarrow i} = \frac{\cos\phi_k\cos\phi_i}{\pi r^2_{ki}} = K_{i\rightarrow k} \equiv K_{i\,k}$$

Resultant correlations $F_{k\rightarrow i} = A_i\,K_{i\,k}$ and $F_{i\rightarrow k} = A_k\,K_{i\,k}$ are valid *only* (!) for *differential* facets. Thereby VF must be a *distribution* on nondifferential surface facets which thus identifies that radiosity is also a distribution. In homogeneous form, for the n finite surface facet domain boundary segments $\partial\Omega_e$ of $\partial\Omega^h$, the GWS^h FE p basis algorithm element matrix for prediction of the radiosity $\mathrm{R}^h(x_i)$ distribution DOF populating $\{\mathrm{R}\}_e$ is

$$\text{GWS}^h\text{:}\ \int_{\Omega_{e=i}}\{N\}\left(\begin{array}{c}\left(\{N\}^T\{\mathrm{R}\}_i - \varepsilon_i\sigma\{N\}^T\{Q^4\}_i\right)dA_i\\ -(1-\varepsilon_i)\left[\sum\limits_{k=1}^{\mathrm{n}}\int_{\Omega_{e=k}}K_{ik}\{N\}^T\{\mathrm{R}\}_k dA_k\right]dA_i\end{array}\right)\equiv\{0\}$$

Completion of the embedded integration generates the surface element Ω_e matrix

$$\text{GWS}^h\text{:}\ \det_i[\text{M200}]\{\mathrm{R}\}_i - (1-\varepsilon_i)\sum_{k=1}^{n}K_{i\,k}\det_k\det_i[\text{M1010}]\{\mathrm{R}\}_k = \varepsilon_i\sigma\det_i[\text{M200}]\{Q^4\}_i$$

which is a *strictly algebraic* (!) statement, hence "[A]{x} = {b}" direct solvable for prediction of the DOF in the $R^h(x_i)$ distribution.

The sequential integration process required in forming the second term therein generates the FE p basis column matrix outer product $\{\text{M10}\}\{\text{M10}\}^T$ labeled [M1010] to distinguish its unique formulation process. For the $n = 2$ bilinear FE $p = 1$ basis, each element of [M1010] is the constant 1/16 with $4\det_k$ and $4\det_i$ the respective element surface areas.

To quantify the fundamental consequence of thus derived GWS^h direct solve algebraic formulation, visually alter the differential facet geometry in Figure 8.4 to finite dimension facets with one facet discretized into two elements Ω_e, Replace the GWS^h matrix statement kernels $K_{i\,k}$ with view factors $F_{i\rightarrow k}$, also the element matrix data 4

($\det_k$, $\det_i$) by integer subscripts on each matrix. Then denoting the rank of each direct solvable matrix statement by a subscript, the resultant GWS^h direct solve matrix statement for prediction of the radiosity DOF distribution for this M = 3 mesh is with visual clarity symbolized as

$$\begin{bmatrix} [\text{M200}]_1, & -(1-\varepsilon_1)F_{1\to2}[\text{M1010}]_{12}, & -(1-\varepsilon_1)F_{1\to3}[\text{M1010}]_{13} \\ -(1-\varepsilon_2)F_{2\to1}[\text{M1010}]_{12}, & [\text{M200}]_2, & -(1-\varepsilon_2)F_{2\to3}[\text{M1010}]_{23} \\ -(1-\varepsilon_3)F_{3\to1}[\text{M1010}]_{13}, & -(1-\varepsilon_3)F_{3\to2}[\text{M1010}]_{23}, & [\text{M200}]_3 \end{bmatrix}_{12\times12}$$

$$\begin{Bmatrix} \{\text{R}\}_1 \\ \{\text{R}\}_2 \\ \{\text{R}\}_3 \end{Bmatrix} = \sigma \begin{Bmatrix} \varepsilon_1[\text{M200}]_1\{Q^4\}_1 \\ \varepsilon_2[\text{M200}]_2\{Q^4\}_2 \\ \varepsilon_3[\text{M200}]_3\{Q^4\}_3 \end{Bmatrix}_{12\times1}$$

For the tensor product FE $p = 1$ basis implementation, the plate discretized into two quadrilateral elements Ω_e share a common pair of geometric nodes. This GWS^h matrix solution statement thereby clearly visualizes that the radiosity $R^h(x_i)$ solution DOF at any such shared node in $\{\text{R}\}_1$ will *not be identical* (!) with the same node radiosity $R^h(x_i)$ solution DOF in $\{\text{R}\}_2$. In fact, augmented by VF and emissivity distinctions, the $R^h(x_i)$ DOF in $\{\text{R}\}_e$ for geometric nodes shared by adjacent elements Ω_e will *always* (!) be unique.

Thereby, for finite dimension facets the weak formulation GWS^h FE $p = 1$ basis algorithm generated $R^h(x_i)$ distribution will always be ***piecewise discontinuous*** (!). Furthermore, GWS^h radiosity algorithm implementation with higher completeness degree FE $p = 2,3$ bases will not alter this observation!.

8.6 Validation, asymptotic convergence, differential radiosity weak formulation GWS^h FE $p = 1$ basis algorithm

For an isolated isothermal plane finite facet, a semianalytical theory predicts that the $R(x_i)$ distribution for steady conduction DE with radiosity NBC is a *paraboloidal surface* with extremum at the centroid, Gould [12]. For an originally isothermal uniform plane facet M = 8 × 8 discretization, the GWS^h FE $p = 1$ basis-generated distribution of $R^h(x_i)$ is perspective projection graphed with mesh overlay in Figure 8.7a.

Noting that the vertical surfaces connecting each plane facet uniform $R^h(x_i)$ distribution therein are a graphics package plotting anomaly, the discontinuous planar union of $R^h(x_i)$ *uniform* $\{\text{R}\}_e$ DOF horizontal distributions is indeed a piecewise discontinuous approximation to a paraboloidal surface with extremum at the centroid. Thereby, the weak formulation steady conduction NS DE + NBC GWS^h FE $p = 1$ basis algorithm generated piecewise discontinuous $R^h(x_i)$ distribution is *quantitatively validated.*

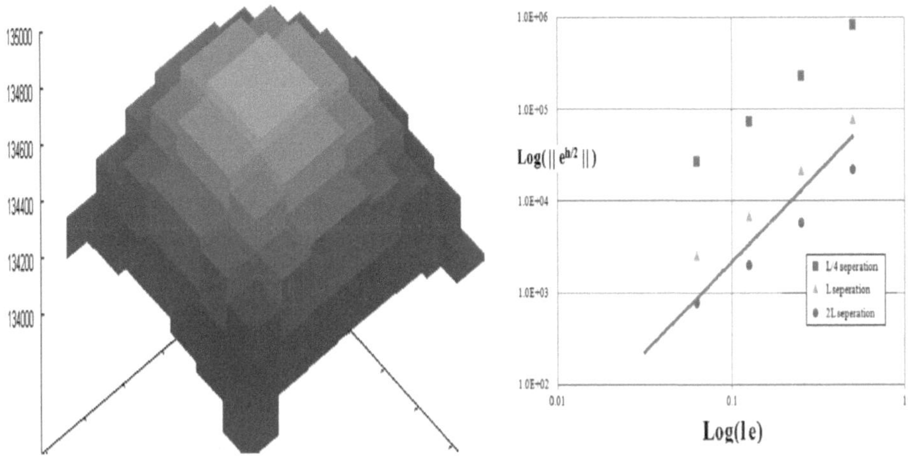

a) isolated plate $R^h(x_i)$ distribution

b) facing plates, $R^h(x_i)$ $O(h^2)$ convergence

Figure 8.7: Validation, differential radiosity weak formulation, steady DE + NBC GWS^h FE $p = 1$ basis algorithm, isolated isothermal plane facet M = 8 × 8 uniform mesh, $\varepsilon = 0.9$: (a) $R^h(x_i)$ piecewise discontinuous distribution, (b) for isothermal plane parallel facing plates, offsets L/4, L, 2L, asymptotic convergence in $(H^0)^2$, red line quadratic slope (adapted from Freeman et al. [10]).

As the Fredholm integral equation for radiosity is a statement of interpolation, continuous weak formulation theory predicts that the *asymptotic convergence* rate for the conduction NS DE + NBC GWS^h FE $p = 1$ basis algorithm solutions

$$
\begin{aligned}
\|\mathrm{R}^h\|_{H^0} \equiv \sum_{\Omega_e} \|\mathrm{R}_e^h\|_{H^0} &= \frac{1}{2}\sum_{\Omega_e}\left(\{\mathrm{R}\}_e^T \int_{\Omega_e} \{N\}\{N\}^T \, d\tau \{\mathrm{R}\}_e\right)^{1/2} \\
&= \frac{1}{2}\left(\sum_{e=1}^{\mathrm{M}} \det_e \{\mathrm{R}\}_e^T [\mathrm{M200}]\,\{\mathrm{R}\}_e\right)^{1/2}
\end{aligned}
$$

will be $O(h)$ in the Hilbert norm H^0, the space of all functions square integrable and finite on Ω. Kolmogorov and Fomin [14]. For an isolated isothermal *normal facing* plane plate pair, each of measure L and emissivity $\varepsilon = 0.9$, conduction NS DE + NBC GWS^h FE $p = 1$ basis uniform mesh refinement sequence (Figure 8.7b) *validates* $R^h(x_i)$ asymptotic convergence is indeed $O(h^2)$ in the H^0 norm squared for plate offsets L/4, L, and 2L. The convergence rate alteration to H^0 squared correlates with weak formulation preferred error estimation in the NS DE PDE energy norm (Appendix B).

8.7 Validation, long time stability, weak formulation gray body unsteady NS DE + NBC GWS^h + θTS FE $p = 1$ basis trapezoidal rule algorithm

For a normal facing plane plate pair each of measure L, with offset distance L and IC the identical uniform temperature, unsteady NS DE + NBC GWS^h + θTS FE $p = 1$ basis algorithm predicted long time decay of each plate temperature on $0 \le n\Delta t \le$ E05 s for grey body *emissivity* range $0.5 \le \varepsilon \le 0.9999$ is semilog scaled in Figure 8.8. The repository for the energy "lost" by each plate during this process is the surrounding transparent half-space of infinite lateral extent within which the plate pair resides.

The long time plate temperature decay rate history is essentially indistinguishable for emissivity ranging $0.5 \le \varepsilon \le 0.8$. The temperature decay rate then progressively decreases for emissivity specifications $\varepsilon = 0.9$, 0.99, 0.9999. As definition $\varepsilon_i = 0.9999$ is the *essential* statement of black body radiation, that the plate temperatures remain uniform at the IC specification for computed evolution period of order 10^5 s validates that the GWS^h + θTS FE $p = 1$ basis algorithm/trapezoidal rule code is long time *stable* for *O*(E+04) the magnitude of the Fredholm integral equation emissivity factor $\varepsilon_i/(1 - \varepsilon_i)$.

Of historical pertinence, this weak formulation GWS^h + θTS algorithm facing plate geometry prediction also serves to validate the pre-CFD era *classic* graphical solution technique detailed in Bird, Stewart, and Lightfoot ([5], Figure 14.5-2).

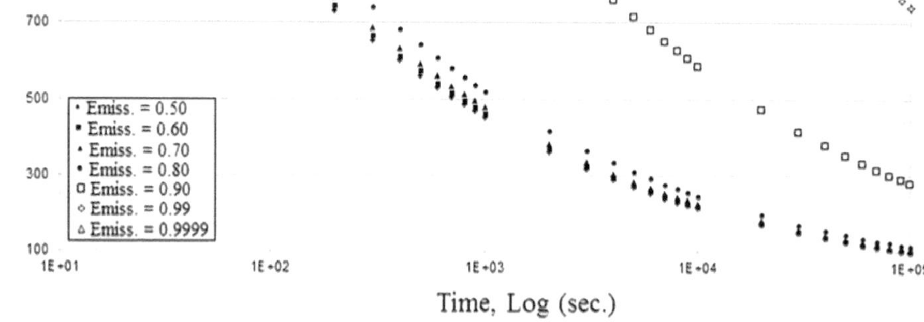

Figure 8.8: Validation, differential radiosity theory stability, unsteady conduction DE + NBC GWS^h + θTS FE $p = 1$ basis algorithm, initially uniform equal temperature normal facing plate pair of measures L, separation offset L, emissivity range $0.5 \le \varepsilon_i \le 0.9999$, plate temperature decay rate ordinate semilog scaled on $0 < n\Delta t \le \mathrm{E} + 05$ s (adapted from Freeman et al. [10]).

8.8 Conclusion, NS/RaNS PDE system *error-freed* CFD *mathematics*

This chapter concludes monograph derivation/validation of NS/RaNS PDE systems weak formulation continuous Galerkin FE p basis algorithms written on *error-freed* CFD *mathematics* alteration of the textbook appearance of incompressible and compressible transonic to hypersonic aerothermodynamic fluid mechanics conservation principles. Appendix C fully details the $\mathrm{GWS}^h + \theta\mathrm{TS}$ FE p-basis algorithm formulation particulars omitted to this juncture to keep theory/validation presentation clarity uncompromised by distracting details.

The remaining topic for weak formulation *error-freed* CFD *mathematics* continuous Galerkin FE p basis algorithms is *analytical closure* of space-filtered NS PDE systems which follows as Part B of this monograph. Fully validated therein is that analytically closed NS PDE system algorithms are quantified $\forall$Re pertinent to prediction of $n = 3$ thermal-velocity vector fields for Reynolds numbers ranging $100 < \mathrm{Re} < 12{,}000$. In particular, Part B content for the extremum specification validates *rigorous mathematical principles* prediction of laminar profile BL velocity insipient transition, separation, turbulent profile BL velocity reattachment then relaminarization in the *total absence* of any form of artificial stabilization or modeling. Thereby, the CFD theorization maturity required to predict hypersonic shock layer BL transition does indeed exist!

References

[1] Anderson, J.D. (1989). *Hypersonic and High Temperature Gas Dynamics*, McGraw-Hill, NY.

[2] Argyris, J., Szimmat, J. (1992). "An analysis of temperature radiation interchange problems," *J Comp Mtd Appl Mech Engr*, 94.

[3] Baker, A.J., Freels, J.D., Iannelli, G.S., Manhardt, P.D. (1994). "A Finite Element Code for Real Gas Aerodynamics Simulation (PRaNS), Vol. I. Algorithm & Theory," US Air Force, Ballistic Missiles Organization Tech. Rpt. BMO:TR-91–39

[4] Baldwin, B.S., Lomax, H. (1979). "Thin-layer approximation and algebraic model for separated turbulent flows," Tech. Rpt. AIAA:78:257

[5] Bird, R.B., Stewart, W.E., Lightfoot, E.N. (1960). *Transport Phenomena*, Wiley, NY

[6] Blair, T.R. (2022). personal communication

[7] Button, K. (2019). "Decoding the Boundary Layer at Hypersonic Speeds," *Aerospace America*, 57, 9.

[8] Candler, G. (2019). "Rate Effects in Hypersonic Flow," *Ann Rev Fluid Mechanics*, 51, 379.

[9] Duffa, G., (2013). *Ablative Thermal Protective Systems Modeling*, AIAA Education Series, ISBN-10: 1624101717

[10] Freeman, E.L., Baker, A.J., Orzechowski, J.A., Wong, K.L. (2014). "Validation of a Time-Accurate Weak Form Algorithm for Convection-Greybody Radiation Dominated Unsteady Heat Transfer," *J Heat Mass Transfer*, 72.

[11] Gnoffo, P.A. (1999). "Planetary-Entry Gas Dynamics," *Ann Rev Fluid Mechanics*, 31, 459.

[12] Gould, N., (2000). "Radiation heat transfer between diffuse-grey surfaces using high order finite elements," AIAA Tech. Rpt. AIAA:2000-2371

[13] Gonzalez, D., Medina, A. (2019). "Experimental-numerical campaigns give insight into complex fluid phenomena," *Aerospace America*, 57, 11.

[14] Kolmogorov, A.N., Fomin, S.V. (1975). *Introductory Real Analysis*, Dover, NY.
[15] Martellucci, A., Weinberg, S. (1982). "MAT Program Test Summary Report – Biconic Body with Slice/ Flap," USAF Technical Report BMO: TR-82–28.
[16] Miller, F.P., Vandome, A.F., Brewster, J.M. (2010). *Gebhart Factors*, VDM Publishing House Ltd,, England.
[17] Tannehill, J.C., Mugge, P.H. (1974). "Improved Curve Fits for the Thermodynamic Properties of Equilibrium Air Suitable for Numerical Computation using Time-dependent or Shock Capturing Methods," NASA Technical Report CR-2470

Part B: **Analytical closure of spatially filtered NS PDE systems**

Chapter 9
Mathematically rigorous closure of spatially filtered NS PDE systems

Libretto: Error freed CFD mathematics alterations of NS/RaNS/LES PDE systems *annihilate* space-time discretization-generated truncation error algebraic instability and scalar member phase aliasing.

Herein rigorously derived *analytical closure* of spatially filtered NS PDE systems is the *transformational* alteration to, "... **stagnation** of CFD capability *advancement* ... exists."

As the derivation never invokes the word *turbulent*, analytically closed spatially filtered NS PDE systems are pertinent for ***arbitrary*** Reynolds number (Re) specification.

This *error-freed* CFD mathematics *advancement* renders obsolete the host of LES algorithms that replace resolved-unresolved scale interaction *and* unresolved scales ***mathematical rigor*** with heuristic "physics-based" subgrid scale tensor models.

Keywords: Navier-Stokes, spatial filtering, analytical closure, Gaussian filter, ∀Re pertinence

9.1 Legacy "physics-based" closure models for spatially filtered NS (LES) PDE systems

From Chapter 3, spatial filtering of incompressible/compressible NS PDE systems inserts the resolved-unresolved scale interaction state variable tensor and vector *quadruples* into thus generated resolved scale $O(1)$ PDE systems

$$\overline{u_j m_i} \equiv \overline{(\bar{u}_j + u'_j)(\bar{m}_i + m'_i)} = \overline{\bar{u}_j \bar{m}_i} + \overline{\bar{u}_j m'_i} + \overline{u'_j \bar{m}_i} + \overline{u'_j m'_i}$$

$$\overline{u_j u_i} \equiv \overline{(\bar{u}_j + u'_j)(\bar{u}_i + u'_i)} = \overline{\bar{u}_j \bar{u}_i} + \overline{\bar{u}_j u'_i} + \overline{u'_j \bar{u}_i} + \overline{u'_j u'_i}$$

$$\overline{u_j q} \equiv \overline{(\bar{u}_j + u'_j)(\bar{q} + q')} = \overline{\bar{u}_j \bar{q}} + \overline{\bar{u}_j q'} + \overline{u'_j \bar{q}} + \overline{u'_j q'}$$

for q a state-variable generic scalar member. That no terms therein correlate with incompressible or compressible spatially filtered NS *resolved scale* $O(1)$ state variables

$$\overline{\mathrm{NS}}\colon \{\bar{q}(x_j, t)\} \equiv \{\bar{\varphi}, \bar{u}_i, \overline{\Theta}, \overline{Y} : \overline{P}\}^T,\ \{\bar{\rho}, \overline{m_i}, \overline{E}, \overline{\rho Y} : \bar{p}\}^T$$

generates the requirement to establish *closure* of the $O(1)$ PDE systems.

https://doi.org/10.1515/9783112206430-010

The original space-filtered incompressible NS **DP** CFD algorithm is Smagorinsky [19]. On a rectangular Cartesian finite difference (FD) mesh, the *box-filtered* NS **DP** closure model is of RaNS *eddy viscosity* mixing length genre with the *single* model constant C_S. Closure model simplicity led to inaccurate predictions, as well anticipated (!), which prompted replacement of the "Smagorinsky constant" C_S with distributions. Such closure alterations include *dynamic-subgrid scale* (SGS) solution-adapted mesh refinement [5, 13], the kinetic energy of the SGS [24], and the mixed scale [16]. The voluminous LES practice is thoroughly archived (cf., [1, 4, 6, 7, 8, 14, 16, 23]).

With very few exceptions legacy LES closure practices replace the NS **DP** *resolved-unresolved scale interaction* tensor pair, also the *unresolved scale* tensor, with heuristically argued "physics-based" SGS tensor models that are *solely* resolved scale $O(1)$ velocity functionally dependent. For example, SGS tensor model identification via the *double decomposition* hypothesis [16]

$$\tau_{ij} \equiv \overline{u_j u_i} - \bar{u}_j \bar{u}_i \Rightarrow (\overline{\bar{u}_j u'_i} + \overline{u'_j \bar{u}_i}) + \overline{u'_j u'_i} \equiv C_{ij} + R_{ij}$$

generates the *cross-stress tensor* pair C_{ij} and a Reynolds SGS tensor R_{ij}, both of which are then modeled. The *triple decomposition* hypothesis assumes all tensors are $O(1)$ state-variable expressible, hence forming $\overline{\bar{u}_j \bar{u}_i}$ requires a second filtering. The remedy

$$\overline{\bar{u}_j \bar{u}_i} = (\overline{\bar{u}_j \bar{u}_i} - \bar{u}_j \bar{u}_i) + \bar{u}_j \bar{u}_i \equiv L_{ij} + \bar{u}_j \bar{u}_i$$

generates the *Leonard* stress tensor L_{ij} [12], wherein the generated space-filtered $O(1)$ velocity tensor product is assumed to be the nonlinear convection term in $O(1)$ **DP**.

Legacy LES closure hypotheses thereby place the entire *burden* of prediction fidelity on *physical insight*, i.e., "physics-based" modeling, which is the consequence of the *inability* to mathematically predict the filtering generated tensor triple $\overline{\bar{u}_j u'_i} + \overline{u'_j \bar{u}_i} + \overline{u'_j u'_i}$. As occurs for all RaNS *scalar* coefficient Reynolds stress tensor closure models, the principle coordinate system for SGS tensor closure models is thereby forced to be identical with that of the filtered $O(1)$ velocity Stokes isotropic strain rate tensor $\bar{S}_{ij}$ which leads to the representative SGS tensor functional form

$$\tau_{ij} \cong f\left(C_S(\mathbf{x}, t), \delta^2, |\bar{S}_{ij}|\right)\bar{S}_{ij}$$

Therein the SGS tensor model scalar coefficient $f\left(C_S(\mathbf{x}, t), \delta^2, |\bar{S}_{ij}|\right)$ is indeed a distributed "turbulent eddy viscosity," recall Chapter 3.2, with $C_S(\mathbf{x}, t)$ the placeholder for the litany of the replacements for the Smagorinsky constant C_S. $|\bar{S}_{ij}|$ is a norm of the Stokes $O(1)$ velocity *isotropic* tensor $\bar{S}_{ij}$. Finally, note the selected filter of uniform measure δ, or distributed measure $\delta(x_j)$, impacts thus closed LES $O(1)$ PDE system CFD prediction *only* in its turbulent eddy viscosity definition.

One alternative to legacy LES physics-based SGS tensor model closure is named the *variational multiscale method* (VMM) [9]. This theory admits resolved scale velocity partitioning into "larger to smaller" scales and closure remains reliant on an SGS tensor model. In clear distinction to legacy LES modeling practice, VMM theorizations recognize (!) that SGS tensor models also act as the CFD algorithm numerical stabilization, hence indeed constitute artificial dissipation. A VMM archive is due to Rasthofer and Gravemeier [15].

9.2 Rigorous analytical closure of spatially filtered NS PDE systems

Alternatives to SGS/VMM tensor physics-based closure modeling for the spatially filtered **DP** tensor quadruple $\overline{u_j u_i}$ involve polynomial interpolation of the inverse of the Fourier transform of a uniform measure δ Gaussian filter. This operation enables *analytical* completion of the Fourier transform operations intrinsic to space filtering.

The second-order Taylor series (TS) interpolation (Figure 9.1a) is truly inaccurate (!) and its poor performance is well documented in the LES literature. The Pade' second and fourth-order *rational* interpolation polynomials (Figure 9.1b,c) generate approximations possessing significantly improved fidelity for the Gaussian filter.

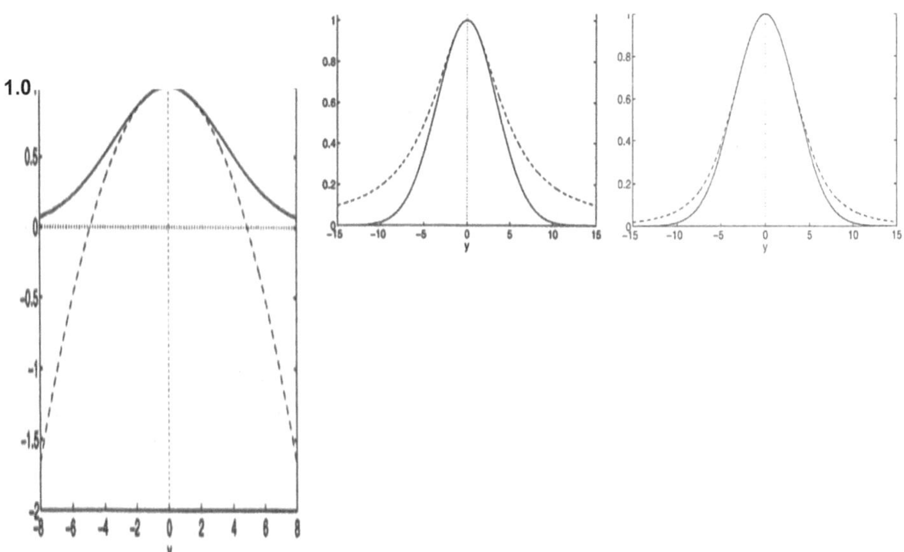

a) 2nd order Taylor series   b) 2nd order rational Pade   c) 4th order rational Pade

Figure 9.1: Polynomial interpolations (the dashed lines) of the Fourier transform inverse of a Gaussian measure δ filter (the solid lines): (a) second-order Taylor series, (b) Pade′ second-order rational polynomial, (c) Pade′ fourth-order rational polynomial.

The interpolation definitions for the inverse of the Gaussian Fourier transform in n-D are

$$\text{second-order Taylor: } e^{ax_i} = 1 + ax_i + O\left(a^2x_i^2\right)$$

$$\text{second-order Pade': } e^{ax_i} = \frac{1}{1 + ax_i} + O\left(a^2x_i^2\right)$$

$$\text{fourth-order Pade': } e^{ax_i} = \frac{1}{1 + ax_i + a^2x_i^2} + O\left(a^4x_i^4\right)$$

and the Gaussian filter of measure δ and shape factor γ is defined

$$g_\delta(x_i) = \left(\frac{\gamma}{\pi\delta^2}\right)\exp\left(-\frac{\gamma}{\delta^2}x_i^2\right)$$

The Fourier transforms for the lead three tensors in the spatially filtered quadruple $\overline{u_ju_i} \equiv \overline{(\bar{u}_j + u'_j)(\bar{u}_i + u'_i)} = \overline{\bar{u}_j\bar{u}_i} + \overline{\bar{u}_ju'_i} + \overline{u'_j\bar{u}_i} + \overline{u'_ju'_i}$ are

$$F\left(\overline{\bar{u}_j\bar{u}_i}\right) = F(g_\delta)F\left(\bar{u}_j\bar{u}_i\right)$$
$$F\left(\overline{\bar{u}_ju'_i}\right) = F(g_\delta)\left(F(\bar{u}_j) * F(u'_i)\right)$$
$$F\left(\overline{u'_j\bar{u}_i}\right) = F(g_\delta)\left(F(u'_j) * F(\bar{u}_i)\right)$$

Since $F(g_\delta)$ is non-zero, their rearrangements define the Fourier transforms for u_i and u'_i

$$F(u_i) = \frac{F(g_\delta)F(u_i)}{F(g_\delta)} = \frac{F(\bar{u}_i)}{F(g_\delta)}, \quad F(u'_i) = \left(\frac{1}{F(g_\delta)} - 1\right)F(\bar{u}_i)$$

which enables alteration of the previous Fourier transform statements to

$$F\left(\overline{\bar{u}_j\bar{u}_i}\right) = F(g_\delta)F\left(\bar{u}_j\bar{u}_i\right)$$
$$F\left(\overline{\bar{u}_ju'_i}\right) = F(g_\delta)\left(F(\bar{u}_j)^*\left(\frac{1}{F(g_\delta)} - 1\right)F(\bar{u}_i)\right)$$
$$F\left(\overline{u'j\bar{u}_i}\right) = F(g_\delta)\left(\left(\frac{1}{F(g_\delta)} - 1\right)F(\bar{u}_j)^*F(\bar{u}_i)\right)$$

As the Fourier transform of the resolved scale velocity is known, completion of the derivations rests on identification of approximations of the Gaussian filter Fourier transforms $F(g_\delta)$ and $1/F(g_\delta)$ in these definitions. For the second-order TS interpolation of the transform inverse, a second-order TS approximation for the Gaussian filter Fourier transform inverse generates these $O(\delta^4)$ approximations

$$F(g_\delta(\delta, y_j)) = 1 - \frac{\|y_j\|_2^2}{4\gamma}\delta^2 + O\left(\delta^4\right), \quad \frac{1}{F(g_\delta(\delta, y_j))} = 1 + \frac{\|y_j\|_2^2}{4\gamma}\delta^2 + O\left(\delta^4\right)$$

Their insertion into the Fourier transforms for the lead three tensors in the $\overline{u_j u_i}$ quadruple then identify their Fourier transform approximations to be

$$F(\overline{\bar{u}_j\bar{u}_i}) = F(\bar{u}_j\bar{u}_i) + \frac{\delta^2}{4\gamma}F\left(\frac{\partial^2(\bar{u}_j\bar{u}_i)}{\partial x_k \partial x_k}\right) + O(\delta^4)$$

$$F(\overline{\bar{u}_j u'_i}) = -\frac{\delta^2}{4\gamma}F\left(\bar{u}_j\frac{\partial^2\bar{u}_i}{\partial x_k \partial x_k}\right) + O(\delta^4)$$

$$F(\overline{u'_j\bar{u}_i}) = -\frac{\delta^2}{4\gamma}F\left(\frac{\partial^2\bar{u}_j}{\partial x_k \partial x_k}u_i\right) + O(\delta^4)$$

Then multiplying through by the Fourier inverse transformation identifies the second-order TS interpolation approximation for the lead three tensors in $\overline{u_j u_i}$ to be

$$\overline{\bar{u}_j\bar{u}_i} = \bar{u}_j\bar{u}_i + \frac{\delta^2}{4\gamma}\frac{\partial^2(\bar{u}_j\bar{u}_i)}{\partial x_k \partial x_k} + O(\delta^4)$$

$$\overline{\bar{u}_j u'_i} = -\frac{\delta^2}{4\gamma}\bar{u}_j\frac{\partial^2\bar{u}_i}{\partial x_k \partial x_k} + O(\delta^4)$$

$$\overline{u'_j\bar{u}_i} = -\frac{\delta^2}{4\gamma}\frac{\partial^2\bar{u}_j}{\partial x_k \partial x_k}\bar{u}_i + O(\delta^4)$$

Tensor field theory coalescence of this Cartesian resolution quantifies that the second-order TS approximation of the uniform measure δ Gaussian filter Fourier transform inverse identifies the lead three tensors in the triple to $O(\delta^4)$ to be

$$\text{second-order TS:}\quad \overline{\bar{u}_j\bar{u}_i} + \overline{\bar{u}_j u'_i} + \overline{u'_j\bar{u}_i} = \bar{u}_j\bar{u}_i + \frac{\delta^2}{2\gamma}\frac{\partial\bar{u}_j}{\partial x_k}\frac{\partial\bar{u}_i}{\partial x_k} + O(\delta^4)$$

The LES literature has labeled this analytical closure as the *gradient* model. Comparison of CFD predictions with the experimentally measured decay of isotropic turbulence yielded a quality correlation [22]. The algebraic instability associated with these gradient model closed CFD codes was traced to insufficient dissipation [21]. Code artificial stabilization via augmentation with a Smagorinsky closure model then led to excess artificial diffusion [2].

For the Pade' second-order rational polynomial interpolation of the Fourier transform inverse of a uniform measure δ Gaussian filter, retaining the second-order TS approximation of the generated Gaussian Fourier transforms truncated at $O(\delta^4)$ generates

$$F(g_\delta(\delta, y_j)) = \frac{\delta^2}{1 + \frac{\|y_j\|_2^2}{4\gamma}} + O(\delta^4), \qquad \frac{1}{F(g_\delta(\delta, y_j))} = 1 + \frac{\|y_j\|_2^2}{4\gamma}\delta^2 + O(\delta^4)$$

Clearing via the inverse Fourier transform then identifies the Cartesian resolution

$$\overline{\bar{u}_j\bar{u}_i} = \left[\delta_{ij} - \frac{\delta^2}{4\gamma}\nabla^2\right]^{-1}(\bar{u}_j\bar{u}_i) + O(\delta^4)$$

$$\overline{\bar{u}_j u'_i} = -\frac{\delta^2}{4\gamma}\left[\delta_{ij} - \frac{\delta^2}{4\gamma}\nabla^2\right]^{-1}\left(\bar{u}_j\frac{\partial^2\bar{u}_i}{\partial x_k \partial x_k}\right) + O(\delta^4)$$

$$\overline{u'_j\bar{u}_i} = -\frac{\delta^2}{4\gamma}\left[\delta_{ij} - \frac{\delta^2}{4\gamma}\nabla^2\right]^{-1}\left(\frac{\partial^2\bar{u}_j}{\partial x_k \partial x_k}\bar{u}_i\right) + O(\delta^4)$$

which via tensor field theory identifies the Pade' second-order rational interpolation approximation to $O(\delta^4)$ to be [3]

$$\text{second-order Pade: } \overline{\bar{u}_j\bar{u}_i} + \overline{\bar{u}_j u'_i} + \overline{u'_j\bar{u}_i} = \bar{u}_j\bar{u}_i + \frac{\delta^2}{2\gamma}\left[I - \frac{\delta^2}{4\gamma}\nabla^2\right]^{-1}\left(\frac{\partial\bar{u}_j}{\partial x_k}\frac{\partial\bar{u}_i}{\partial x_k}\right) + O(\delta^4)$$

The repeating of this sequence for Gaussian filter Fourier transform inverse interpolation via the Pade' fourth-order rational polynomial identifies the tensor field definition for the lead three tensors in $\overline{u_j u_i}$ as [10]

$$\text{fourth-order Pade: } \overline{\bar{u}_j\bar{u}_i} + \overline{\bar{u}_j u'_i} + \overline{u'_j\bar{u}_i} = \bar{u}_j\bar{u}_i + \frac{\delta^2}{2\gamma}\left[I - \frac{\delta^2}{4\gamma}\nabla^2 + \frac{\delta^4}{32\gamma^2}\nabla^4\right]^{-1}$$

$$\times\left(\frac{\partial\bar{u}_i}{\partial x_k}\frac{\partial\bar{u}_j}{\partial x_k} - \frac{\delta^4}{16\gamma^2}\left(\nabla^2\bar{u}_j\nabla^2\bar{u}_i + \frac{\partial\nabla^2\bar{u}_i}{\partial x_k}\frac{\partial\bar{u}_j}{\partial x_k} + \frac{\partial\bar{u}_i}{\partial x_k}\frac{\partial\nabla^2\bar{u}_j}{\partial x_k} + \nabla^2\left(\frac{\partial\bar{u}_i}{\partial x_k}\frac{\partial\bar{u}_j}{\partial x_k}\right)\right)\right) + O(\delta^6)$$

The symbols "I" in the $\overline{u_j u_i}$ lead triple tensor approximation generated by second and fourth-order Pade' rational polynomial interpolations denotes the identity matrix. The CFD algorithm implementation *challenge* for thus identified $O(\delta^4)$ and $O(\delta^6)$ analytical closures resides in the matrix inverse differential operator therein with the $\delta^2/4\gamma$ multiplier

$$[A] \equiv \left[I - \frac{\delta^2}{4\gamma}\nabla^2\right]^{-1}, [A] \equiv \left[I - \frac{\delta^2}{4\gamma}\nabla^2 + \frac{\delta^4}{32\gamma^2}\nabla^4\right]^{-1}$$

The rational LES literature labels this issue the *auxiliary problem* for which a variety of convolution-based resolutions have been reported to generate unsatisfactory results [10].

The second CFD algorithm implementation *challenge* for the second-order Pade' rational polynomial $O(\delta^4)$ closure is that the analytical derivation determines that the unresolved scale tensor

$$\text{second-order Pade: } \overline{u'_j u'_i} = \frac{\delta^4}{16\gamma^2}\left[\nabla^2\bar{u}_j\nabla^2\bar{u}_i\right] + O(\delta^6)$$

possesses *negligible* (!) significance as the lead term multiplier δ^4 is the truncation error order of the $\overline{u_j u_i}$ triple approximation. Of true fundamental theoretical impact, this second-order Pade' interpolation result has rigorously determined that the *significance* of the unresolved scale tensor is bounded $O(\delta^4) < O(\overline{u'_j u'_i}) < O(\delta^2)$. Thereby, the second-order Pade' identified unresolved tensor $\overline{u'_j u'_i}$ has been labeled the *subfilter scale* (SFS) tensor to distinguish it from the range of legacy LES identified SGS tensor models of universal $O(\delta^2)$ significance.

9.3 Resolution of the auxiliary problem, *analytical* FaNS theory PDE systems

The *verity* in fluid mechanics theoretical analysis that is clearly expressed by classic perturbation theory [20]

> . . . Most useful approximations (in fluid mechanics theory) are valid when one or more of the parameters or variables in the problem is small (or large).

correlates *precisely* with the second-order Pade rational polynomial interpolation generated lead three tensors $\overline{u_j u_i}$ an even order TS in Gaussian filter measure δ to $O(\delta^4)$. Of fundamental *mathematics of physics of fluids* pertinence to analytical closure of spatially filtered NS PDE systems, thereby enabled is *rigorous* resolution of the Pade' interpolation theory generated "auxiliary problem", Sekachev et al [17].

The non-D significance of the resolved scale velocity vector $\bar{u}_i$ in space-filtered NS **DP** is by definition $O(1)$ as is that of the second filtering generated **DP** convection term $\overline{\bar{u}_j \bar{u}_i}$. For CFD algorithm discrete implementation on a mesh of measure h, Figure 4.1 clearly quantifies that NS state variable Fourier spectral content of $O(\leq 2h)$ is not resolvable. Thereby, the significance of the non-D *unresolved* scale velocity vector u'_i cannot exceed $O(2h)$, which via the filtered NS adequate energetic capture constraint $\delta \geq 2\,h$, [14], quantifies its significance is indeed $O(\delta)$. The second-order Pade' rational interpolation predicted constraint $O(\delta^4) < O(\overline{u'_j u'_i}) < O(\delta^2)$ thereby quantifies that the significance of $\overline{u'_j u'_i}$ is *twice* less than that of u'_i. Hence, as $O(\delta)$ and $O(\delta^3)$ rigorously bound the *resolved-unresolved scale interaction* pair tensor, the significance of tensor pair $\overline{\bar{u}_j u'_i} + \overline{u'_j \bar{u}_i}$ cannot be other than $O(\delta^2)$.

Thereby, classic fluid mechanics perturbation theory predicts that the significance of the Pade' second-order rational interpolation generated $\overline{u_j u_i}$ lead triple tensor to $O(\delta^4)$ is

$$\overline{\bar{u}_j \bar{u}_i} = \bar{u}_j \bar{u}_i + O(\delta^2) + \ldots.$$

which interestingly (*thankfully* !?) is consistent with legacy LES practice formulated $O(\delta^2)$ SGS tensor models. However, totally *unique* (!) to the Pade' second-order interpolation $O(\delta^4)$ TS prediction is that this $O(\delta^2)$ term is *continuum* calculus identified to be

$$\text{second-order Pade:}\quad \overline{\bar{u}_j u'_i} + \overline{u'_j \bar{u}_i} = \frac{\delta^2}{2\gamma}\left[I - \frac{\delta^2}{4\gamma}\nabla^2\right]^{-1}\left(\frac{\partial \bar{u}_j}{\partial x_k}\frac{\partial \bar{u}_i}{\partial x_k}\right) + O(\delta^4)$$

Clearing the matrix differential operator inverse and common multipliers, then labeling the resolved-unresolved scale interaction $O(\delta^2)$ tensor pair $\tau_{ij}(\mathbf{x},t) \equiv \overline{\bar{u}_j u'_i} + \overline{u'_j \bar{u}_i}$, the classic fluid mechanics perturbation theory identified rigorous *resolution* of the Pade' second-order rational interpolation auxiliary problem *is* the linear harmonic EBV PDE

$$\mathrm{D}\tau_{ij}: -\delta^2\nabla^2\tau_{ij} + 4\gamma\tau_{ij} - 2\delta^2\frac{\partial \bar{u}_i}{\partial x_k}\frac{\partial \bar{u}_j}{\partial x_k} = 0 + O(\delta^4)$$

That $\mathrm{D}\tau_{ij}$ is linear and constant coefficient *guarantees* that all thus predicted $O(\delta^2)$ significance resolved-unresolved scale interaction tensor pairs $\tau_{ij}(\mathbf{x},t) \equiv \overline{\bar{u}_j u'_i} + \overline{u'_j \bar{u}_i}$ will be *symmetric, realizable,* and both translation and Galilean *invariant*. Thereby, this resolution of the "auxiliary problem" rigorously satisfies *all necessary* mathematical requirements of NS tensor field theory [14].

Direct extensions of this analysis identify resolution of the auxiliary problem for all filtered NS state variable member $O(\delta^2)$ resolved-unresolved scale interaction *vector* pairs. Specifically, for the $O(\delta^2)$ thermal *vector* pair $v_j^{\Theta} \equiv \overline{\bar{u}_j\Theta'} + \overline{u'_j\bar{\Theta}}$

$$\mathrm{D}v_j^{\Theta}: -\delta^2\nabla^2 v_j^{\Theta} + 4\gamma v_j^{\Theta} - 2\delta^2\frac{\partial \bar{u}_j}{\partial x_k}\frac{\partial \bar{\Theta}}{\partial x_k} = 0 + O(\delta^4)$$

and the species mass fraction $O(\delta^2)$ *vector* pair $v_j^{Y\alpha} \equiv \overline{\bar{u}_j Y'_\alpha} + \overline{u'_j \bar{Y}_\alpha}$

$$\mathrm{D}v_j^{Y\alpha}: -\delta^2\nabla^2 v_j^{Y\alpha} + 4\gamma v_j^{Y\alpha} - 2\delta^2\frac{\partial \bar{u}_j}{\partial x_k}\frac{\partial \bar{Y}_\alpha}{\partial x_k} = 0 + O(\delta^4)$$

Thus rigorously identified is the *O*(1;δ^2) significance state variable for ***F****iltered* ***a****nalytically* ***N****avier*-***S****tokes* $O(h^4, m^4, \delta^4)$ PDE system, which hereon is referenced as "FaNS theory", a *pronounceable* acronym that rhymes with RaNS. As the FaNS theory $O(1;\delta^2)$ scaled state variable PDE system is uniformly elliptic boundary value (EBV), to be bounded domain well-posed, i.e., of *practical use* for physics of fluids CFD quantification, requires that non-homogeneous boundary conditions (BCs) be *a priori* identified $\forall$ state variable member resolutions on the complete enclosure $\partial\Omega \in \mathbf{R}^{n-1}$ of the NS PDE system domain Ω. Rigorous analytical resolution of these *critical* requirements is derived in Chapter 10.

Classic perturbation theory resolution of the Pade' fourth-order rational interpolation generated $O(\delta^6)$ auxiliary problem generates the linear constant coefficient *bi-harmonic* EBV PDE

$$\mathrm{D}\tau_{ij}\text{: } \delta^4\nabla^4\tau_{ij} - 8\gamma\delta^2\nabla^2\tau_{ij} + 32\gamma^2\tau_{ij} = 16\gamma\delta^2\frac{\partial\bar{u}_i}{\partial x_k}\frac{\partial\bar{u}_j}{\partial x_k}$$

$$- 2\delta^4\left(\nabla^2\bar{u}_j\nabla^2\bar{u}_i + \frac{\partial\nabla^2\bar{u}_i}{\partial x_k}\frac{\partial\bar{u}_j}{\partial x_k} + \frac{\partial\bar{u}_i}{\partial x_k}\frac{\partial\nabla^2\bar{u}_j}{\partial x_k} + \nabla^2\left(\frac{\partial\bar{u}_i}{\partial x_k}\frac{\partial\bar{u}_j}{\partial x_k}\right)\right) = 0 + O(\delta^6)$$

Clearly evident therein is the $O(\delta^2)$ distinction in significance that exists between the first and the second highly differentiated $O(1)$ velocity source terms.

9.4 Compilation, RaNS, LES, SA, FaNS theory tensor closures

Comparison of the FaNS theory $O(\delta^2)$ resolved-unresolved scale interaction tensor and vectors with RaNS/LES PDE system Reynolds stress tensor closure models is very informative. Recalling the time-averaged and/or space-filtered Stokes isotropic strain rate tensor definition

incompressible RaNS **DP**: $\tau_{ij} \equiv -2\nu^T\bar{S}_{ij}, \quad \nu^T \equiv C_4 k^2/\varepsilon$

$$\bar{S}_{ij} \equiv \left(\partial\bar{u}_i/\partial x_j + \partial\bar{u}_j/\partial x_i\right)/2$$

incompressible FaNS **DP**: $-\nabla^2\tau_{ij} + \dfrac{4\gamma}{\delta^2}\tau_{ij} \equiv -2\dfrac{\partial\bar{u}_i}{\partial x_k}\dfrac{\partial\bar{u}_j}{\partial x_k}$

legacy LES **DP**: $\tau_{ij} \cong f\left(C_S(\mathbf{x},t), \delta^2, |\bar{S}_{ij}|\right)\bar{S}_{ij}$

compressible SA **DP**: $\tau_{ij} \equiv -2\rho(\nu+\tilde{\nu})\left[\bar{S}_{ij} - \dfrac{2}{3}\bar{S}_{kk}\delta_{ij}\right]$

$$\tilde{\nu} \equiv \nu^T/(\tilde{\nu}/\nu)^3\left((\tilde{\nu}/\nu)^3 + c_{v1}^3\right)$$

compressible RaNS **DP**: $\tau_{ij} \equiv -2\mu^T\left[\bar{S}_{ij} - \dfrac{2}{3}\left(\bar{S}_{kk} + \bar{\rho}k\right)\delta_{ij}\right]$

$$\mu^T \equiv \rho\nu^T$$

compressible FaNS **DP**: $-\nabla^2\tau_{ij} + \dfrac{4\gamma}{\delta^2}\tau_{ij} \equiv -2\dfrac{\partial\bar{u}_i}{\partial x_k}\dfrac{\partial\bar{m}_j}{\partial x_k}$

compressible FaNS DE: $-\nabla^2 v_j^T + \dfrac{4\gamma}{\delta^2}v_j^T \equiv -2\dfrac{\partial\bar{u}_j}{\partial x_k}\dfrac{\partial\bar{T}}{\partial x_k}$

compressible FaNS DY: $-\nabla^2 v_j^{Y_\alpha} + \dfrac{4\gamma}{\delta^2}v_j^{Y_\alpha} \equiv -2\dfrac{\partial\bar{u}_j}{\partial x_k}\dfrac{\partial\bar{Y}_\alpha}{\partial x_k}$

Clearly evident is that the FaNS theory $O(\delta^2)$ PDE identifications are totally unique in being ***nonlinearly sourced*** by the *tensor product* of $O(1)$ velocity vector gradients or the $O(1)$ velocity-momentum vector gradients tensor products. That the FaNS theory $O(1)$ DE and DY PDE $O(\delta^2)$ state variable closure is *nonlinearly* sourced by product of $O(1)$ velocity and scalar member gradients is totally distinct from RaNS/LES DE turbulent Prandtl number Pr^T closure models an algebraic alteration of the **DP** SGS tensor model. Furthermore, that the FaNS theory closure tensor is *anisotropic* is totally distinct from closures secured using the linear *isotropic* Stokes tensor.

9.5 Derivation, FaNS theory $O(\delta^3)$ unresolved scale tensor and vectors

Pade' second-order rational polynomial interpolation theory bounds the significance of the FaNS theory *unresolved scale* tensor $\overline{u'_j u'_i}$ and *unresolved scale* vectors $\overline{u'_j \Theta'}$, $\overline{u'_j Y \alpha'}$ by $O(\delta^3)$. Legacy CFD theorizations have historically interpreted these functionals as dissipation mechanisms operational at the unresolved scale *threshold* as the "replacement" for NS viscous dissipation. The rational LES literature documents the less than satisfactory performance of SFS tensor $O(\delta^3)$ closure models theorized on classic fluid dynamics "physics-based" hypotheses [11].

Factually, NS/RaNS CFD user base *literati* fully realize, as this monograph content has fully quantified, that *the phenomenon* required altered at the unresolved scale threshold $O(h) < O(\delta)$ is the $O(h^2, m^2, m^3)$ truncation errors induced by space-time discretization. TEA theorized nonlinear tensor product functionals operation at $O(h)$ precisely serves the SFS $h < \delta$ requirement.

Thereby spatially filtered NS **DP** *analytical* closure accrues to the TEA-theorized $O(\delta^3)$ SFS tensor

$$-\overline{u'_j u'_i} \equiv \frac{C_S h^2 Re}{12} \bar{u}_j \bar{u}_k \frac{\partial \bar{u}_i}{\partial x_k}$$

Recalling the energetic capture adequacy requirement $\delta \geq 2h$, the $O(\delta^2)$ significance intrinsic to TEA-theorized SFS tensor $\overline{u'_j u'_i}$ must then be reduced one order, hence the significance of coefficient C_S must be $O(\delta)$, i.e., $C_S \Rightarrow C_S(\delta)$

Interpretation of this TEA-theorized SFS tensor $\overline{u'_j u'_i}$ to be of eddy viscosity genre with diffusion coefficient $\left(h^2 \mathrm{Re} \bar{u}_j \bar{u}_k / 12\right)$ is totally *flawed*, as the $O(1)$ velocity tensor product $\bar{u}_j \bar{u}_k$ is *non-positive definite*. Thereby, the action of this SFS tensor $\overline{u'_j u'_i}$ can be dispersive, i.e., *anti-dissipative* (!), which intrinsically admits the space-filtered NS theory requirement of *backscatter*, the energetic cascade from smaller to larger scales [14].

By direct extension, the TEA theorized $O(\delta^3)$ SFS *vectors*

$$-\overline{u'_j \Theta'} \equiv \frac{C_S(\delta) h^2 \,\mathrm{Re}\,\mathrm{Pr}}{12} \left[\bar{u}_j \bar{u}_k \frac{\partial \bar{\Theta}}{\partial x_j}\right], \quad -\overline{u'_j Y'_\alpha} \equiv \frac{C_S(\delta) h^2 \,\mathrm{ReSc}}{12} \left[\bar{u}_j \bar{u}_k \frac{\partial \bar{Y}_\alpha}{\partial x_j}\right]$$

provide completely analytical identification of the FaNS theory $O(h^4, m^4, \Delta t^4, \delta^4)$ PDE system state variable. That $C_S(\delta)$ modification of TEA theorized SFS tensor/vectors does not compromise FaNS theory $O(1)$ PDE system $O(h^2)$ TE *annihilation* will be validated $\forall \mathrm{Re} \leq O(\mathrm{E}+04)$.

For an incompressible fluid, the FaNS theory $O(h^4,m^4,\Delta t^4,\delta^4)$ PDE system state variable

$$\text{FaNS: } \{\overline{q}(\mathbf{x},t)\} \equiv \left\{\overline{\varphi}, \overline{u}_i, \overline{\Theta}, \overline{Y_a} : \overline{P}; \tau_{ij}, v_j^{\Theta}, v_j^{Y_a}; \overline{u'_j u'_i}, \overline{u'_j \Theta'}, \overline{u'_j Y'_a}\right\}^T$$

involves the complete compliment of the $O(\delta^2, \delta^3)$ tensors and vectors. The semicolons therein denote the FaNS theory state variable $O(1;\delta^2;\delta^3)$ significance partitioning. Extending the detailed incompressible FaNS theory PDE system to a compressible fluid simply involves replacing $\overline{u_j u_i}$ with $\overline{u_j m_i}$ and recognizing that analytically space filtered $\mathrm{D\overline{M}}$ introduces no new correlations. Compressible FaNS theory $O(h^4, m^4, \Delta t^4, \delta^4)$ PDE system state-variable $O(1;\delta^2;\delta^3)$ significance semicolon partitioned is

$$\text{FaNS: } \{\overline{q}(\mathbf{x},t)\} \equiv \left\{\bar{\rho}, \overline{m}_i, \overline{E}, \overline{\rho Y_a}, \overline{p}; \tau_{ij}, v_j^{E}, \rho v_j^{Y_a}; \overline{u'_j m'_i}, \overline{u'_j T'}, \overline{u'_j \rho Y'_a}\right\}^T$$

9.6 Conclusion, analytical FaNS theory $O(h^4,m^4,\Delta t^4,\delta^4)$ PDE systems

Incompressible/compressible spatially filtered NS analytically closed FaNS theory $O(h^4,m^4,\Delta t^4,\delta^4)$ PDE systems are the *mathematically rigorous* replacement for legacy LES NS PDE systems closure reliance on heuristic "physics-based:" models solely $O(1)$ state variable-dependent. That FaNS PDE system derivation never invoked the word *turbulent* portends that FaNS theory algorithms/codes will be pertinent ∀Re fluid-thermal-multispecies applications across the complete spectrum of computational fluid dynamics.

Thereby, this fundamentally *transformative* FaNS theory advance is ***precisely*** responsive to the absolutely *key* CFD capability *unstagnation* requirement [18]

> . . .*single* most critical issue . . adequately predict viscous turbulent flows, ... boundary layer transition . . .

References

[1] Berselli, L.C., Iliescu, T., Layton, W.J. (2006). *Mathematics of Large Eddy Simulation of Turbulent Flows*. Springer, Heidelberg.

[2] Clark, R.A., Ferziger, J.H., Reynolds, W.C. (1979). "Evaluation of subgrid scale models using an accurately simulated turbulent flow," *J Fluid Mech*, 91.

[3] Galdi, G.P., Layton, W.J. (2000). "Approximation of the larger eddies in fluid motion, II. A model for space filtered flow," *J Math Modeling Mtd App Science*, 10.

[4] Garnier, E., Adams, N., Sagaut, P. (2009). *Large Eddy Simulation for Compressible Flows*

[5] Germano, M., Piomelli, U., Moin, P., Cabot, W. (1991). "A dynamics subgrid scale eddy viscosity model," *Physics of Fluids*, A, V. 3.

[6] Ghosal, S., Moin, P. (1995). "The basic equations for the large eddy simulation of turbulent flows in complex geometry," *J Comp Physics*, 118.
[7] Grinstein, F.F., Fureby, C. (2004). "From Canonical to Complex Flows: Recent Progress on Monotonically Integrated LES," *Comp Science Engineering*, 6.
[8] Grinstein, F.F., Margolin, L.G., Rider, W.F. (Eds.) (2007). Implicit Large Eddy Simulation: Computing Turbulent Fluid Dynamics, Cambridge University Press, England.
[9] Hughes, T.R.J., Feijdo, G.R., Mazzei, L., Quincy, J.-B. (1998). "The variational multi-scale method–a paradigm for computational mechanics," *Comp. Mtd. Appl. Mech. Engr*, 166.
[10] John, V. (2004). *Large Eddy Simulation of Turbulent Incompressible Flows, Analytical and Numerical Results for a Class of LES Models*. in Lecture Notes in Computational Science and Engineering, vol. 34, Springer, Berlin.
[11] John, V. (2005). "An assessment of two models for the subgrid scale tensor in the rational LES model," *J Comp Appl Math*, 163.
[12] Leonard, A. (1974). "Energy cascade in large eddy simulations of turbulent fluid flows," *Advances in Geophysics*, A, V18.
[13] Lily, D.K. (1992). "A proposed modification of the Germano subgrid scale closure method," *Physics of Fluids*, A,V.4.
[14] Pope, S.B. (2000). *Turbulent Flows*. Cambridge University Press, NY.
[15] Rasthofer, U., Gravemeier, V. (2017). Recent Developments in Variational Multi-Scale Methods for Large Eddy Simulation of Turbulent Flow. in *Archives Comp Mtd Engr*, Springer, Berlin.
[16] Sagaut, P. (2006). *Large Eddy Simulation for Incompressible Flows*, 3rd ed., Springer, Berlin.
[17] Sekachev, M.A., Baker, A.J., Wong, K.L. (2016a). "Totally Analytical Closure of Space Filtered Navier-Stokes for Arbitrary Reynolds Number, Part I. Theory, Resolutions," *J Num Heat Trans, Part B*, 70(4).
[18] Slotnick, J., Khodadoust, A., Alonso, J., Darmofal, D., Gropp, W., Lurie, E., Mavriplis, D., (2014). "CFD Vision 2030 Study: A Path to Revolutionary Computational Aerosciences," NASA Tech Report. CR-2014-218178
[19] Smagorinsky, J. (1963). "General Circulation Experiments with the Primitive Equations, I. The Basic Experiments," *Monthly Weather Revue*, 91.
[20] Van Dyke, M. (1975). *Perturbation Methods in Fluid Mechanics*. Academic, NY.
[21] Vreman, B. (1995). "Direct and large-eddy simulation of the compressible turbulent mixing layer," PhD thesis, University of Twente, Netherlands.
[22] Winckelmans, G.S., Wray, A.A., Vasilyev, G.V., Jeanmart, H. (2001). "Explicit filtering large-eddy simulations using the tensor-diffusivity model supplemented by a dynamic Smagorinsky term," *Physics Fluids*, 13.
[23] Yang, Z. (2015). "Large-Eddy Simulation: Past, Present and Future," *Chinese Journal of Aeronautics*, 28(1).
[24] Yoshizawa, A., Horiuti, K. (1985). "A Statistically-Derived Subgrid Scale Kinetic Energy Model for the Large-eddy Simulation of Turbulent flow," *J Phys Soc Japan*, 54, 2834.

Chapter 10
FaNS theory $O(1;\delta^2)$ PDE system rendered bounded domain well-posed

Libretto: For FaNS $O(1{:}\delta^2)$ EBV PDE systems to be bounded domain well-posed requires *a priori* definition of boundary conditions (BCs) on the encompassing $\partial\Omega$ of NS PDE domain Ω.

For NS spatial filtering to be rigorous requires that the filter measure δ be uniform. FaNS Gaussian filter will thus penetrate *all* NS PDE boundaries $\partial\Omega$ with the NS no-slip BC.

Rigorous resolution accrues to FaNS $O(1)$ PDE system augmentation with the *boundary commutation error* (BCE) integrals which require *convolution* of the *unfiltered* (!) NS state-variable member normal *efflux* on surfaces $\partial\Omega$ with NS no-slip BC specifications.

The definition of $O(\delta^4)$ *approximate deconvolution* (AD) enables GWS prediction of BCE identified nonhomogenous Neumann BC (NBC) $\forall O(1)$ state-variable members.

Validation of fully implicit $O(1)$:AD:BCE^h:NBC^h GWS^h algorithm quantifies that the *optimal* filter/mesh measure ratio for FaNS theory is $\delta/h \equiv 4$.

A nonhomogeneous DBC weak formulated AD:DBC^h GWS^h convolution algorithm is *a priori* hypothesized for FaNS $O(\delta^2)$ state-variable members.

Resultant FaNS $O(1)$ $O(h^4,m^4,\Delta t^3)$ PDE:AD:BCE^h:NBC^h:$O(\delta^2)$ $O(h^2,\delta^4)$ PDE:AD:DBC^h: $O(\delta^3)$SFS tensor/vector GWS^h + θTS algorithm is thus validated to be bounded domain well-posed.

Keywords: space-filtered NS, analytical closure, FaNS theory, BCE integral, approximate deconvolution (AD), $O(1)$ Neumann BCs, $O(\delta^2)$ Dirichlet BCs, optimal measure ratio, bounded domain well-posed

10.1 Incompressible FaNS theory $O(1)$ $\mathrm{D}^m\overline{P}$ completion for pressure

For an incompressible fluid, the NS DM^h differential constraint pressure kinetic impact parameter P^* (Chapter 5) is replaced with spatially filtered $\overline{P}^*$ in the kinetic flux vector for FaNS theory $O(\delta^4)$ $\mathrm{D}^m\overline{\mathbf{P}}$. Thus altered $\mathrm{D}^m\overline{\mathbf{P}}$ remains equated to $0 + O(h^4)$

https://doi.org/10.1515/9783112206430-011

$$\mathrm{D}^m\overline{\mathbf{P}}\colon \frac{\partial \overline{u}_i}{\partial t} + \frac{\partial}{\partial x_j}\left(\overline{u}_j\overline{u}_i + \tau_{ij} + \overline{P}^*|_{n+1}^p \delta_{ij}\right) + \frac{\mathrm{Gr}}{\mathrm{Re}^2}\overline{\Theta}\,\hat{g}_i$$
$$-\frac{\partial}{\partial x_j}\left[\frac{1}{\mathrm{Re}}\left(\frac{\partial \overline{u}_i}{\partial x_j} + \frac{\partial \overline{u}_j}{\partial x_i}\right) - \frac{\mathrm{C_S}(\delta)h^2\mathrm{Re}}{12}\overline{u}_j\overline{u}_k\frac{\partial \overline{u}_i}{\partial x_k}\right] = 0 + O(h^4)$$

as the $O(\delta^3)$ constraint on TEA theorized SFS tensor will be validated to *not compromise* $O(h^2)$ TEA functionality for FaNS $O(1)$ PDE $\mathrm{GWS}^h + \theta\mathrm{TS}$ algorithm predictions. The absolutely *key* observable in incompressible FaNS theory $\mathrm{D}^m\overline{\mathbf{P}}$ PDE is resolved-unresolved scale interaction tensor $\tau_{ij}(\mathbf{x},t) \equiv \overline{\overline{u}_j u'_i} + \overline{u'_j \overline{u}_i}$ disposition in the *kinetic* flux vector which renders its functionality *modulo* divergence. This is totally distinct from legacy LES D**P** algorithms wherein the $O(\delta^2)$ SGS tensor model resides in the *dissipative* flux vector.

The definition for space-filtered pressure kinetic impact variable $\overline{P}^*$ for incompressible $\mathrm{D}^m\overline{\mathbf{P}}\colon \mathrm{D}\overline{\mathrm{M}}^h$ PDE system GWS^h algorithm is essentially unaltered

$$\overline{P}^*|_{n+1}^p \equiv \overline{P^h}(x_i, t_n) + \frac{1}{\theta\Delta t}\sum_{\alpha=1}^{p}\delta\overline{\phi}^h(x_i, t_{n+1})|^\alpha$$

where $\overline{u_j^h}(x_i, t_{n+1})$ remains *any* approximation to divergence-free $\overline{u_j}(x_i, t_{n+1})$. As with NS and RaNS PDE system algorithms, the ***sole admissible*** solution to iteratively converged space-filtered $\mathrm{D}^m\overline{\mathbf{P}}\colon \mathrm{D}\overline{\mathrm{M}}^h$ PDE system with pure Neumann BC

$$\mathrm{D}\overline{\mathrm{M}}^h\colon \frac{{}^2\partial\overline{\phi}}{\partial x_k^2} + \frac{\partial \overline{u_j^h}}{\partial x_j} = 0,\ \ \overline{\mathrm{NBC}}\colon \left(\frac{\partial\overline{\phi}}{\partial x_j} - \left(\overline{u_j} - \overline{u_j^h}\right)\right)\widehat{n}_j = 0$$

must possess $O(\varepsilon)$ *significance*, which equates to the iterative convergence requirement, typically $\varepsilon \le$ E-04. Thereby the $O(h^4)$ asymptotic convergence rate accorded to NS/RaNS $O(\varepsilon)$ *divergence-free* velocity GWS^h FE $p = 1$ basis algorithms remains pertinent for any tridiagonal stencil equivalent $\mathrm{D}^m\overline{\mathbf{P}}\colon \mathrm{D}\overline{\mathrm{M}}^h$ PDE system algorithm under uniform and/or regular solution adapted nonuniform mesh refinement.

10.2 Resolution, FaNS uniform measure δ filter penetration of NS no-slip BC $\partial\Omega$

FaNS theory Gaussian filter δ measure defined uniform *guarantees* that convolution and differentiation *commute*, a *necessary* requirement for NS PDE system spatial filtering to be rigorous (Dunca et al., 2003). The consequence is that the uniform measure δ filter will penetrate *all* NS domain Ω bounding surfaces $\partial\Omega$ on which the NS velocity BC is no-slip (Figure 10.1).

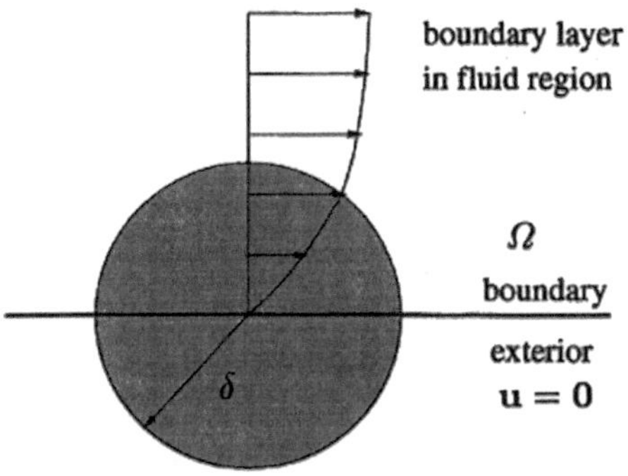

Figure 10.1: Penetration of uniform measure δ filter on a NS no-slip BC boundary ∂Ω.

Rigorous accounting of this fundamental issue is derived via mathematical properties assessments of the non-D incompressible NS **DP** in vector form:

$$\mathbf{DP}\text{:}\ \frac{\partial \mathbf{u}}{\partial t} + \nabla \cdot \mathbf{uu} + \nabla P - \frac{1}{\text{Re}} \nabla \cdot \nabla \mathbf{u} = \mathbf{f}$$

wherein *data* $\mathbf{f} \equiv -\text{Gr}\,\Theta\,\widehat{\mathbf{g}}/\text{Re}^2$ is state-variable member-dependent. The rigorous resolution of $\partial\Omega$ filter penetration requires that the NS PDE domain Ω be enlarged to $\Omega^+ \subset \text{R}^\text{n}$ such that it encompasses all Ω bounding surfaces $\partial\Omega$ upon which *a priori* defined NS Dirichlet data exist.

Recalling that the Hilbert space H^m contains *all* functions with m spatial derivatives in all tensor product combinations that are square integrable and finite on Ω, NS **DP** resident first-order weak derivatives $\partial\mathbf{u}/\partial t$, $\nabla\mathbf{u}$, $\nabla \cdot \mathbf{u}$, $\nabla \cdot \mathbf{uu}$ each possess the *regularity* required to be well defined on NS domain enlargement $\Omega^+ \subset \text{R}^\text{n}$ since

$$\mathbf{u} \in \left(H_0^1(\text{R}^\text{n})\right)^n,\ P \in L^2(\text{R}^\text{n}),\ \textit{for } t \in [0, T]$$

$$\mathbf{u} \in \left(H^1(0,\ t)\right)^n,\ \textit{for } \mathbf{x} \in \text{R}^\text{n}$$

Conversely, the third and fourth terms in NS **DP** lack the differentiability required to exist on enlarged domain $\Omega^+ \subset \text{R}^\text{n}$ as $\mathbf{u} \notin \left(H^2(\text{R}^\text{n})\right)^n$ and $P \notin H^1(\text{R}^\text{n})$.

The rigorous resolution for domain enlargement engendered differentiability requirements requires that the *total* stress tensor in NS **DP**

$$\text{D}(\mathbf{u}, \text{P}) \equiv -\nabla\text{P} + (2/\text{Re})\nabla \cdot \text{S}(\mathbf{u})$$

be defined in the sense of *distributions* [3], wherein $\text{S}(\mathbf{u})$is the dyadic form of the Stokes tensor. It then follows that NS state-variable $\{\mathbf{u}, P\}$ will fulfill NS PDE differentiability requirements on extended domain $\Omega^+ \subset \text{R}^\text{n}$ upon insertion of the *boundary commutation error* (BCE) integrals into NS **DP**

$$\text{D}^{\text{BCE}}\mathbf{P}\text{:}\ \frac{\partial \mathbf{u}}{\partial t} + \nabla \cdot \mathbf{uu} + \nabla P - \frac{1}{\text{Re}} \nabla \cdot \nabla \mathbf{u} = \mathbf{f} + \int_{\partial\Omega} \text{D}(\mathbf{u}, \text{P})(\text{s})\mathbf{n}(\text{s})\phi(\text{s})\text{ds}$$

This theorization is valid for all piecewise continuous functions $\phi(s) \in C_0^\infty(\Omega) \subset H^0$ and arbitrary *data* $\mathbf{f}$, for coordinate s spanning $\partial\Omega$ with outward pointing unit normal vector $\mathbf{n}(s)$.

The BCE integral augmented FaNS theory $O(h^4)$ $D^m\overline{\mathbf{P}}$ in tensor notation is

$$D^m\overline{\mathbf{P}}: \frac{\partial \bar{u}_i}{\partial t} + \frac{\partial}{\partial x_j}\begin{bmatrix} \overline{u}_j\overline{u}_i + \tau_{ij} & -\dfrac{1}{\mathrm{Re}}\left(\dfrac{\partial \bar{u}_i}{\partial x_j} + \dfrac{\partial \bar{u}_j}{\partial x_i}\right) \\ +\overline{P}^*\big|_{n+1}^{p}\delta_{ij} - & \dfrac{C_S(\delta)h^2\mathrm{Re}}{12}\bar{u}_j\bar{u}_k\dfrac{\partial \bar{u}_i}{\partial x_k} \end{bmatrix} + \frac{\mathrm{Gr}}{\mathrm{Re}^2}\bar{\Theta}\hat{g}_i - A_\delta\left(D_{ij}(u_i, P)\right)$$
$$= 0 + O(h^4)$$

for $D^m\overline{\mathbf{P}}$ BCE definition

$$\mathrm{BCE}(D^m\overline{\mathbf{P}}):\ A_\delta\left(D_{ij}(u_i, P^*)(x_k, t)\right) \equiv \int_{\partial\Omega} g_\delta(x_k - s_k) * \left[-P^*\delta_{ij} + \frac{1}{\mathrm{Re}}\left(\frac{\partial u_i}{\partial x_j} + \frac{\partial u_j}{\partial x_i}\right)\right]\hat{n}_j ds$$

Note specifically that BCE integral augmentation requires *convolution* (denoted *) of ***unfiltered*** (!) NS velocity and pressure efflux with filter function $g_\delta(x_k - s_k)$.

Recalling how Green-Gauss divergence theorem enforcement on the weak formulation continuous Galerkin $D^m\overline{\mathbf{P}}$ inserts nonhomogeneous Neumann boundary conditions (NBCs) into the GWS^h statement, completion of this convolution operation thereby ***inserts*** the wall-normal nonhomogeneous Neumann BC (!) for space-filtered $O(1)$ velocity into the FaNS theory-resolved scale $O(h^4)$ $D^m\overline{\mathbf{P}}$. Thus rigorously resolved is requisite *a priori* identification of encompassing BCs for the FaNS theory $O(1)$ $D^m\overline{\mathbf{P}}$ EBV PDE system to be bounded domain well-posed.

BCE integral augmentation of incompressible space-filtered $D\overline{M}^h$ is not required as NS Dirichlet (DBC) data do not exist! Following convergence of thus identified FaNS theory $O(h^4, h^2)$ $D^m\overline{\mathbf{P}}: D\overline{M}^h$ GWS^h algorithm at time t_{n+1}, and for any genuine pressure DBC data extended onto $\Omega^+ \subset R^n$, prediction of space-filtered NS genuine kinematic pressure is the post convergence operation on spatially filtered $D\overline{P}$ Poisson PDE with exact Neumann NBC

$$D\overline{P}: \frac{\partial^2\overline{P}}{\partial x_k^2} + \frac{\partial}{\partial x_j}\left[\frac{\partial\overline{u_j u_k}}{\partial x_k} + \frac{\mathrm{Gr}}{\mathrm{Re}^2}\bar{\Theta}\hat{g}_j\right] - A_\delta(\nabla_k P)\hat{n}_k = 0,\quad \mathrm{NBC}: \left(\frac{\partial \overline{P}}{\partial x_j} - \frac{1}{\mathrm{Re}}\frac{\partial^2\bar{u}_j}{\partial x_k^2}\right)\hat{n}_j = 0$$

for $D\overline{P}$ source term definition $\overline{u_j u_k} \equiv \bar{u}_j\bar{u}_k + \tau_{jk} + \overline{u'_j u'_k}$ and $D\overline{P}$ BCE integral

$$\mathrm{BCE}(D\overline{P}):\ A_\delta(\nabla P(x_k, t)) \equiv \int_{\partial\Omega} g_\delta(x_k - s_k) * \left[\frac{\partial P}{\partial x_j}\right]\hat{n}_j\, ds$$

Recall that $D\overline{P}$ PDE BCE integral evaluation is required *only* on the surface $\partial\Omega$ of NS domain Ω with inflow as $(\partial\overline{P}/\partial x_j)\,\hat{n}_j$ vanishes everywhere else on NS PDE bounding surfaces $\partial\Omega$.

FaNS theory incompressible BCE integral augmented $O(h^4,\Delta t^3)$ $\mathrm{D}^m\overline{\mathrm{E}}$ is

$$\mathrm{D}^m\overline{\mathrm{E}}:\ \frac{\partial\overline{\Theta}}{\partial t}+\frac{\partial}{\partial x_j}\left[u_j\overline{\Theta}+\overline{v_j^{\Theta}}-\frac{1}{\mathrm{RePr}}\frac{\partial\overline{\Theta}}{\partial x_j}-\frac{\mathrm{RePr}\,h^2}{12}\overline{u}_j\overline{u}_k\frac{\partial\overline{\Theta}}{\partial x_k}\right]$$

$$-\mathrm{A}_\delta(\nabla\Theta(x_k,t))=0+O(h^4)$$

with requisite nonhomogeneous NBC

$$\mathrm{BCE}(\mathrm{D}^m\overline{\mathrm{E}}):\ \mathrm{A}_\delta(\nabla\Theta(x_k,t))\equiv\int_{\partial\Omega}g_\delta(x_k-s_k)*\left[\frac{1}{\mathrm{Re\,Pr}}\frac{\partial\Theta}{\partial x_j}\right]\hat{n}_j\,ds$$

From Chapter 7, the FaNS $O(h^4,m^4,\Delta t^4)$ $\mathrm{D}^m\overline{\mathrm{Y}}$ BCE integral augmentation is

$$\mathrm{D}^m\overline{\mathrm{Y}}:\ \left(1-\frac{\gamma\Delta t^2}{6}\right)\frac{\partial}{\partial x_j}\left(\overline{u_j u_k}\frac{\partial}{\partial x_k}\right)\frac{\partial\overline{Y_a}}{\partial t}+\left(1+\frac{\mathrm{Re}h^2}{6\Delta t}\right)\frac{\partial}{\partial x_j}\left(\overline{u}_j\overline{Y}_a+\overline{v_j^{Y_a}}\right)$$

$$-\mathrm{A}_\delta(\nabla Y_a(x_k,t))-\frac{\partial}{\partial x_j}\left[\left(\frac{1}{\mathrm{ReSc}}+\frac{h^2}{6\Delta t}\right)\frac{\partial\overline{Y_a}}{\partial x_j}+\frac{C_S(\delta)h^2\mathrm{ReSc}}{12}u_j\,u_k\frac{\partial\overline{Y_a}}{\partial x_k}\right]$$

$$=0+O(h^4,m^4,\Delta t^4)$$

for *optimal* $\gamma\equiv-0.5$, $\theta\equiv0.5$, and $\mathrm{D}^m\overline{\mathrm{Y}}$ nonhomogeneous NBC

$$\mathrm{BCE}(\mathrm{D}^m\overline{\mathrm{Y}}):\ \mathrm{A}_\delta(\nabla Y_a(x_k,t))\equiv\int_{\partial\Omega}g_\delta(x_k-s_k)*\left[\frac{1}{\mathrm{ReSc}}\frac{\partial Y_a}{\partial x_j}\right]\hat{n}_j\,ds$$

For a compressible fluid, that spatially filtered $\mathrm{D}\overline{\mathrm{M}}$ is a pure initial value PDE confirms that BCE integral augmentation is not required. For NS density Dirichlet BC (DBC) data extended onto $\Omega^+\subset\mathrm{R}^n$

$$\mathrm{D}^m\overline{\mathrm{M}}:\ \frac{\partial\overline{\rho}}{\partial t}+\frac{\partial}{\partial x_j}\left[\overline{m}_j-\frac{h^2\mathrm{Re}}{12}\overline{u}_k\overline{\overline{u}}_j\frac{\partial\overline{\rho}}{\partial x_k}\right]=0+O(h^4)$$

BCE integral augmented non-D compressible $O(1)$ FaNS theory $O(h^4)$ $\mathrm{D}^m\overline{\mathbf{P}}$ is

$$\mathrm{D}^m\overline{\mathbf{P}}:\ \frac{\partial\overline{m_i}}{\partial t}$$

$$+\frac{\partial}{\partial x_j}\left[\begin{array}{l}\overline{u}_j\overline{m_i}+\tau_{ij}+\mathrm{Eu}\overline{p}\,\delta_{ij}\\[4pt] -\dfrac{\mu^*}{\mathrm{Re}}\left(\dfrac{\partial\overline{u}_i}{\partial x_j}+\dfrac{\partial\overline{u}_j}{\partial x_i}-\dfrac{\partial\overline{u}_k}{\partial x_k}\left(1-\dfrac{2}{3}\delta_{ij}\right)\right)-\dfrac{C_S(\delta)h^2\mathrm{Re}}{12\mu^*}\left(\overline{u}_j\left(\overline{u}_k-\dfrac{\partial\overline{u}_k}{\partial x_l}\underline{\delta_{kl}}\right)\dfrac{\partial\overline{m_i}}{\partial x_k}\right)\end{array}\right]$$

$$-\mathrm{A}_\delta\left(\mathrm{D}_{ij}(u_i,p)\right)=0+O(h^4)$$

$$\text{BCE}(\text{D}^m\overline{\mathbf{P}}):\ \text{A}_\delta\left(\text{D}_{ij}(u_i,p)(x_k,t)\right)$$
$$\equiv \int_{\partial\Omega} g_\delta(x_k-s_k)*\left[\begin{array}{l}-\text{Eu}\,p\delta_{ij}\\ +\dfrac{\mu^*}{\text{Re}}\left(\dfrac{\partial u_i}{\partial x_j}+\dfrac{\partial u_j}{\partial x_i}-\dfrac{\partial u_k}{\partial x_k}\left(1+\dfrac{2}{3}\delta_{ij}\right)\right)\end{array}\right]\hat{n}_j ds$$

for Eu the Euler number, μ^* the non-D temperature-dependent absolute viscosity. The underbar denoted Kronecker delta remains indicators only, no summation.

BCE integrals augmented compressible FaNS theory $O(h^4,\Delta t^3)$ $\text{D}^m\overline{\text{E}}$ and $O(h^4,m^4,\Delta t^3)$ $\text{D}^m\overline{\rho Y}$ are

$$\text{D}^m\overline{\text{E}}:\ \frac{\partial\overline{E}}{\partial t}+\frac{\partial}{\partial x_j}\left[\begin{array}{l}\overline{u}_j\left(\overline{E}+\overline{p}\right)+\overline{v_j^E}-\dfrac{\kappa^*}{\text{Re Pr}}\dfrac{\partial\overline{T}}{\partial x_j}-\dfrac{h^2\text{RePr}}{12\kappa^*}\left(\overline{u}_k\left(\overline{u}_j-\dfrac{\partial\overline{u}_l}{\partial x_l}\delta_{\underline{jl}}\right)\dfrac{\partial\overline{T}}{\partial x_k}\right)\\ -\dfrac{\mu^*\overline{u}_i}{\text{Re}}\left(\dfrac{\partial\overline{u}_i}{\partial x_j}+\dfrac{\partial\overline{u}_j}{\partial x_i}-\dfrac{\partial\overline{u}_k}{\partial x_k}\left(1+\dfrac{2}{3}\delta_{ij}\right)\right)\end{array}\right]$$
$$-\text{A}_\delta(\nabla T(x_k,t))=\ 0+O(h^4)$$

$$\text{BCE}(\text{D}^m\overline{\text{E}}):\ \text{A}_\delta(\nabla T(x_k,t))\equiv\int_{\partial\Omega} g_\delta(x_k-s_k)*\left[\frac{\kappa^*}{\text{Re Pr}}\frac{\partial T}{\partial x_j}\right]\hat{n}_j\,ds$$

$$\text{D}^m\overline{\rho\text{Y}}_\alpha:\ \left(1-\frac{\gamma\Delta t^2}{6}\right)\frac{\partial}{\partial x_j}\left(\overline{u_j u_k}\frac{\partial}{\partial x_k}\right)\frac{\partial\overline{\rho Y_\alpha}}{\partial t}+\left(1+\frac{\text{Re}h^2}{6\Delta t}\right)\frac{\partial}{\partial x_j}\left(\overline{u}_j\overline{\rho Y_\alpha}+\overline{v_j^{\rho Y_\alpha}}\right)$$
$$-\text{A}_\delta(\nabla\rho Y_\alpha(x_k,t))-\frac{\partial}{\partial x_j}\left[\left(\frac{\text{D}^*}{\text{ReSc}}+\frac{h^2}{6\Delta t}\right)\frac{\partial\overline{\rho Y_\alpha}}{\partial x_j}+\frac{\text{C}_\text{S}(\delta)h^2\text{ReSc}}{12\,\text{D}^*}\overline{u}_j\overline{u}_k\frac{\partial\overline{\rho Y_\alpha}}{\partial x_k}\right]$$
$$=\ 0+O(h^4,m^4,\Delta t^4)$$

$$\text{BCE}(\text{D}^m\overline{\rho\text{Y}}):\ \text{A}_\delta(\nabla\rho Y_\alpha(x_k,t))\equiv\int_{\partial\Omega} g_\delta(x_k-s_k)*\left[\frac{\text{D}^*}{\text{ReSc}}\frac{\partial\rho Y_\alpha}{\partial x_j}\right]\hat{n}_j\,ds$$

for D^* the non-D binary diffusion coefficient, κ^* the non-D temperature-dependent thermal diffusivity. Compressible space-filtered dynamic pressure of the form $\overline{p}=\overline{p}(\gamma,\overline{E})$ is computed algebraically or as altered for aerothermodynamics impact on a hypersonic flow field.

To conclude, incompressible/compressible fluid FaNS theory state variables

$$\{\overline{q}(\mathbf{x},t)\}=\left(\begin{array}{l}\left\{\overline{\phi},\ \overline{u}_i,\overline{\Theta},\overline{Y_\alpha},\overline{P};\ \phi,\ u_i,\Theta,\ Y_\alpha,\ P;\ \tau_{ij},v_j^\Theta,v_j^{Y_\alpha};\ \overline{u_j'u_i'},\ \overline{u_j'\Theta'},\ \overline{u_j'Y'_\alpha}\right\}^T\\ \left\{\overline{\rho},\ \overline{m}_i,\ \overline{E},\overline{\rho Y_\alpha},\ \overline{p};\ \rho,m_i,E,\ \rho Y_\alpha,\ p;\ \tau_{ij},v_j^E,v_j^{\rho Y_\alpha};\ \overline{u_j'u_i'},\ \overline{u_j'T'},\ \overline{u_j'Y'_\alpha}\right\}^T\end{array}\right)$$

involve the ***full*** complement of $O(\delta^2)$ tensor and vector quadruples ***plus*** (!) the unfiltered NS state variable, hence can number up to several dozen members.

10.3 Derivation, weak formulation BCE integral convolution GWS^h algorithm

An algorithm is required derived to execute BCE integral convolution of each unfiltered NS state-variable member efflux on NS domain Ω boundary closure segments ∂Ω. Derivation of the weak formulation GWS^h algorithm design addresses the $D^m\overline{\mathbf{P}}$ BCE integral absent the pressure term as representative of the requirement

$$\mathrm{BCE}(\mathrm{D}^m\overline{\mathbf{P}}):\ \mathrm{A}_\delta\left(\mathrm{D}_{ij}(u_i,\hat{n}_j)(x_k,t)\right) \Rightarrow \frac{1}{\mathrm{Re}}\int_{\partial\Omega} g_\delta(x_k - x_s) * \frac{\partial u_i(x_s,t)}{\partial x_j}\,\hat{n}_j ds$$

The BCs for unfiltered NS velocity $u_i(x_i, t)$ are nonhomogeneous Dirichlet on ∂Ω surfaces with inflow, homogeneous Dirichlet (no-slip) on all domain Ω solid surface segments ∂Ω, and homogeneous Neumann on domain bounding outflow surfaces. Therefore, the $\mathrm{BCE}(\mathrm{D}^m\overline{\mathbf{P}})$ integrand is required evaluated on *all* NS PDE domain Ω bounding surfaces ∂Ω excepting those at domain outflows.

The mathematical correlation between the filtered and unfiltered NS velocity to FaNS theory PDE pertinent significance $O(\delta^4)$ is expressed by the statement of *approximate deconvolution* (AD)

$$\mathrm{AD}:\ u_i(x_k,t) \ = \ \overline{u}_i(x_k,t) - \frac{\delta^2}{4\gamma}\nabla^2\overline{u}_i(x_k,t) + O(\delta^4)$$

No doubt should now exist that the *optimal* solution strategy is a weak formulation continuous Galerkin algorithm. For AD definition altered to its *differential statement* (DS),

$$\mathrm{DS}:\ \mathrm{D}(u_i(x_k,t)\,) \equiv u_i - \overline{u}_i + \frac{\delta^2}{4\gamma}\nabla^2\overline{u}_i = 0 + O(\delta^4)$$

the weak formulation continuous Galerkin GWS^h statement

$$\mathrm{GWS}^h(t):\ \int_\Omega \{N\}\mathrm{D}\left(u_i^h(x_k,t)\right)\,\mathrm{d}\tau = \int_\Omega \{N\}\left(u_i^h - \overline{u}_i^h + \frac{\delta^2}{4\gamma}\nabla^2\overline{u}_i^h\right)\mathrm{d}\tau$$

when Green-Gauss divergence theorem altered is the GWS^h FE basis $\{N\}$ element matrix statement

$$\mathrm{GWS}^h(t):\ [\mathrm{M200}]_e\{\mathrm{UI}\}_e = [\mathrm{M200}]_e\left\{\overline{\mathrm{U}I}\right\}_e + \frac{\delta^2}{4\gamma}\,[\mathrm{M2}LL]_e\left\{\overline{\mathrm{U}I}\right\}_e - \frac{\delta^2}{4\gamma}\,[\mathrm{m20}L]_e\hat{n}_l\left\{\overline{\mathrm{U}I}\right\}_e$$

Theorem-generated surface normal derivative matrix $[\mathrm{m20}L]_e$ is now deleted as NS $\mathrm{D}^m\overline{\mathbf{P}}$ homogeneous Neumann BC data transition directly to space-filtered $O(1)$ velocity. The FE p basis element interpolation (mass) matrix $[\mathrm{M200}]_e$ and the Laplacian equivalent element matrix $[\mathrm{M}2LL]_e$ for $1 \le L \le n$ are of course evaluated using the appropriate order Gauss quadrature.

As differential statement $\mathrm{D}(u_i(x_k, t))$ is *not* a PDE, there exist no *a priori* given BCs. Compensation for this absence is the requirement that the NS velocity zero *magnitude* DOF at each no-slip BC node $\mathbf{x}_b \in \partial\Omega_{e,\hat{n}_j}$ must be enforced prior to algebraic processing of the globally assembled GWS$^h(t)$ matrix statement.

The compute *efficient* alternative to the standard practice of global [M200] row/column elimination is to multiply the diagonal entry in [M200] by a large number, e.g., E+10, for each no-slip BC node DOF. This "informs" the algebraic process that these DOF are *inoperable* in unaltered matrix [M200], hence their requisite *nil alteration* is enforced to this digit significance in the linear algebra solution process. The miniscule error thus generated is then removed by simply not updating the data for these no-slip BC DOF.

This $\mathrm{D}(u_i(x_k, t))$ GWS$^h(t)$ FE $p = 1$ basis algorithm generates the *piecewise continuous* function $u_i^h(\mathbf{x}, t) \equiv \cup_e \left(u_i(\mathbf{x}, t)\right)_e$, a differentiable distribution that enables analytical prediction of $\partial u_i^h(x_s, t)/(\partial x_j)\hat{n}_j$ at each no-slip BC node $\mathbf{x}_b \in \partial\Omega_{e,\hat{n}_j}$. A second-order accurate endpoint TS is appropriate for a domain discretization with wall normal nodal disposition.

The alternative for a mesh absent such nodal strings is a GWS$^h(t)$ algorithm written on the differential statement for NS velocity efflux definition

$$\mathrm{DS:\ D}\left(\frac{\partial u_i(\mathbf{x}, t)}{\partial x_j}\hat{n}_j\right) \equiv \nabla_n u_i(t) - \frac{\partial u_i(\mathbf{x}, t)}{\partial x_j}\hat{n}_j = 0$$

which generates the element matrix statement

$$\mathrm{GWS}_e^h(t)\text{: } [\mathrm{M}200]_e\{\nabla_n \mathrm{U}I(t)\}_e = [\mathrm{M}20J]_e\{\mathrm{U}I\}_e\,\hat{n}_J,\ for\ 1 \le (I,J) \le n$$

As before, the *nil* magnitude of each NS no-slip BC DOF must be enforced prior to algebraic processing of the globally assembled [M200] matrix statement.

Via either procedure the GWS$^h(t)$:AD:BCE^h algorithm-generated *piecewise continuous* approximation to $\mathrm{BCE}(\mathrm{D}^m\overline{\mathbf{P}})$ integrand NS velocity efflux DOF $\nabla_n UI$ is

$$\mathrm{BCE}(\mathrm{D}^m\overline{\mathbf{P}})\text{: } \mathrm{A}_\delta\left(\mathrm{D}_{ij}(u_i, \hat{n}_j)\right) = \frac{1}{\mathrm{Re}}\int_{\Omega^+} g_\delta(x_k - x_s) * \left(\{N(x_s)\}^T\{\nabla_n \mathrm{U}I(t)\}_e\right)\mathrm{ds},\ \mathbf{x}_b \in \partial\Omega_{\hat{n}_j}$$

Therein, the FE basis argument x_s spans the NS domain $\partial\Omega$ surface which lies *interior* (!) to domain extension $\Omega^+ \subset \mathrm{R}^n$. The shaded region in Figure 10.2 illustrates the planar quadrature rule footprint for uniform filter/mesh measure ratio $\delta/h = 2$.

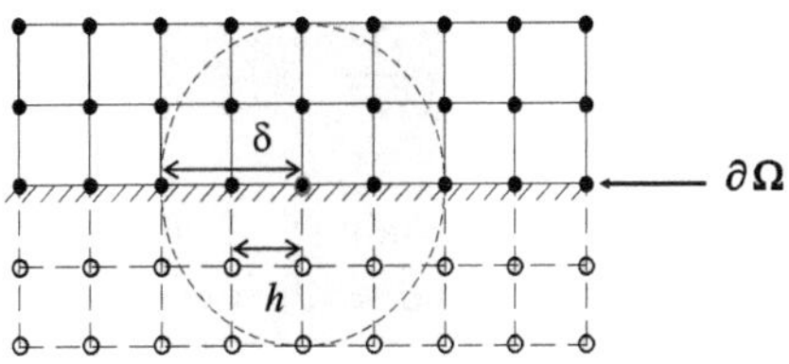

Figure 10.2: Planar illustration of GWS^h(*t*) AD-BCE^h algorithm convolution numerical quadrature footprint for $\delta/h = 2$ on $\Omega^+ \subset R^3$ (adapted from Sekachev [4]).

Convolution is executed via numerical quadrature

$$\nabla_n \overline{\mathrm{U}}I(t)|_{\mathbf{x}_b \in \partial\Omega_{e,\hat{n}_j}} \cong \sum_{l=1}^{L_X}\sum_{m=1}^{M_Y}\sum_{n=1}^{N_Z}\begin{pmatrix} w_l w_m w_n g_\delta(x_b - x_l, y_b - y_m, z_b - z_n) \\ \times \left(\{N(x_l, y_m, z_n)\}^T\{\nabla_n \mathrm{U}I(t)\}_e\right)\end{pmatrix}_{\forall \mathbf{x}_b \in \partial\Omega_{e,\hat{n}_j}}$$

for n_x, n_y, n_z quadrature rule with corresponding weights w_l, w_m, w_n appropriate for the completeness degree p of the selected FE basis.

FaNS theory $O(1)$ velocity efflux nonhomogeneous Neumann BC DOF identification at *each* NS velocity no-slip BC node $\mathbf{x}_b \in \partial\Omega_{e,\hat{n}_j}$ is complete. Minor notation alterations in this AD-BCE($\mathrm{D}^m\overline{\mathbf{P}}$) algorithm statement enable prediction of the nonhomogeneous Neumann BC DOF at no-slip nodes $\mathbf{x}_b \in \partial\Omega_{e,\hat{n}_j}$ for each FaNS theory $O(1)$ state-variable scalar member.

Alteration of the GWS^h(*t*) AD:BCE^h algorithm for the kinematic pressure P DOF in BCE($\mathrm{D}^m\overline{\mathbf{P}}$) NS simply omits TS prediction of the wall normal gradient prior to convolution. Alteration of this detailed incompressible NS GWS^h(*t*) BCE($\mathrm{D}^m\overline{\mathbf{P}}$) algorithm for a compressible fluid amounts essentially to replacement of 1/Re with μ^*/Re and P with the product Eu p.

10.4 Characterization, incompressible FaNS theory $\mathrm{D}^m\overline{\mathrm{E}}$:AD-BCE^h GWS^h(*t*) algorithm

Validation of the GWS^h(*t*) algorithm for AD-BCE^h predictions addresses FaNS BCE integral-augmented $\mathrm{D}^m\overline{\mathrm{E}}$ PDE reduced to unsteady conduction

$$\mathrm{D}^m\overline{\mathrm{E}}: \frac{\partial\overline{\Theta}}{\partial t} - \frac{1}{\mathrm{Re\,Pr}}\frac{\partial^2\overline{\Theta}}{\partial x_j^2} - \overline{s}_\Theta - \int_{\partial\Omega} g_\delta(x_k - s_k) * \left[\frac{1}{\mathrm{Re\,Pr}}\frac{\partial\Theta}{\partial x_j}\right]\hat{n}_j\, ds = 0$$

The nonhomogeneous data *distributions* defined for this manufactured statement are DBC $\Theta_b(x, y=0) = \sin(\pi x)\,/2\pi$ on one segment of domain closure $\partial\Omega$ and the domain source term $s(x,y) = 2\pi^2\sin(\pi x)\sin(\pi y)$.

For these definitions, classic separation-of-variables complimentary-particular EBV PDE practice enables analytical derivation of the NS unsteady non-D temperature distribution on the $n = 2$ unit square [4]

$$\text{DE: } \Theta(\mathbf{x},\ t) = \frac{\sin(\pi x)\ \sinh(\pi - \pi y)}{2\pi\ \sinh(\pi)} + \sin(\pi x)\ \sin(\pi y)\left(1 + \left(\frac{1}{2\pi^2} - 1\right) e^{-2\pi^2 t}\right)$$

Thereby, for *any* discretization of Ω projected onto enlarged domain $\Omega^+ \cup \partial\Omega \Leftrightarrow \Omega^{+h} \cup \partial\Omega^h$, and for *any* Gaussian uniform measure filter/mesh ratio $\delta \equiv nh$, n an integer, analytical convolution of this NS DE exact solution $\Theta(\mathbf{x}, t)$ generates the ***exact*** space-filtered temperature solution $\overline{\Theta}_{exact}(\mathbf{x}, t, \delta)$ for *exact* space-filtered $\overline{s}(\mathbf{x}, \delta)$ and $\overline{\Theta}_b(x, y = 0, \delta)$ data distributions. Thus enabled is *independent* quantification of the *distribution* of error

$$\overline{e}^h(\mathbf{x}, \delta, t) \equiv \overline{\Theta}_{exact}(\mathbf{x}, \delta, t) - \bar{\Theta}^h(\mathbf{x}, \delta, t)$$

resulting from operational decisions made on the FaNS $O(h^4)$ $D^m\overline{E}$:AD-BCE^h GWS^h + θTS FE $p = 1$ basis algorithm.

The base data specifications are an M = 20^2 uniform FE $p = 1$ basis discretization of the unit square and *minimum* energetic capture filter/mesh ratio $\delta \equiv 2\ h$ with IC the NS exact solution $\Theta(\mathbf{x}, t = 0)$ projected onto the nodalization of mesh M. Accuracy assessment of FaNS theory $O(h^4, \delta^4, \Delta t^3) D^m\overline{E}$:AD-BCE^h GWS^h + θTS unsteady algorithm predictions for uniform Δt is executed for the solution process stopped at non-D time $n\Delta t = 0.5 \equiv T$.

Algorithm initial evaluations utilize analytical convolution, the exact filtered data, and the *exact* NS solution efflux $\nabla\Theta(\mathbf{x}_b, t) \bullet \hat{n}|_{\partial\Omega_e}$. The error resultant from total omission of the AD-BCE^h algorithm in the FaNS $O(h^4, \Delta t^3)$ $D^m\overline{E}$ GWS^h + θTS prediction

$$\overline{e}^h(\mathbf{x}, \delta, t = \mathrm{T}) \equiv \overline{\Theta}_{exact}(\mathbf{x}, \delta, t = \mathrm{T}) - \overline{\Theta}^h_{\Omega^+}(\mathbf{x}, \delta, t = \mathrm{T})$$

is equidistributed over *all* surfaces $\partial\Omega$ bounding Ω, perspective projection (Figure 10.3a), with significance of the negative *extremum* equal to O(E-02).

Now including the AD-BCE^h algorithm but omitting the projection of the DBC data onto domain extension $\Omega^+ \subset \mathrm{R}^n$, the error $\overline{e}^h(\mathbf{x}, \delta, t = \mathrm{T})$ generated by the FaNS $O(h^4, \delta^4, \Delta t^3) D^m\overline{E}$:AD-BCE^h GWS^h + θTS FE $p = 1$ basis algorithm is restricted to the $\partial\Omega$ surface segment with the DBC data specification (Figure 10.3b), with *extremum* remaining of O(E-02).

Finally, extending the *exact* filtered DBC data onto domain extension $\Omega^+ \subset \mathrm{R}^n$, the distribution of error $\overline{e}^h(\mathbf{x}, \delta, t = \mathrm{T})$ for the *as-designed* FaNS $O(h^4, \delta^4, \Delta t^3) D^m\overline{E}$:AD-BCE^h GWS^h + θTS algorithm transitions to *nil magnitude* everywhere on $\partial\Omega$ (Figure 10.3c). The domain Ω centered symmetric distribution of error $\overline{e}^h(\mathbf{x}, \delta, t = \mathrm{T})$, at two orders smaller extremum O(E-04), is that inherent to FaNS theory $O(h^4, \delta^4, \Delta t^3)$ $D^m\overline{E}$:AD-BCE^h GWS^h + θTS FE $p = 1$ basis solution $\overline{\Theta}^h_{\Omega^+}(\mathbf{x}, \delta(h), t = \mathrm{T})$ on the uniform M = 20^2 mesh.

With FaNS theory $O(h^4, \delta^4, \Delta t^3)$ $D^m\overline{E}$:AD-BCE^h GWS^h + θTS algorithm design thus validated, independent assessment of errors induced by numerical quadrature and

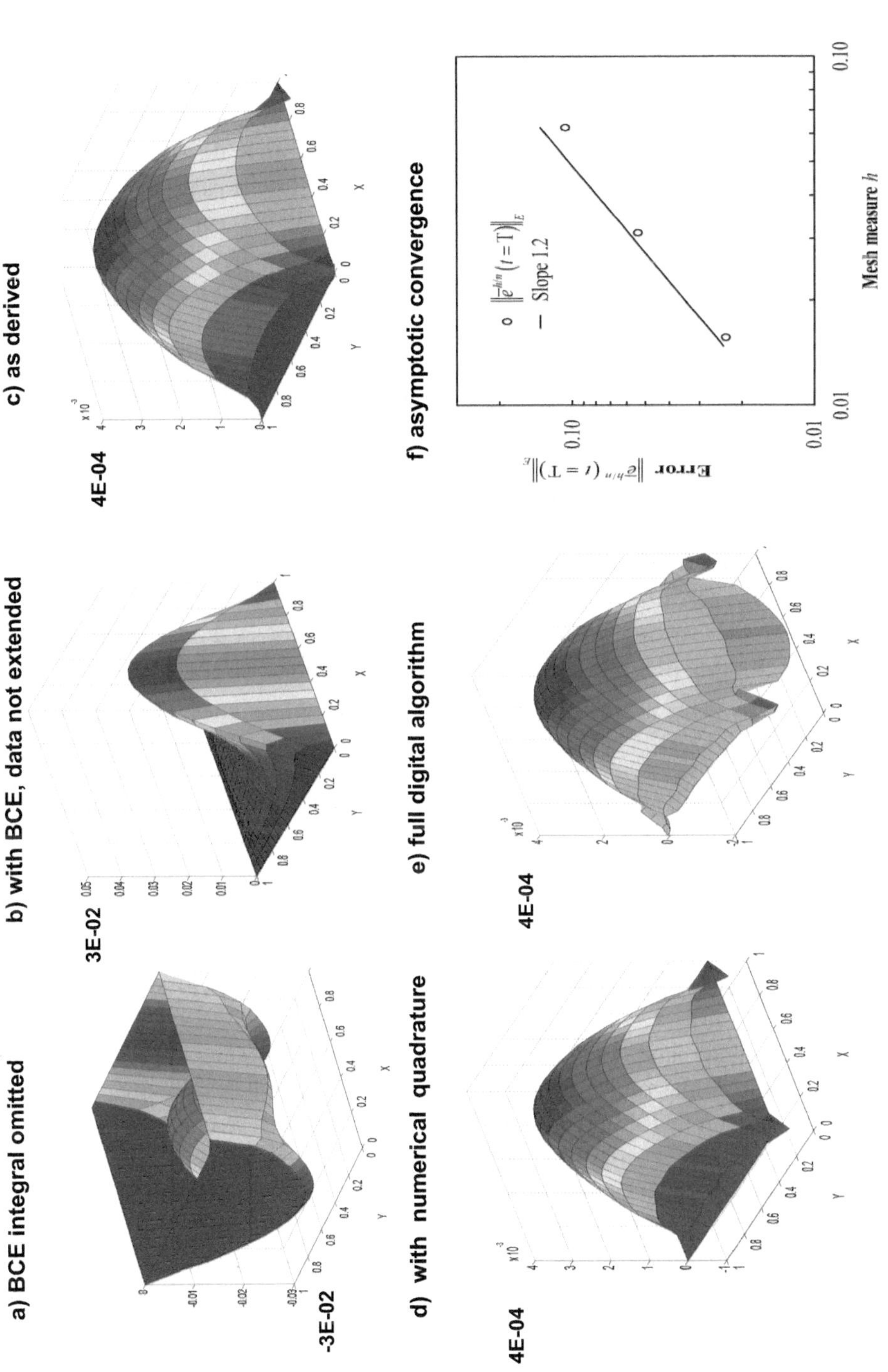

Figure 10.3: Synopsis, error $\bar{e}^h(\mathbf{x}, \delta, t = \mathrm{T})$ distributions, FaNS $O(h^4,\delta^4,\Delta t^3)$ $\mathrm{D}^m\bar{E}$:AD-BCE^h GWS^h + θTS FE $p = 1$ basis predicted $\overline{\Theta}^{h/n}_{\Omega^+}(\mathbf{x}, \delta(h), t = \mathrm{T})$, M = 20^2 uniform mesh, $n\Delta t \equiv \mathrm{T} = 0.05$ non-D, convergence validation.

replacement of exact NS efflux $\nabla\Theta(\mathbf{x}_b,t)\bullet\hat{n}|_{\partial\Omega_e}$ with AD-BCE^h generated NS efflux $\nabla\Theta^h(\mathbf{x}_b,t)\bullet\hat{n}|_{\partial\Omega_e}$ is enabled. Quadrature selection is the Newton-Cotes composite $O(h^6)$ 5-point closed "Boole's rule" [1], on sequential five node subsets centered on each $\mathbf{x}_b\in\partial\Omega_e$ (Figure 10.2). For DOF located nearer than $4h$ to the corners of domain Ω, Boole's rule is replaced with Simpson's $O(h^4)$ 4- and 3-node rules, and the $O(h^2)$ 2-node end point trapezoidal rule for each DOF at the domain Ω corner nodes.

Retaining exact NS $\nabla\Theta(\mathbf{x}_b,t)\bullet\hat{n}|_{\partial\Omega_e}$ the error $\bar{e}^h(\mathbf{x},\delta,t=\mathrm{T})$ induced by numerical quadrature in the $O(h^4,\delta^4,\Delta t^3)\mathrm{D}^m\bar{\mathrm{E}}$:AD-BCE^h GWS^h + θTS solution possesses *nil* significance everywhere except at each end point $O(h^2)$ trapezoidal rule influenced Ω corner node DOF (Figure 10.3d). Finally, the error $\bar{e}^h(\mathbf{x},\delta,t=\mathrm{T})$ resultant from replacing the exact NS efflux $\nabla\Theta(\mathbf{x}_b,t)\bullet\hat{n}|_{\partial\Omega_e}$ with AD-BCE^h algorithm-generated NS efflux $\nabla\Theta^h(\mathbf{x}_b,t)\bullet\hat{n}|_{\partial\Omega_e}$ possesses comparatively nil significance on all non-DBC boundaries $\partial\Omega$ (Figure 10.3e). The *extremum,* which is located on the DBC boundary $\partial\Omega$, is half that intrinsic to the FE $p=1$ basis uniform M = 20^2 mesh solution $\bar{\Theta}^h_{\Omega^+}(\mathbf{x},\delta(h),t=\mathrm{T})$ at the center of domain Ω (Figure 10.3c).

The requisite terminal assessment is to *validate* that asymptotic convergence is *monotone* for FaNS theory $O(h^4,\delta^4,\Delta t^3)\mathrm{D}^m\bar{\mathrm{E}}$:AD-BCE^h GWS^h + θTS algorithm. For the factor of two uniform mesh refinement sequence $\Omega^{h/\mathrm{n}}$, $8\le\mathrm{n}\le 64$, the asymptotic convergence of FaNS $O(h^4,\delta^4,\Delta t^3)\mathrm{D}^m\bar{\mathrm{E}}$:AD-BCE^h algorithm solutions $\bar{\Theta}^{h/\mathrm{n}}(\mathbf{x},\delta(h),t=\mathrm{T})$ in the DE PDE energy error norm $\|\bar{e}^{h/n}(\delta(h),\mathrm{T})\|_E$ is validated *monotone* (Figure 10.3f). The apparent slope ~1.2 is again erroneous as error bounding norms $\|\bar{s}(\mathbf{x},\delta)\|_{L2,\Omega}$, $\|\bar{\Theta}_b(x,y=0,\delta)\|_{L2,\partial\Omega}$ and $\|\bar{\Theta}^h(\delta,t_0)\|^2_{H^1(\Omega)}$ are each *explicitly* mesh $\Omega^{h/\mathrm{n}}$ dependent, cf., Appendix B.

10.5 Characterization, FaNS theory mesh resolution requirement, turbulent/laminar BL profile *O*(1) velocity efflux NBC

FaNS theorization required no assumption on the near wall profiles of $O(1)$ $\bar{u}_i(x_k,t)$ velocities. $\mathrm{D}^m\bar{\mathbf{P}}$ BCE^h integrand convolution of unfiltered NS velocity wall normal nonhomogeneous Neumann BC (NBC) requires AD:BCE^h algorithm mesh resolution assessment of *near wall profile resolution* AD:BCE^h:NBC^h algorithm requirements.

Reynolds ordering of an incompressible NS PDE non-D system quantifies the comparative significance of the wall normal resolution of the boundary layer (BL) velocity is $O(\mathrm{Re}^{-1/2})$. Thereby deemed appropriate for AD:BCE^h:NBC^h algorithm design validation is the $O(\delta^4)$ AD differential statement altered to the NS $O(1)$ BL velocity wall parallel resolution $u_{\mathrm{AD}}(x_k,t)$

$$\mathrm{DS:}\ \mathrm{D}(u_{\mathrm{AD}}(x_k,t))\equiv u_{\mathrm{AD}}(x_k,t)-\bar{u}(x_k,t)+\frac{\delta^2}{4\gamma}\nabla^2\bar{u}(x_k,t)=0+O(\delta^4)$$

Green-Gauss divergence theorem enforcement generates the $\mathrm{GWS}_e^h(t)$ matrix statement,

$$\mathrm{GWS}_e^h(t):\ [\mathrm{M200}]_e\{UAD\}_e = [\mathrm{M200}]_e\{\bar{\mathrm{U}}\}_e - \frac{\delta^2}{4\gamma}[\mathrm{M2}LL]_e\{\bar{\mathrm{U}}\}_e + \frac{\delta^2}{4\gamma}[\mathrm{m20}L]_e\{\bar{\mathrm{U}}\}_e\hat{n}_L$$

Note that the DOF in $O(1)$ velocity efflux matrix $[\mathrm{m20}L]_e\{\bar{\mathrm{U}}\}_e\hat{n}_L$ are not predictable by solution of globally assembled AD:BCE^h:NBC^h GWS$^h(t)$ algorithm as these DOF are the *sought output* (!) at NS no-slip BC nodes $\mathbf{x}_b \in \partial\Omega_{e,\hat{n}_j}$. Resolution of this AD:BCE^h:NBC^h algorithm impasse requires replacement of the element efflux matrix $[\mathrm{m20}L]_e\{\bar{\mathrm{U}}\}_e\hat{n}_L$ with the element matrix $[\mathrm{M200}]_e\{\mathrm{d}\bar{\mathrm{U}}/\mathrm{d}n\}_e$, wherein the DOF in $\{\mathrm{d}\bar{\mathrm{U}}/\mathrm{d}n\}_e$ are the wall normal derivative of the wall parallel $O(1)$ velocity distribution at no-slip BC nodes $\mathbf{x}_b \in \partial\Omega_{e,\hat{n}_j}$.

The obvious AD:BCE^h:NBC^h $\mathrm{GWS}_e^h(t)$ algorithm requirement is to predict the DOF in $\{\mathrm{d}\bar{\mathrm{U}}/\mathrm{d}n\}_e$ with accuracy sufficient to *extremize* the difference between algorithm-generated DOF in $\{UAD\}$ and the NS no-slip BC $u_{\mathrm{AD}}(\mathbf{x}_b) \equiv 0$ at $\mathbf{x}_b \in \partial\Omega_{\hat{n}_i}$. This requires quantification of the *distribution of* error generated by uniform measure δ Gaussian filtering of a turbulent BL velocity profile, which for reference is graphed in BL similarity coordinates in Figure 10.4a. AD:BCE^h:NBC^h $\mathrm{GWS}_e^h(t)$ algorithm assessment must then be extended to the range of Re-dependent laminar BL velocity profiles that remain similarity coordinates correlated $u+ = y+$ to well past laminar profile BL velocity transition to fully turbulent at $y+ \approx 30$ for sufficiently large Re.

Selected turbulent BL velocity *validation* profile is the "hard" ($\mathrm{Re}_\tau = 180$) turbulent BL distribution that was independently validated via direct numerical simulation (DNS), cf., [6]. Fortunately, this profile is CFD predictable using the RaNS:TKE EDDYBL FD algorithm/code detailed in [6] (appendix D). This code provides the *critical* support for this validation objective as its execution generates detailed resolution of the $u_{\mathrm{W}}^+(y^+)$ DOF distribution in the BL sublayer reach $6 < y^+ < 30$ as well as in the viscosity-dominated wall adjacent sublayer reach $0.1 \le y^+ \le 6$. With reference to this requisite sublayer span, the $y^+ = 1$ origin of the reference BL profile in Figure 10.4a is one order larger (!) than validation requisite $y^+ = 0.1$.

As laminar and turbulent BL velocity profile distributions are correlated in the BL similarity coordinate system (u_{W}^+, y^+), the *similarity variable* definition for the Gaussian uniform measure δ filter is $\delta^+ \equiv u_\tau\delta/v$ [4]. For a BL wall-normal solution adapted 501 node mesh M with $y^+ = 0.1$ the similarity coordinate of the first off wall node, the EDDYBL code-generated *validation* turbulent BL $u_{\mathrm{W}}^+(y^+)$ distribution is similarity coordinates compared with its *analytically* convolved distribution, for filter definitions $36 \ge \delta^+ \ge 4.5$, is labeled UW+ in Figure 10.4b.

The distribution of convolution injected *error* into the space-filtered turbulent BL velocity validation profile is therein quantified to possess *nil* significance $\forall y^+ > 30$, the sublayer coordinate at which a laminar profile velocity transition to turbulent is completed. Within the BL sublayer profile transition reach $6 < y^+ \le 30$, the distribution of profile *error* induced by space filtering is progressively moderated by reduction of simi-

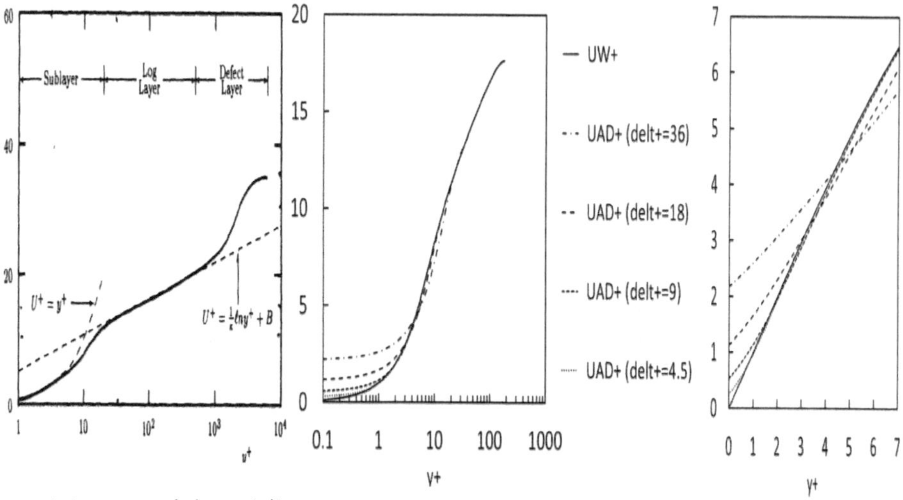

a) turbulent BL correlation $u^+_w(y^+)$ b) log $u^+_w(y^+, \delta^+)$, $36 \geq \delta^+ \geq 4.5$ c) linear $u^+_w(y^+ < 7, \delta^+)$

Figure 10.4: Synopsis, uniform measure δ Gaussian space-filtered $O(1)$ BL velocity similarity variable $\delta^+ \equiv u_\tau \delta / \nu$ parameterization of *error* distribution in $O(1)$ velocity $\overline{u^+_W}(y^+, \delta^+)$, analytical convolution of DNS validated $Re_\tau = 180$ turbulent velocity $u^+_W(y^+)$ BL distribution.
(a) NS turbulent BL velocity $u^+_W(y^+)$ similarity variables correlation on $1 \leq y^+ \leq 10^3$
(b) NS turbulent $u^+_W(y^+)$validation profile comparison with analytically convolved $\overline{u^+_W}(y^+, \delta^+)$profile, $36 \geq \delta^+ \geq 4.5$
(c) viscous sublayer $\overline{u^+_W}(y^+, \delta^+)$ distributions linear in y^+ comparison with $u^+_W(y^+)$ validation profile, $36 \geq \delta^+ \geq 4.5$

larity variable δ^+. The mechanism clearly visible in the scaled linear in y^+ graphic Figure 10.4c, is "rotation" of the $\overline{u^+_W}(y^+, \delta^+)$ distribution about the coordinate $y^+ = 4\pm$ in the viscosity-dominated sublayer.

These data quantify that BL profile distinctions between the $Re_\tau = 180$ turbulent validation BL $u^+_W(y^+)$ profile and its *analytically* convolved spatially filtered $\overline{u^+_W}(y^+, \delta^+)$ approximation persist ***solely*** within the near wall viscosity-dominated BL sublayer $y^+ \leq 6$ for *any* $\delta^+ \leq 9$. Improvement in convolution fidelity correlates therein with reduction of δ^+ terminating at $y^+ \approx 0.25\delta^+$, which corresponds with wall normal non-D coordinate $y/\delta \leq 4$. Thus quantified is that the AD:BCE^h:NBC^h GWS^h algorithm *must* function within the viscosity-dominated BL sublayer reach $y^+ \leq 6$, which of course requires *significant* (!) wall normal mesh resolution to support sufficiently accurate prediction of the wall normal distribution of the DOF in $\{\overline{U}\}_e$.

Resolution of this wall normal mesh requirement accrues to the embedding of four sub-elements, each of wall normal span $h \Rightarrow \Delta y = \delta/16$, in the reach $0 \leq y/\delta \leq 4$ of each wall adjacent FE $p = 1$ basis spanned element Ω_e (Figure 10.5). The GWS^h AD:BCE^h:NBC^h

algorithm will then generate $\{\bar{U}\}_e$ DOF at the wall normal nodal string 2 ↔ 5, the latter of which coincides with energetic capture minimum $h = \delta/2$. Creation of this $\{\bar{U}\}_e$ DOF distribution will then enable prediction of the DOF in wall normal efflux $\{d\bar{U}/dn\}$ via the $O(h^3)$ accurate end point TS at each NS no-slip BC node $\mathbf{x}_b \in \partial\Omega_{\hat{n}_i}$.

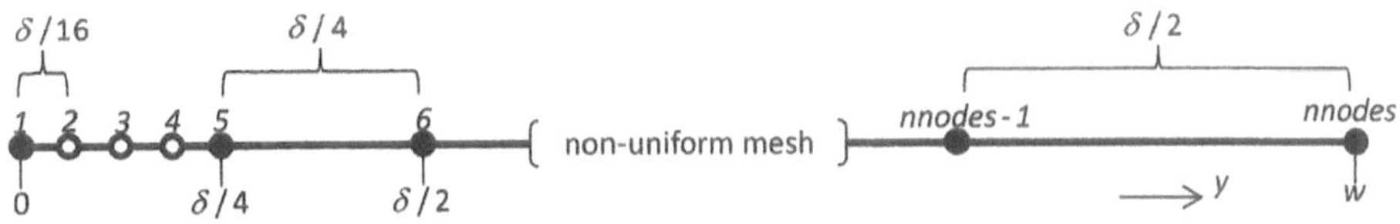

Figure 10.5: AD:BCE^h:NBC^h GWS^h algorithm wall normal mesh node embedding strategy to enable end point $O(h^3)$ TS prediction of DOF $\{d\bar{U}/dn\}$ at NS no-slip BC nodes $\mathbf{x}_b \in \partial\Omega_{\hat{n}_i}$.

10.6 Validation, FaNS theory $O(h^4)$ D^mP:AD-BCE^h-NBC^h GWS^h algorithm

For a mirror symmetric turbulent BL velocity pair in a channel of transverse half-span L = E-02 m, a uniform mesh refinement sequence enabled validation of FaNS theory $O(h^4)$ $D^m\mathbf{P}$:AD-BCE^h-NBC^h GWS^h + θTS algorithm prediction of $O(1)$ velocity wall efflux at NS no-slip nodes $\mathbf{x}_b \in \partial\Omega_{\hat{n}_i}$. The channel wall normal (subscript n) uniform mesh refinement sequence $20 \leq M_n \leq 320$ generates wall normal non-D mesh measures ranging $0.1 \geq h/L \geq 0.003125$, hence half-channel $\{\bar{U}\}$ DOF nodal resolutions ranging 11 ⇔ 161.

The design optimization parameter for FaNS theory $O(h^4)$ $D^m\mathbf{P}$:AD-BCE^h-NBC^h GWS^h algorithm is filter/mesh measure ratio $\delta/h \equiv$ n, for integers $n \geq 2$. The normalized *error* $\left|(d\bar{U}/dn)/(du^+_W/dn) - 1\right|$ in the algorithm predicted distributions of wall normal efflux DOF $\{d\bar{U}/dn\}$ at nodes $\mathbf{x}_b \in \partial\Omega_{\hat{n}_i}$ for n = 2, 4, 8, and 16 are logarithm scaled in percent (Figure 10.6).

The energetic capture minimum definition n = 2 is unusable as the error ranging 20–40% is totally nonresponsive to mesh refinement. Definition n = 8 reduces the error extremum to ~1% but these data do not exhibit monotone asymptotic convergence. Essentially nil error with extremum ~0.2% results for definition $\delta/h \equiv 16$, which is an impractically large filter/mesh measure ratio.

Finally, for n = 4 the $O(h^4)$ $D^m\mathbf{P}$:AD-BCE^h-NBC^h GWS^h algorithm-generated error in DOF $\{d\bar{U}/dn\}$ is ~10% on the extremely coarse half-channel $M_n = 10$ mesh ($h/L = 0.1$). Algorithm error then decreases to ~1% for the $M_n = 80$, $h/L = 0.0125$ mesh and then to ~0.2% for the $M_n = 160$, $h/L = 0.00625$ mesh. Of theory requisite rigor, the *asymptotic* convergence for FaNS theory $O(h^4)$ $D^m\mathbf{P}$:AD-BCE^h- NBC^h GWS^h generated $\{d\bar{U}/dn\}$ DOF at $\mathbf{x}_b \in \partial\Omega_{\hat{n}_i}$ for filter/mesh measure ratio $\delta/h = 4$ is indeed *monotone* (Figure 10.6). The apparent rate of convergence is again *erroneous* as quantification is possible *iff* theory error bounding norms are independent of the sequence M_n (Appendix B). Of *critical*

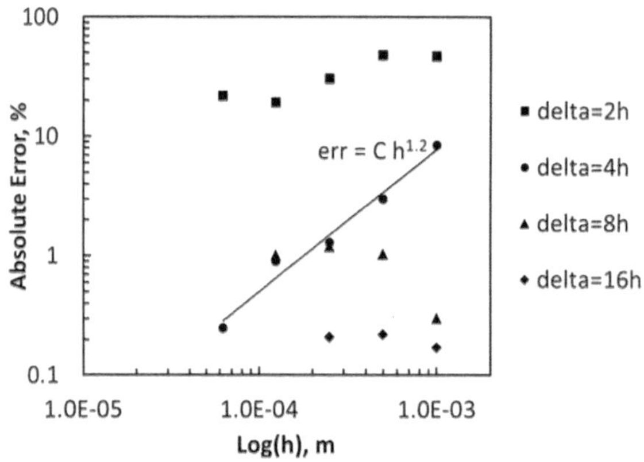

Figure 10.6: Dependence of FaNS $O(h^4)$ D^m**P**:AD-BCE^h-NBC^h GWS^h algorithm normalized error $\left|(d\overline{U}/dn)/\left(d\overline{u_W^+}/dn\right)-1\right|$ in $\{d\overline{U}/dn\}$ DOF at nodes $\mathbf{x}_b \in \partial\Omega_{\hat{n}_i}$, filter/mesh measure ratios $\delta/h \equiv n$ for $2 \le n \le 16$, wall normal uniform mesh refinement sequence $20 \le M_n \le 320$, mirror symmetric fully turbulent BL pair in a channel of half-span L = E-02m.

pertinence to physics of fluid prediction *utility*, thus validated FaNS theory *optimal* filter/mesh ratio $\delta/h \equiv 4$ is *functionally practical.*

Continuing, turbulent BL validation of the $O(h^4)$ D^m**P**:AD-BCE^h-NBC^h GWS^h optimal $\delta/h \equiv 4$ algorithm must be extended to Re-dependent *laminar* BL velocity profile $\{d\overline{U}/dn\}$ DOF predictions at nodes $\mathbf{x}_b \in \partial\Omega_{\hat{n}_i}$. The Blasius *analytical* flat plate laminar BL velocity solution, [2] (Ch. VIII), is the validation selection. These BL velocity distributions for E + 04 ≤ Re ≤ E + 06, similarity coordinates graphed in Figure 10.7a, quantify profile adherence to universal correlation $u+ = y+$ to well past laminar BL profile transition to turbulent for sufficiently large Re.

FaNS $O(h^4)$ D^m**P**:AD-BCE^h-NBC^h GWS^h + θTS algorithm/code alteration to NS simply involves deletion of the $O(\delta^2)$ state variable and NBC alteration to the NS no-slip BC. Validation statement is the velocity distribution at the trailing edge of an infinite lateral span, finite axial span flat plate for wall normal geometric progression ratio (pr) solution adapted $M_n = 68$ nonuniform mesh (Sekachev et al., 2014).

At Re = E+06, omission of the AD-BCE^h-NBC^h algorithm leads to $O(1)$ velocity trailing edge BL $\overline{u^+}^h(x_{TE}, y^+, z_{SYM}, T)$ distribution *disagreement* with Blasius $\forall y^+ \le 15$, data labeled U+(AD BC) in Figure 10.7b. Thereafter, for $\delta/h \equiv 4$ the $O(h^4)$ D^m**P**:AD-BCE^h-NBC^h algorithm departure of the $\overline{u^+}^h(x_{TE}, y^+, z_{SYM}, T)$ profile from Blasius is restricted to $y+ \le 1\pm$, data labeled U+(AD BC;BCE) in Figure 10.7b.

The repeating of this exercise with inclusion of the FaNS theory $O(\delta^2)$ state variable generates the NS laminar BL velocity $\overline{u^+}^h(x_{TE}, y^+, z_{SYM}, T)$ profile labeled U+(AD BC;BCE;RLES) in Figure 10.7b. That these data exactly overly that generated for omission of the $O(\delta^2)$ state variable data further *quantifies* that the resolved-unresolved scale interaction state variable exerts nil impact on laminar BL velocity prediction at large Re.

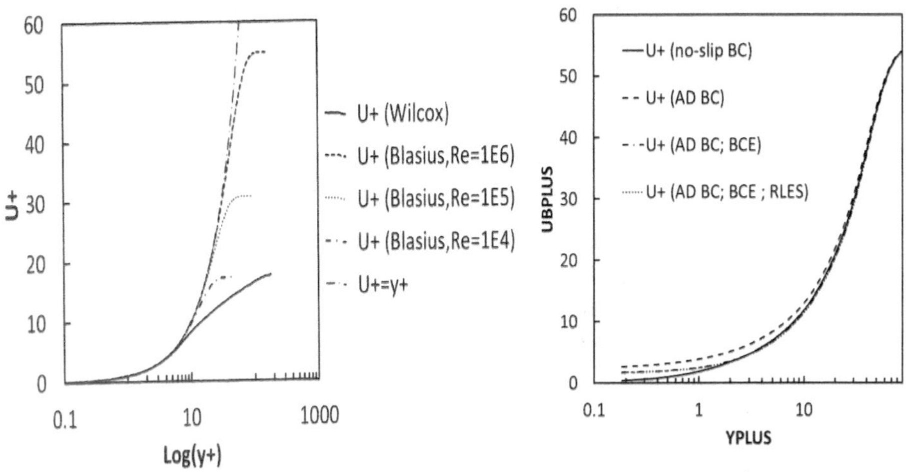

a) laminar BL $u_w^+(y^+)$ as $f(\text{Re})$

b) trailing edge $u^{+\,h}(x_{\text{TE}}, y^+, z_{\text{SYM}}, \text{T})$

Figure 10.7: Validation, laminar BL profile prediction via FaNS theory $O(h^4)$ $\text{D}^m\mathbf{P}$:AD-BCE^h-NBC^h GWS^h + θTS algorithm: (a) Blasius analytical laminar BL similarity coordinates distributions for E + 04 ≤ Re ≤ E + 06, (b) $\overline{u^+}^{\,h}(x_{\text{TE}}, y^+, z_{\text{SYM}}, \text{T})$ distributions compared with Blasius, trailing edge symmetry plane of semi-infinite span $n = 3$ finite axial span flat plate, Re = E+06 channel, $\delta/h \equiv 4$.

To conclude, thus detailed laminar/turbulent BL *validation* of the FaNS theory $O(h^4)$ $\text{D}^m\mathbf{P}$:AD-BCE^h-NBC^h GWS^h + θTS algorithm will not be unaffected by farfield pressure gradients as their influence is thoroughly diminished within the viscosity-dominated BL sublayer reach $y^+ \leq 6$ (Yang, 2015). Elementary notation alterations in the $\text{D}^m\mathbf{P}$:AD-BCE^h-NBC^h algorithm define its extension $\forall O(1)$ state-variable scalar member near wall NBC predictions.

10.7 Derivation, FaNS theory $O(\delta^2)$ state-variable nonhomogeneous Dirichlet BC

Any resolution of the FaNS theory $O(1;\delta^2)$ PDE system $O(\delta^2)$ tensor/vector state variable are solutions to a set of linear constant coefficient harmonic PDEs. For this transformative identification to be bounded domain well-posed, hence of practical impact to current practice CFD unstagnation, requires a nonhomogeneous DBC be *a priori* defined on all bounding surfaces $\partial\Omega$ of NS domain Ω, in particular those with the NS no-slip BC.

The $O(\delta^4)$ AD statement written for the nonphysical, i.e., *unfiltered* (!), resolved-unresolved scale interaction tensor $\tau_{ij}(\mathbf{x}, t)$ is [4]

$$\text{AD: } \tau_{ij} = \overline{\tau}_{ij} - \frac{\delta^2}{4\gamma}\nabla^2\overline{\tau}_{ij} + O(\delta^4)$$

Inserting appropriate overbar notation into the definition for $O(\delta^4)\mathrm{D}\tau_{ij}$ (Chapter 9.3), then multiplying through by $\delta^2/4\gamma$ and subtracting this from overbar altered $\mathrm{D}\tau_{ij}$ predicts that the definition for *non-physical* unfiltered $O(\delta^2)$ tensor $\tau_{ij}(\mathbf{x},t)$ is

$$\tau_{ij}(\mathbf{x},t) = \frac{\delta^2}{2\gamma}\left(\frac{\partial \bar{u}_i}{\partial x_k}\frac{\partial \bar{u}_j}{\partial x_k}\right) + O(\delta^4)$$

It is perhaps of some theoretical merit that this definition is identical to that generated by the second-order TS interpolation of the Gaussian Fourier transform inverse *space-filtered* $O(\delta^2)$ tensor (Chapter 9.2).

The nonhomogeneous DBC for spatially filtered $O(\delta^2)$ $\bar{\tau}_{ij}$ DOF at NS no-slip BC nodes $\mathbf{x}_b \in \partial\Omega_{\hat{n}_i}$ is thereby predictable via convolution

$$\bar{\tau}_{ij}(\mathbf{x}_b \in \partial\Omega_{\hat{n}_k}) \equiv \left(g_\delta{}^*\tau_{ij}\right) = \frac{\delta^2}{2\gamma}\int_\Omega g_\delta(x_m - x_l)^*\left(\frac{\partial \bar{u}_i}{\partial x_k}\frac{\partial \bar{u}_j}{\partial x_k}\right)_{x_l,t} d\tau$$

Expressing the *integrand* in the convolution tensor product as a differential statement

$$\mathrm{DS}{:}\left(d\bar{u}_i d\bar{u}_j\right) \equiv d\bar{u}_i d\bar{u}_j - \frac{\partial \bar{u}_i}{\partial x_k}\frac{\partial \bar{u}_j}{\partial x_k} = 0$$

and labeling thus generated DOF as $\{\mathrm{D\bar{U}}I\mathrm{D\bar{U}}J\}_e$, the weak formulation GWS^h algorithm element matrix for this convolution integrand function is

$$\mathrm{GWS}^h_e(t){:}\ [\mathrm{M200}]_e\{\mathrm{D\bar{U}}I\mathrm{D\bar{U}}J\}_e = \{\mathrm{\bar{U}}I\}_e^T[\mathrm{M3}K\mathrm{0}K]_e\{\mathrm{\bar{U}}J\}_e$$

for $1 \le (I, J, K) \le n$.

Gauss quadrature-generated DOF-independent FE p basis-dependent matrix $[\mathrm{M3}K\mathrm{0}K]_e$, $1 \le K \le n$, is a newly identified *hypermatrix*. Assembling each of these GWS^h $1 \le (I, J) \le n$ element matrices to global, the subsequent algebraic processes generate at any time t the $1 \le (I, J) \le n$ set of FE $p = 1$ basis piecewise continuous distributions

$$\mathrm{GWS}^h(t){:}\ \left(\frac{\partial \bar{u}_i}{\partial x_k}\frac{\partial \bar{u}_j}{\partial x_k}\right)^h_{x_k,t} = \cup_e\left(\{N(x_k)\}^T\{\mathrm{D\bar{U}}I\mathrm{D\bar{U}}J(t)\}_e\right)$$

Insertion of these distributions into the DBC convolution statement, the FE basis appropriate order numerical quadrature

$$\tau^h_{ij}(t)|_{\mathbf{x}_b\in\partial\Omega_{\hat{n}_i}} = \frac{\delta^2}{2\gamma}\sum_{l=1}^{\mathrm{L}x}\sum_{m=1}^{\mathrm{M}y}\sum_{n=1}^{\mathrm{N}z}\left(\begin{array}{c} w_l w_m w_n g_\delta(x_b - x_l, y_b - y_m, z_b - z_n) \\ \times\left(\{N(x_l, y_m, z_n)\}^T\{\mathrm{D\bar{U}}I\mathrm{D\bar{U}}J(t)\}_e\right)\end{array}\right)$$

then generates the nonhomogeneous DBC DOF for $\tau_{ij}(\mathbf{x},t)$ at nodes $\mathbf{x}_b \in \partial\Omega_{\hat{n}_i}$

Continuing, for the FaNS theory "unfiltered" $O(\delta^2)$ *vector* functional identifications

$$v_j^{\Theta}(\mathbf{x},t)=\frac{\delta^2}{2\gamma}\frac{\partial\bar{u}_j}{\partial x_k}\frac{\partial\overline{\Theta}}{\partial x_k}+O(\delta^4),\ v_j^{Y_a}(\mathbf{x},t)=\frac{\delta^2}{2\gamma}\frac{\partial\bar{u}_j}{\partial x_k}\frac{\partial\overline{Y_a}}{\partial x_k}+O(\delta^4)$$

minor notation alterations to the detailed $\tau_{ij}(\mathbf{x},t)$ GWS^h algorithm identify numerical quadrature prediction of the DBC DOF set at $\mathbf{x}_b\in\partial\Omega_{\hat{n}_i}$ for the $O(\delta^2)$ vectors

$$\begin{aligned}v_j^{\Theta h}(t)|_{\mathbf{x}_b\in\partial\Omega_{\hat{n}_i}} &= \frac{\delta^2}{2\gamma}\int_{\Omega^+} g_\delta(x_m-x_l)\left(\frac{\partial\overline{\Theta}}{\partial x_k}\frac{\partial\overline{u}_j}{\partial x_k}\right)^h d\tau,\ \mathbf{x}_b\in\partial\Omega_{\hat{n}_k}\\ &= \frac{\delta^2}{2\gamma}\sum_{l=1}^{\mathrm{L}x}\sum_{m=1}^{\mathrm{M}y}\sum_{n=1}^{\mathrm{N}z}\left(\begin{array}{l} w_l w_m w_n g_\delta(x_b-x_l, y_b-y_m, z_b-z_n)\\ \times\left(\{N(x_l,y_m,z_n)\}^T\{\mathrm{D\overline{\Theta}D\overline{U}}J(t)\}_e\right)\end{array}\right)\end{aligned}$$

$$\begin{aligned}v_j^{Y_a h}(t)|_{\mathbf{x}_b\in\partial\Omega_{\hat{n}_i}} &= \frac{\delta^2}{2\gamma}\int_{\Omega^+} g_\delta(x_m-x_l)\left(\frac{\partial\overline{Y_a}}{\partial x_k}\frac{\partial\overline{u}_j}{\partial x_k}\right)^h d\tau,\ \mathbf{x}_b\in\partial\Omega_{\hat{n}_k}\\ &= \frac{\delta^2}{2\gamma}\sum_{l=1}^{\mathrm{L}x}\sum_{m=1}^{\mathrm{M}y}\sum_{n=1}^{\mathrm{N}z}\left(\begin{array}{l} w_l w_m w_n g_\delta(x_b-x_l, y_b-y_m, z_b-z_n)\\ \times\left(\{N(x_l,y_m,z_n)\}^T\{\mathrm{D\overline{Y_a}D\overline{U}}J(t)\}_e\right)\end{array}\right)\end{aligned}$$

Following iteration p convergence of FaNS theory $O(1;\delta^2)$ PDE system GWS^h + θTS algorithm at time t_{n+1}, these GWS^h algorithm-generated solutions will *a priori* identify the requisite set of $O(\delta^2)$ state-variable DBC DOF in the global matrices $\{\mathrm{TAUIJ}(t)\}$, $\{\mathrm{V\Theta}J(t)\}$, and $\{\mathrm{VY}J(t)\}$ for use in iteration $p+1$ of the FaNS theory $O(1;\delta^2)$ PDE system at time t_{n+1}.

10.8 Commentary, *a priori* identified FaNS theory $O(\delta^2)$ state-variable DBC

As the $O(\delta^2)$ state variable members in FaNS theory are strictly mathematical entities identified in securing bounded domain closure of space-filtered NS PDE systems, generation of experimental data to support *validation* of the DBC algorithm set is likely not possible. Thereby, assessment of FaNS $O(\delta^2)$ state-variable AD-BCE^h-DBC^h GWS^h algorithm-generated $O(\delta^2)$ member DBC predictions must be inferred from computational data.

The next two monograph chapters document this critical assessment for $O(\delta^2)$ state-variable member near wall and farfield Cartesian resolution multiscale distributions for Reynolds numbers ranging 100 < Re < 12,000. For the extremum Re, $O(\delta^2)$ state-variable data mining in the FaNS $O(1)$PDE:AD-BCE^h-NBC^h:$O(\delta^2)$PDE:AD-DBC^h:$O(\delta^3)$ SFS tensor/vector system GWS^h + θTS FE $p=1$ basis algorithm/code-generated characterization of unsteady, periodic wall attached laminar profile BL $O(1)$ velocity *insipient* transition, separation, turbulent profile reattachment then relaminarization, supports detailed computational data assessment of these $O(\delta^2)$ state-variable AD-BCE^h-DBC^h algorithms.

References

[1] Boole, G., Moulton, J.F. (1960). *A treatise on the Calculus of Finite Differences*, Dover, NY.

[2] Schlichting, H. (1979). *Boundary Layer Theory*, McGraw-Hill, NY.

[3] Schwartz, L. (1966). *Theorie des distributions*, Hermann, Paris.

[4] Sekachev, M.A., (2013). "Essentially analytical theory closure for thermal-incompressible space filtered Navier-Stokes PDE system," PhD dissertation, University of Tennessee

[5] Sekachev, M.A., Baker, A.J., Wong, K.L. (2016). "Totally Analytical Closure of Space Filtered Navier-Stokes for Arbitrary Reynolds Number, Parts I., II, III," *J Numerical Heat Transfer*, Part B, 70(4).

[6] Wilcox, D.C. (2006). *Turbulence Modeling for* CFD, DCW Industries, La Canada, CA.

Chapter 11
Validation, FaNS, NS PDE systems same code/mesh predictions,100 < Re < 4,000

Libretto: FaNS $O(1{:}\delta^2)$ PDE system GWS^h+θTS FE $p = 1$ basis algorithm generates long-time stable, laminar-featured, steady-to-unsteady, multi-scale space filtered $n = 3$ thermal-velocity vector fields for 100 < Re < 4,000.

The $n = 3$ *virtual fluid dynamics laboratory* (VFDL) manufactured statement enables *precise* control of Re specifications for incompressible thermal-velocity vector field predictions.

NS and FaNS PDE systems same code/mesh VFDL predictions *validate* steady transition to unsteady at NS stability theory-predicted *critical* Rayleigh number $\mathrm{Ra_{CR}} = 3.1\mathrm{E}{+}05$, equivalent to Re = 750.

Same code/mesh FaNS and NS PDE thermal-velocity vector field *unsteady* 750 < Re < 4,000 comparisons quantify $\delta/h \equiv 4$ filtering *annihilation* of NS multi-scale $O(\leq 4\ h)$ spectral content.

Keywords: space-filtered NS, FaNS theory, validation, VFDL, critical Ra, ∀Re < 4,000

11.1 Design of FaNS theory validation, the $n = 3$ virtual fluid dynamics laboratory (VFDL)

Analytically closed FaNS $O(h^4)\mathrm{D}^m\mathbf{P}{:}\mathrm{D}^m\mathrm{E{:}DM}^h\mathrm{PDE{:}AD{:}BCE}^h{:}\mathrm{NBC}^h{:}O(h^2, \delta^4)\mathrm{PDE{:}AD\text{-}DBC}^h$ $\mathrm{PDE}{:}O(\delta^3)$ SFS tensor/vector system weak formulation continuous Galerkin GWS^h+θTS FE $p = 1$ basis algorithm/code is required validated for prediction of thermal-velocity vector fields ranging from steady laminar to unsteady multi-scale laminar featured to *transitioned* to turbulent. Meeting this requirement is precisely enabled by extending to three dimensions ($n = 3$) the *virtual fluid dynamics laboratory* (VFDL), an $n = 2$, 8×1 vertical aspect ratio thermal cavity manufactured statement [7].

Temperature difference Dirichlet BC imposition on opposing vertical surfaces enables precise Re specifications for VFDL steady laminar to bounded unsteady laminar, to unbounded unsteady periodic laminar-dominant, to unsteady multi-scale periodic transitioned-to-turbulent velocity vector flowfields for which, ". . . the spatial structure of flow is made of vertical and horizontal boundary layers, corner structures, and a stratified core . . . to give rise to very complex time behaviours (sic) resulting from several instability mechanisms . . . " [5].

NS linear stability theory predicts that the steady thermal-velocity field for air (Pr = 0.71) at Ra = E+05 in the $n = 2$ VFDL is characterized by an infinite set of real and complex conjugate pair eigenvalues [7]. That select eigenvalue pairs cross the imagi-

https://doi.org/10.1515/9783112206430-012

nary axis for the modest increment Ra > E+05 portends the initiation of flowfield bifurcation in the sense of Hopf.

Thereby, CFD prediction of steady velocity transition to unsteady in the vicinity of the linear theory-predicted *critical* Rayleigh number $Ra_{CR} \approx 3.1E+05$ constitutes an opportunity for quantitative validation. Several dozen NS CFD code predictions for unsteady thermal-velocity vector fields at Ra = 3.4E+05 > Ra_{CR} in the $n = 2$ VFDL are coalesced [2]. While voluminous, these data do not quantify the Ra(Re)-dependent transition process nor do they support validation of NS stability theory-predicted Ra_{CR}.

FaNS theory validation geometry is defined the mirror-symmetric half of an $n = 3$, 8 × 1 × 16 VFDL, Figure 11.1. Retaining the Boussinesq assumption on buoyancy, Gebhart et al. [3], NS and FaNS PDE systems are non-dimensionalized assuming that thermal-fluid dynamic processes are in balance, hence Gr/Re ≡ 1. The resultant reference velocity definition is $U = (g\beta\Delta TL^3/\nu^2)^{1/2}$ where g is the gravity vector magnitude, β is the incompressible fluid volumetric expansion coefficient and ΔT is the *a priori* defined Dirichlet BC (DBC) difference in temperature on the opposing 8 × 8 vertical sidewall surfaces.

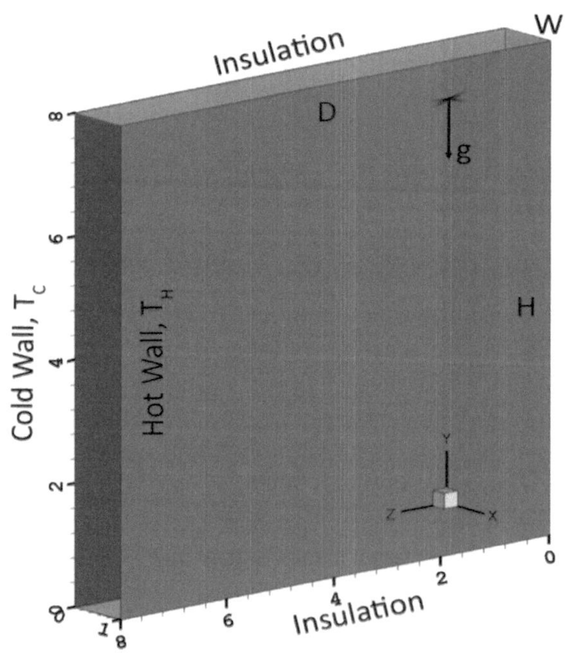

Figure 11.1: Virtual fluid dynamics laboratory (VFDL), $n = 3$, 8 × 1 × 16 symmetric half-domain, Dirichlet BC temperature data T_H ($\boldsymbol{x}_b$), T_C ($\boldsymbol{x}_b$) specified on opposed plane vertical bounding surfaces.

Definitions for the remaining NS non-D groups are Ra $\equiv$ Pr(gβΔTL^3/ν^2) and Re $\equiv$ (Ra/Pr)$^{1/2}$. Thereby, Raleigh number specifications ranging E+04 ≤ Ra ≤ E+08 in air (Pr = 0.71) correspond one-to-one with Reynolds number range E+02 < Re < 2E+04, which encompasses that required for validation of FaNS theory assertion of ∀Re pertinence.

Following Xin and Le Quere [7], the uniform M = 206 × 64 (vertical × horizontal) FE $p = 1$ basis mesh for the $n = 2$ VFDL was quantified, Grubert [4], to provide adequate mesh resolution for NS PDE predictions ∀Re ≤ 2,000. Thereby, the $n = 3$ VFDL symmetric half-domain discretization definition is M = 206 × 64 × 40. The mesh is defined uniform on all transverse planes, and imposition of an axial (z) direction geometric progression factor (pr) admits VFDL end plane region ($z \geq 0+\varepsilon$) adequate resolution while locating the $z = 8$ symmetry plane sufficiently remote therefrom.

The thermal BCs for the $n = 3$ VFDL top, bottom, and end planes are homogeneous Neumann (adiabatic). The sole Dirichlet BC on the symmetry plane is no throughflow, $u_3 = 0$, with homogeneous Neumann BCs defined for all remaining NS and/or FaNS theory $O(1)$ state variable members. The homogeneous Dirichlet BC is specified for the velocity wall-normal resolution, and the wall non-homogeneous Neumann NBC for $O(1)$ state variable members is generated by the AD-BCE^h-NBC^h algorithm. Finally, each NS no-slip BC node on VFDL bounding surface adjacent elements Ω_e is wall-normal direction sub-element enriched (Figure 10.5) to support AD-BCE^h-NBC^h GWS^h algorithm functionality.

FaNS theory mesh parameter h is defined the measure of the uniform discretization on each transverse plane. Thereby, for validated, *optimal* measure ratio $\delta/h \equiv 4$, the gaussian uniform filter measure is $\delta \equiv 0.05$ non-D, $\delta \approx 0.011$ m, 0.0333 ft dimensional. These definitions locate AD-BCE^h-NBC^h algorithm NS no-slip BC enforcement at abscissa coordinate 1E-04 in Figure 10.6, for which the error in NBC prediction is validated to be ≤0.5%. AD-BCE^h-NBC^h GWS^h algorithm $O(1)$ state variable member *efflux* convolution is computed at *every* VFDL domain Ω surface node $\mathbf{x}_b \in \partial\Omega_{e,\widehat{n}_j}$ except those on the symmetry plane.

11.2 Validation, NS and FaNS theory same code/mesh predictions, 100 < Re < 400

As definitions E+05 < Ra < E+07 correlate with 100 < Re < 4,000, FaNS theory and NS PDE systems GWS^h+θTS algorithm/code-generated thermal-velocity vector fields will exhibit strictly laminar features. To ensure the generation of *exactly* comparable FaNS/NS steady-to-periodic unsteady solutions requires that the DOF projection of weak formulation theory initial condition (IC) error bounding norm $\|\{q^h(t_0)\}\|^2_{H^1(\Omega)}$, Appendix B, be mesh-independent. This requirement is ensured by the *sole* non-homogeneous IC specification the linear interpolation of the VFDL bounding surface temperature difference ΔT(Ra) DBC on each transverse plane of mesh M.

Following a sufficient number of algorithm/code time steps nΔt to washout this IC specification *error* in NS $\Theta^h(x, y, z,\ t = 0)$ and FaNS $O(1)$ $\bar{\Theta}^h(x, y, z,\ t = 0)$ DOF distributions, resultant steady-to-unsteady periodic transitioning to multi-scale predictions will be *exactly comparable* ∀Ra specifications. Alteration of the FaNS theory $O(1;\delta^2)$ PDE system algorithm/code to support same code/mesh NS PDE system predictions is elementary via:

bypassing prediction of the $O(\delta^2)$ state variable,

$O(\delta^3)$ SFS tensor/vector coefficient C_δ defined unity,

bypassing the AD-BCE^h-NBC^h algorithm, and

imposing the NS no-slip velocity BC.

Of note, the FaNS predecessor incompressible NS $O(h^2)$ **DP**:$O(h^2)$ DM^h:DP PDE system GWS^h + θTS FE $p = 1$ basis algorithm has been thoroughly *validated* for this range of Reynolds number [1, Chs. 4–7].

Specifications Ra = E+04, Ra = E+05 correlate with Re = 113, Re = 375, hence the resultant thermal-velocity vector fields will be single-scale, resolutely laminar-featured. The FaNS theory and NS $O(h^4)$ PDE system GWS^h+θTS algorithm/Newton jacobian code executions each converged monotonically to steady state for specification $|\{\delta Q(n\Delta t)\}|_{\max} \le$ E−03. Resultant steady FaNS $\bar{\Theta}^h(x, y, z = 8, \mathrm{T})$ and NS $\Theta^h(x, y, z, \mathrm{T})$ *unitized scale* isotherm distribution perspectives on the top plane of the VFDL at Re = 375 (Figure 11.2a, b), confirm the impact of the end wall adiabatic BC, the forefront surfaces therein, is nil. Further, FaNS $\bar{\Theta}^h(x, y, z = 8, \mathrm{T})$ and NS $\Theta^h(x, y, z = 8, \mathrm{T})$ isotherm *true-scale* distributions on the VFDL symmetry plane *exactly* overlie at Ra = E+05 (Figure 11.2c). That this also occurred at Ra = E +04 quantitatively *validates* that the FaNS theory $O(\delta^2)$ state variable exerts *nil* impact on strictly laminar feature predictions ∀Re < 400.

11.3 Validation, NS stability theory-predicted VFDL critical Raleigh number Ra_{CR}

Precise assessment of NS linear stability theory predicted $Ra_{CR} \approx$ 3.1E+05 first addresses the $n = 2$ VFDL with uniform M = 206 × 64 FE $p = 1$ basis discretization. The "S-shaped" isotherm distributions on the $n = 3$ VFDL symmetry plane, clearly detailed in Figure 11.2c, are the precursor to VFDL thermal-velocity field temporal instability that results for a modest increment Ra > E+05. The measure selected to quantify code execution *transition propensity to unsteady* is the NS/FaNS $O(1)$ state variable DOF *infinity norm* $\|\{Q((n+1)\Delta t)\} - \{Q(n\Delta t)\}\|_\infty$, [6]. The IC for code execution of each E+05 increment in Ra in the sequence is linear interpolation of the Ra = E+05 ΔT DBC onto the DOF disposition of $n = 2$ mesh M.

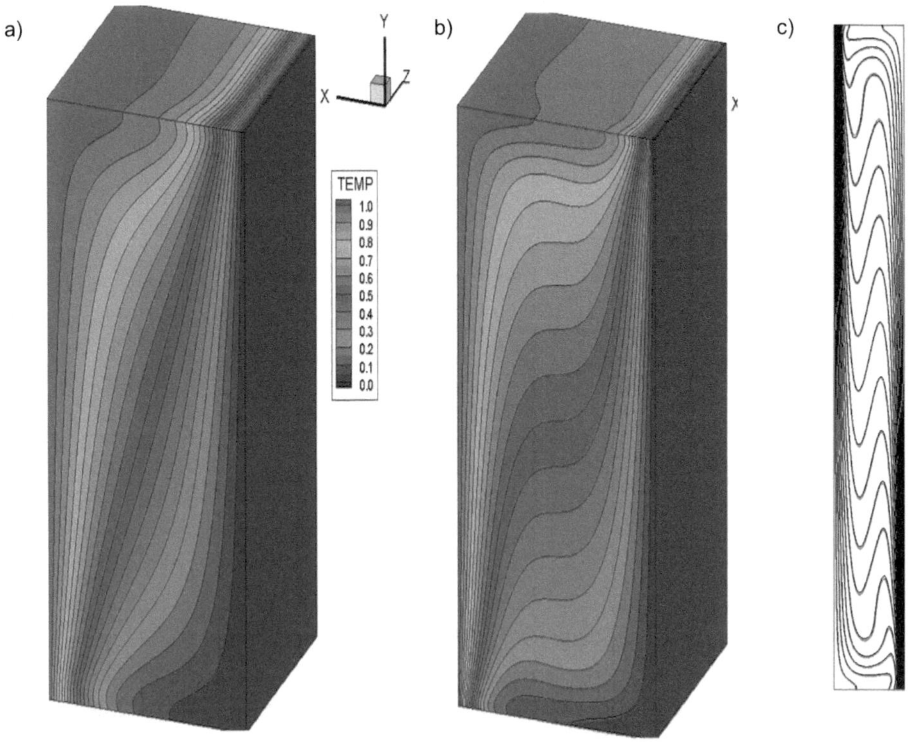

Figure 11.2: NS $O(h^4)$ PDE GWS^h+θTS FE $p = 1$ basis algorithm steady $\Theta^h(\mathbf{x}, \mathrm{T})$ isotherm distributions, symmetric half $n = 3$ VFDL, Pr = 0.71: unitized scale perspective projections; (a) Ra ≡ E+04, Re = 119; (b) Ra ≡ E+05, Re = 375; (c) symmetry plane NS $\Theta^h(x, y, z = 8, \mathrm{T})$ and FaNS $\overline{\Theta}^h(x, y, z = 8, \mathrm{T})$ isotherm true scale overlay, Ra ≡ E+05.

At Ra = 2E+05 both the NS and FaNS theory GWS^h+θTS algorithm same code/mesh converged solutions were *quasi-steady*, true steady state prevented by the *bounded* periodic oscillations in the local extrema in the S-shaped isotherm distributions (Figure 11.2c). For $t \equiv 0$ defined at washout of each Ra increment execution IC error, the NS $O(h^4)$ PDE GWS^h+θTS algorithm infinity norm evolution time histories for increments Ra = (3,4,5,6,7)E+05 on time interval $245 < \mathrm{n}\Delta t < 645$ s past $t = 0$ are comparatively graphed (Figure 11.3a).

The time evolution of the Ra = 3E+05 DOF infinity norm quantifies that oscillations in the S-shaped isotherm extrema are single-period to $\mathrm{n}\Delta t \approx 260$ s, whereafter the Newton iteration algorithm transitioned to single-step convergence. Hence, these bounded *quasi*-steady periodic temperature distribution oscillations will persist for code execution *ad infinitum*. For Ra = 4E+05, NS DOF infinity norm evolution is dual-frequency periodic to $\mathrm{n}\Delta t \approx 350$s (Figure 11.3a), whereupon it transitions to single-frequency periodic until execution termination at $\mathrm{n}\Delta t = 600$ s past $t = 0$. This occurrence is clearly quantified in the norm evolution local zoom (Figure 11.3b).

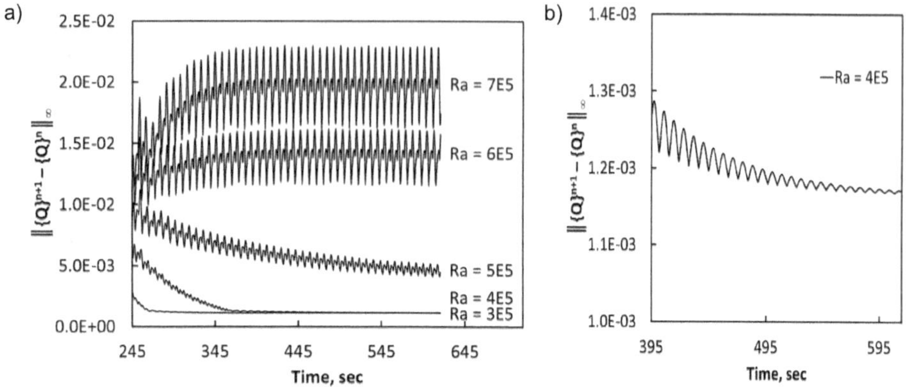

Figure 11.3: Time evolution of the NS DOF infinity norm $\|\{Q((n+1)\Delta t)\} - \{Q(n\Delta t)\}\|_\infty$ for $t = 0$ defined at the washout of the Ra = E+05 IC error, $n = 2$ 8 × 1 VFDL, NS $O(h^4,m^3,\Delta t^3)$ PDE GWS^h+θTS FE $p = 1$ basis algorithm/code, Pr = 0.71: (a) For increments Ra = (2, 3,4,5,6, 7)E+05. (b) Infinity norm long-time asymptote zoom for Ra = 4E+05.

In clear distinction to the Ra = 3E+05 quasi-steady solution, the single-frequency long-time asymptote magnitude of the Ra = 4E+05 infinity norm exceeds the Newton iteration convergence requirement $\varepsilon \le$ E−03. Thus, quantitatively validated is the Ra = 4E+05 thermal-velocity vector field is indeed unsteady ∀time. That the long time asymptotes of NS $O(h^4,h^2,\Delta t^3)$ PDE GWS^h+θTS Ra = 3E+05 and 4E+05 infinity norm bound linear stability theory predicted $Ra_{CR} \approx$ 3.1E+05 quantitatively *validates* the NS PDE GWS^h+θTS algorithm $n = 2$ VFDL stability limit prediction.

Continuing, at Ra = 5E+05 the transition propensity infinity norm evolves into a multi-frequency, time-independent asymptote with magnitude well in excess of Newton convergence requirement $\varepsilon \le$ E−03 (Figure 11.3a). The Ra = (6,7)E+05 long time norm evolution asymptotes are truly distinct from all other Re specifications in that their identical period, multi-frequency long time asymptote magnitudes *exceed* that engendered by the Ra = E+05 ΔT DBC interpolation IC.

These NS $O(h^4,m^4,\Delta t^3)$ PDE system GWS^h+θTS FE $p = 1$ basis algorithm/code data clearly quantify the $n = 2$ VFDL thermal-velocity flowfield monotonic progression from steady to unsteady, to multi-scale unsteady for Ra increments bounding linear stability predicted Ra_{CR}, existence of a critical Rayleigh number Le Quere and Benhia [5]. The $n = 2$ VFDL transition propensity norm $\|\{Q((n+1)\Delta t)\} - \{Q(n\Delta t)\}\|_\infty$ evolution histories generated by FaNS theory $O(h^4,h^2,\Delta t^3,\delta/h \equiv 4)$PDE:AD^h-BCE^h-NBC^h GWS^h+θTS algorithm/code, omitting the $O(\delta^2)$ state variable, are not measurably distinct from those graphed in Figure 11.3 [6].

Finally, repeating the Ra incremented sequence for the symmetric half of the $n = 3$ VFDL, the FaNS theory $O(1)$ D^m**P**:D^{m}E:DM^h:AD-BCE^h-NBC^h;$O(\delta^2)$PDE:AD-DBC^h PDE:$O(\delta^3)$ SFS tensor/vector GWS^h+θTS FE $p = 1$ basis algorithm/code predicted thermal-velocity transition from quasi-steady to unsteady for Ra specifications exactly bounding NS stability theory predicted Ra_{CR}. Thus, *validated* is that the FaNS theory $O(\delta^2)$ state variable exerts nil influence on steady to unsteady, periodic, laminar-featured thermal-velocity vector fields ∀Re < 1,100.

11.4 FaNS filtering annihilation of NS PDE-generated *O*(≤4 *h*) multi-scale content, 2,000 < Re < 4,000

For Ra ≡ 3E+06, equivalent to Re = 2,056, the NS PDE system generates *unsteady periodic* laminar features dominated *multi-scale* thermal-velocity vector fields. Both of the same code/mesh NS and FaNS executions required several hundred time steps to wash out the identical Ra = E+06 ΔT DBC IC interpolation error in $\Theta^h(x, y, z, t=0)$ and $\bar{\Theta}^h(x, y, z, t=0)$ distributions. Execution continuation to $n\Delta t = 600$ s past IC error wash out generates the exactly comparable NS $\Theta^h(\mathbf{x}, n\Delta t)$ and FaNS $\overline{\Theta}^h(\mathbf{x}, n\Delta t)$ unsteady periodic isotherm distributions, Figure 11.4, in the $n = 3$ VFDL. The directly comparable

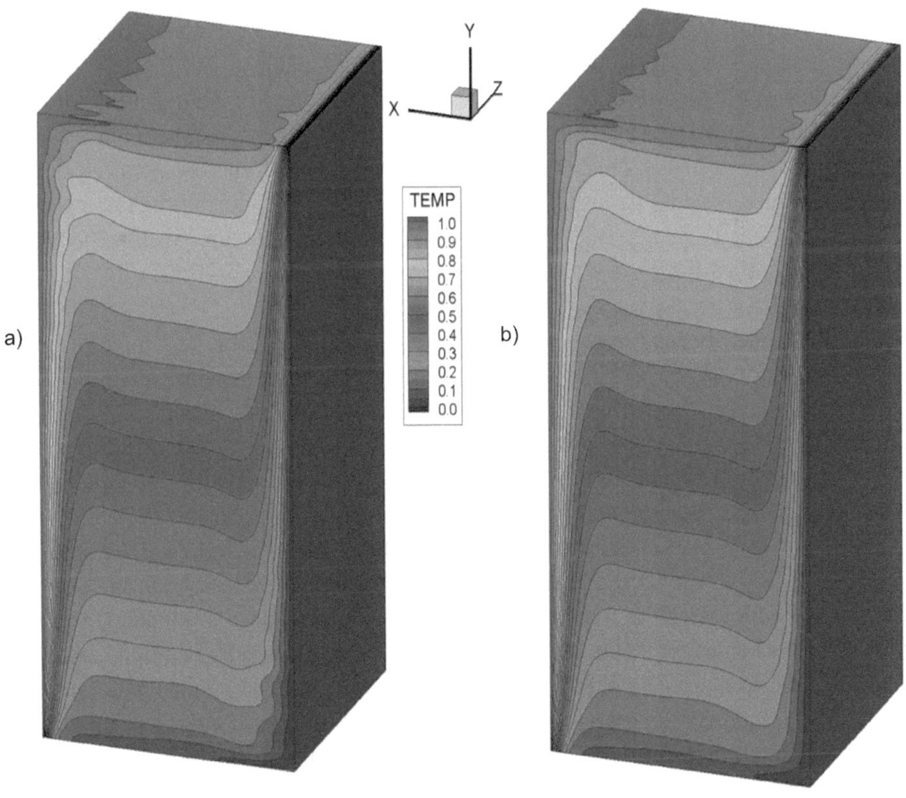

Figure 11.4: Unsteady periodic temperature distribution isotherm unitized scale, $n\Delta t = 600$ s perspective snapshots, view to adiabatic endwall, symmetric half 8 × 1 × 16, $n = 3$ VFDL, Ra ≡ 3E+06, Pr = 0.71, Re = 2,056: (a) NS $\Theta^h(\mathbf{x}, n\Delta t)$, (b) FaNS theory $\overline{\Theta}^h(\mathbf{x}, n\Delta t)$.

unitized scale global perspective snapshots, view to adiabatic endwall, confirm that the z-dependence of the top plane NS solution is *slightly* more noticeable than that of the FaNS solution. The surface isotherm distributions on the adiabatic endwall quantify FaNS theory's *just visible* moderation of the isotherm *undulations* NS predictions at horizontal surface terminations of each wall jet directly adjacent to the VFDL vertical corners.

On the VFDL symmetry plane, the same mesh/code directly comparable. $\Theta^h(x,y,z=8,\mathrm{n}\Delta t)$ and $\overline{\Theta}^h(x,y,z=8,\mathrm{n}\Delta t)$ unitized scale $\mathrm{n}\Delta t = 600$ s snapshots, (Figure 11.5a, b), quantify FaNS theory notable *moderation* of the $O(\mathrm{n}h)$ scale, $\mathrm{n} \le 4$, oscillations in NS isotherm distributions on the ΔT DBC bounding surfaces. This strictly wall-initiated phenomena does not penetrate into the VFDL interior at Re = 2,056.

At Ra ≡ E+07, equivalent to Re = 3,753 and following washout of the Ra ≡ E+06 IC interpolation error, essentially no distinctions exist in the VFDL top plane distributions of NS $\Theta^h(\mathbf{x}, n\Delta t)$ and FaNS $O(1)$ $\overline{\Theta}^h(\mathbf{x}, n\Delta t)$ unsteady unitized scale perspective $\mathrm{n}\Delta t = 600$ s snapshots (Figure 11.6a, b). On the forefront adiabatic wall surface, the NS $\Theta^h(\mathbf{x}, n\Delta t)$ isotherm distribution (Figure 11.6a) quantifies that the wall-adjacent undulations persist throughout the vertical travel of both wall jets. These NS multi-scale laminar features are just barely moderated in the same mesh/code FaNS theory-predicted $O(1)$ $\overline{\Theta}^h(\mathbf{x}, n\Delta t)$ isotherm distribution (Figure 11.6b).

For Re = 3,753, the NS $\Theta^h(x,y,z=8,\mathrm{n}\Delta t)$ periodic unsteady isotherm distribution at $\mathrm{n}\Delta t = 600$ s unitized scale snapshot on the VFDL symmetry plane (Figure 11.7a), fully details the unsteady generation of multi-frequency VFDL core-penetrating significant magnitude thermal fingers ejected normal from the wall during the latter half of wall jet vertical surface travel. The same code/mesh exactly comparable FaNS $O(1)$ $\overline{\Theta}^h(x,y,z=8,\mathrm{n}\Delta t)$ unitized scale $\mathrm{n}\Delta t = 600$ s symmetry plane isotherm distribution snapshot (Figure 11.7b), validates *complete* annihilation of these NS generated $O(\mathrm{n}h)$, $\mathrm{n} \le 4$, mesh scale thermal features by FaNS theory spatial filtering. That no significant alteration occurs in the VFDL core and in the horizontal wall adjacent large scale isotherm distributions quantifies these wall jet vertical surface adjacent multi-scale *dynamic* interactions are truly local.

The undulations present in the NS $\Theta^h(x,y,z=8,\mathrm{n}\Delta t)$ distribution during the latter half of the wall jet's vertical travel, which engender the clearly detailed thermal ejections (Figure 11.8a), are due to the embedding of z-axis *vortex* velocity $\mathbf{u}^h(0 \le x < 1, y, z=8, \mathrm{n}\Delta t)$ distributions (Figure 11.8b). FaNS theory spatial filtering completely suppresses the formation of these vortex embeddings (Figure 11.8c). The functional design of the AD:BCE^h:NBC^h GWS^h algorithm is thereby validated to be *the critical* (!) resolution that renders FaNS theory well-posed on a bounded domain.

The unitized scale perspective/planar graphics that enable data visualization and interpretation grossly distort the *true* dimensionality of NS solution multi-scale features removed by FaNS theory spatial filtering. Centered on the wall jet embedded *extremum* magnitude vortex at the right wall elevation $y \approx 3.6$ on the VFDL symmetry plane (Figure 11.8b), the NS speed flood local snippet with velocity vector $\mathbf{u}^h(0 \le x < 1, 3.0 \le y \le 4.0, z=8, \mathrm{n}\Delta t)$ overlay is side by side compared in *unitized*

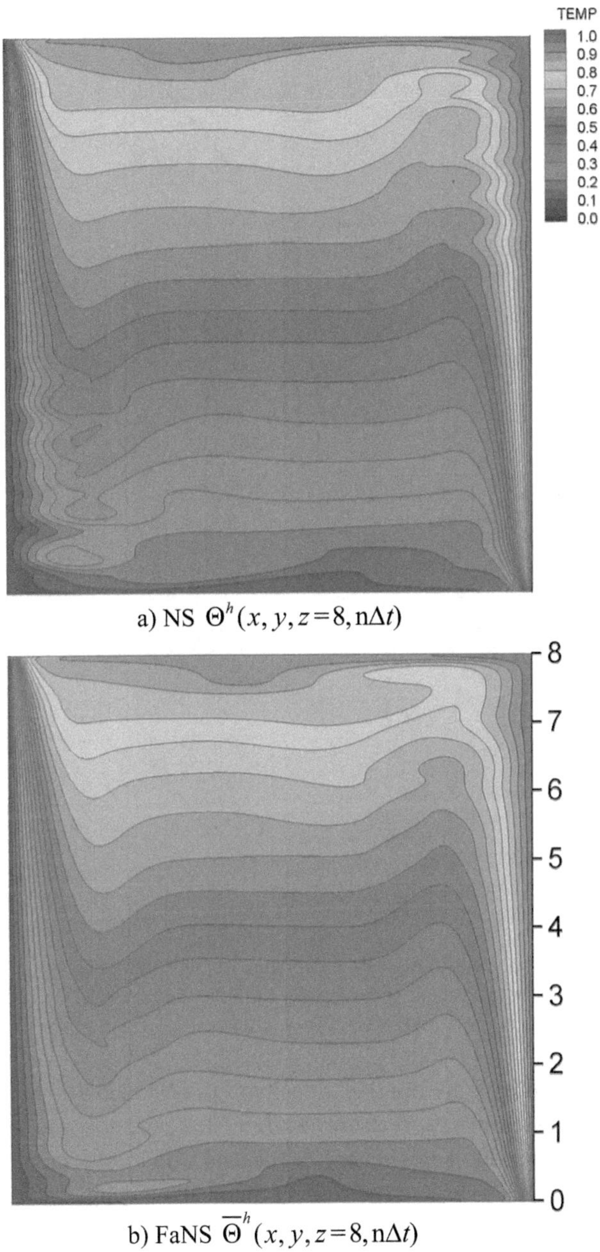

Figure 11.5: FaNS theory and NS unsteady periodic isotherm distributions on the VFDL symmetry plane, unitized scale $\mathrm{n}\Delta t = 600$ s snapshot, Ra $\equiv$ 3E+06, Re = 2,056, Pr = 0.71, $8 \times 1 \times 16$ $n = 3$ VFDL; (a) NS $\Theta^h(x, y, z = 8, \mathrm{n}\Delta t)$, (b) FaNS $\overline{\Theta}^h(x, y, z = 8, \mathrm{n}\Delta t)$.

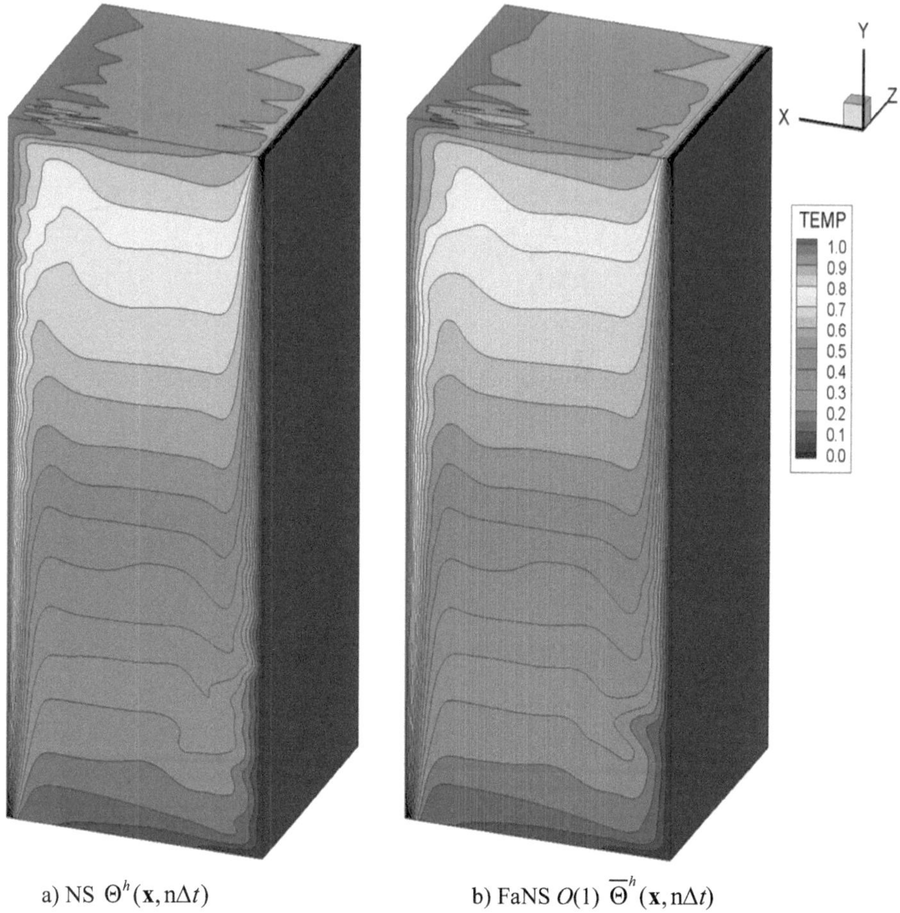

Figure 11.6: Same mesh/code FaNS theory, NS unsteady periodic temperature isotherm unitized scale global perspective, nΔt = 600 s snapshots, view to adiabatic endwall, Ra ≡ E+07, Re = 3,753, Pr = 0.71, symmetric half 8 × 1 × 16, n = 3 VFDL: (a) NS $\Theta^h(\mathbf{x}, n\Delta t)$, (b) FaNS $O(1)\ \overline{\Theta}^h(\mathbf{x}, n\Delta t)$.

and true scales (Figure 11.9 a, b). Judging by the DOF density of the NS velocity vector, the *measure* of this multi-scale feature is precisely $O(4\ h)$.

Same mesh/code exactly comparable FaNS theory $O(1)$ speed flood with vector $\overline{\mathbf{u}}^h(0 \le x < 1,\ 3.0 \le y \le 4.0,\ z = 8,\ n\Delta t)$ overlay true and unitized scale color floods (Figure 11.9c, d), clearly quantify space filtering complete annihilation of the NS generated, dominantly $O(4\ h)$ isotherm structures as well as the additional half dozen thermal structures with measures $O(<4\ h)$ embedded in the NS wall jet at Re = 3,753 (Figure 11.8a, b). These actions correlate precisely with FaNS non-homogeneous NBC near wall mesh resolution identification of the *optimal* filter/mesh measure ratio $\delta/h \equiv 4$, hence validate design of the FaNS theory algorithm.

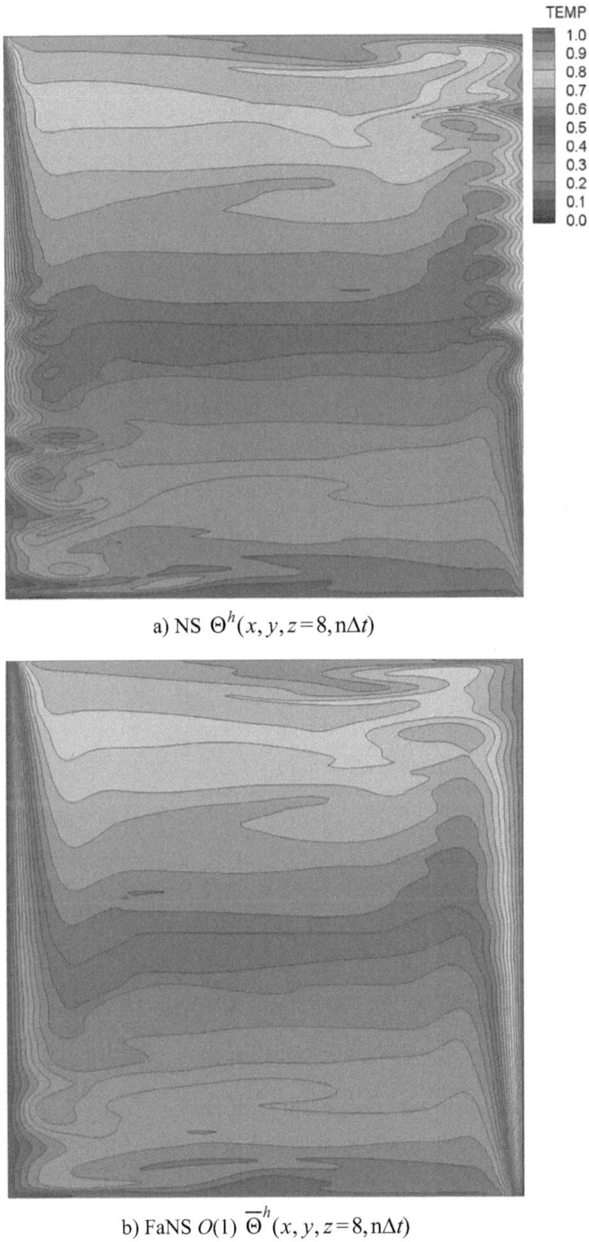

Figure 11.7: Same mesh/code FaNS theory, NS unsteady periodic unitized scale isotherm $n\Delta t = 600$ s snapshot floods on the VFDL symmetry plane, Ra $\equiv$ E+07, Re = 37,538, Pr = 0.71, 8 × 1 × 16 $n = 3$ VFDL: (a) NS $\Theta^h(x, y, z = 8, n\Delta t)$, (b) FaNS $O(1)$ $\overline{\Theta}^h(x, y, z = 8, n\Delta t)$.

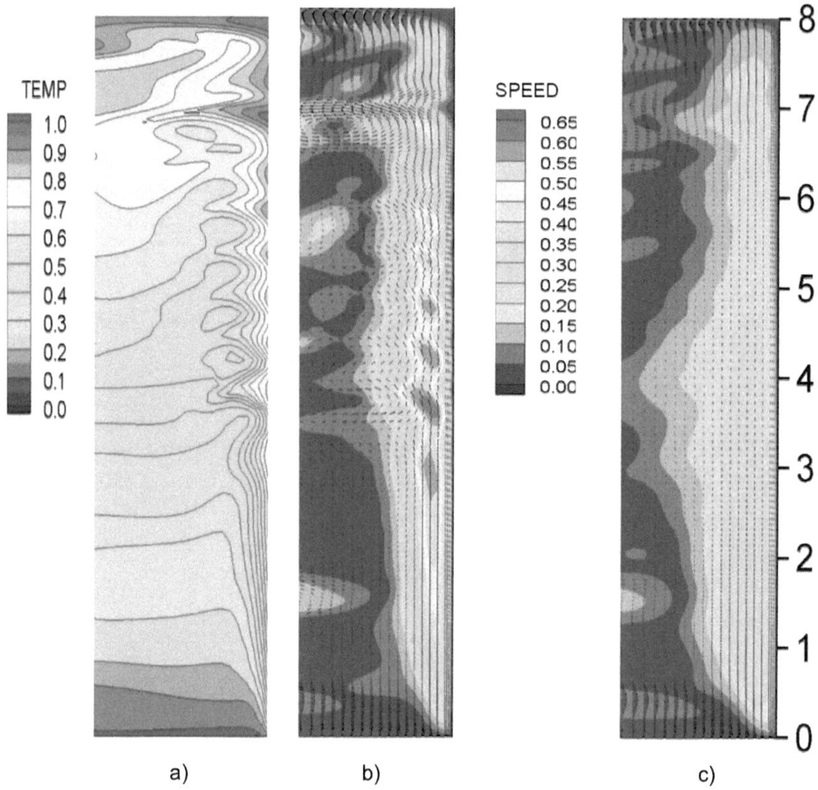

Figure 11.8: FaNS and NS state variable member symmetry plane wall adjacent unsteady periodic unitized scale nΔt = 600 s snapshots, Ra ≡ E+07, Re = 3,753, Pr = 0.71: 8 × 1 × 16 n = 3 VFDL: (a) NS $\Theta^h(x, y, z = 8, \text{n}\Delta t)$, (b) $\mathbf{u}^h(0 \le x < 1, y, z = 8, \text{n}\Delta t)$, (c) FaNS $O(1)\,\overline{\mathbf{u}}^h(0 \le x < 1, y, z = 8, \text{n}\Delta t)$.

11.5 Validation, NS/FaNS state variable non-mirror-symmetric VFDL distributions

A consistency requirement for CFD-predicted $n = 2$ cavity thermal-velocity vector fields is the persistence of *precisely* mirror-symmetric state variable distributions ∀Ra specifications. Mirror symmetry indeed is exhibited in the FaNS and NS PDE generated steady $n = 3$ VFDL symmetry plane isotherm distributions at Ra = E+04, E+05 (Figure 11.2c). However, for Ra = 3E+06 both the FaNS and NS PDE algorithms generate unsteady periodic symmetry plane isotherm distributions that are not mirror symmetric (Figures 11.5 and 11.6). Finally, the asymmetries existent in FaNS and NS state variable distributions for Ra ≡ E+07 are *very pronounced* (!) (Figure 11.7).

These Ra-sensitive distinctions result from the action of the buoyancy body force *vector* in the $n = 3$ VFDL vertical corners formed by the confluence of the adiabatic end-

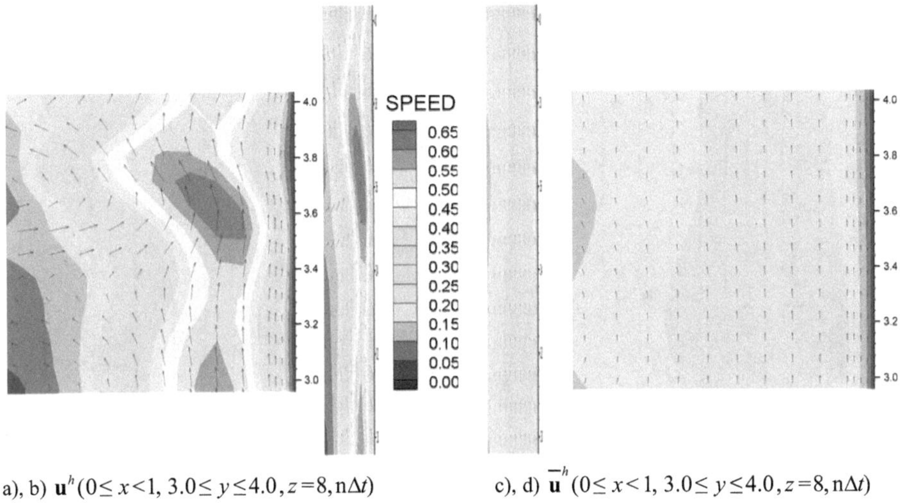

Figure 11.9: FaNS spatial filtering annihilation of NS thermal-velocity periodic $O(\leq 4\ h)$ multi-scale features, VFDL symmetry plane, speed flood with vector overlay, $\mathrm{n}\Delta t = 600$ s snapshot snippets, Ra ≡ E+07, Re = 3,753, Pr = 0.71: 8 × 1 × 16, $n = 3$ VFDL: (a) NS $\mathbf{u}^h(0 \leq x < 1,\ 3.0 \leq y \leq 4.0,\ z = 8,\ \mathrm{n}\Delta t)$ unitized scale, (b) true scale, (c) FaNS $O(1)$ $\overline{\mathbf{u}}^h(0 \leq x < 1,\ 3.0 \leq y \leq 4.0,\ z = 8,\ \mathrm{n}\Delta t)$ true scale, (d) unitized scale.

wall with the ΔT DBC surfaces. For Ra ≡ E+07, Re = 3,753, the DOF distributions for NS velocity vector resolutions $\mathbf{u}^h(x = \delta/2, y, z \leq 2, \mathrm{n}\Delta t) \times \widehat{\mathbf{i}}$ and $\mathbf{u}^h(x = 1 - \delta/2, y, z \leq 2, \mathrm{n}\Delta t) \times \widehat{\mathbf{i}}$ on VFDL transverse planes displaced $|x| \pm \delta/2$ from the ΔT DBC surfaces are unitized scale graphed in Figure 11.10a, b. Clearly quantified is that $n = 3$ VFDL vertical corner flowfield bounding is the cause for the very evident non-symmetry in NS velocity vector distributions on the near endwall transverse planes. Factually, this is a DOF distribution attribute exhibited in all NS state variable member distributions. Further, the NS velocity resolution DOF distributions in Figure 11.10 validate the adequacy of mesh resolution adjacent to the endwall.

Same code/mesh Ra ≡ E+07 FaNS theory $O(1)$ velocity resolution DOF distributions $\overline{\mathbf{u}}^h(x = \delta/2, y, z \leq 2, \mathrm{n}\Delta t) \times \widehat{\mathbf{i}}$ and $\overline{\mathbf{u}}^h(x = 1 - \delta/2, y, z \leq 2, \mathrm{n}\Delta t) \times \widehat{\mathbf{i}}$ on the identical planes, are unitized scale compared in Figure 11.10 c), d). These data corroborate conclusions drawn from Figure 11.10a, b. regarding the $n = 3$ VFDL vertical corners as the cause for non-mirror symmetric FaNS $O(1)$ state variable distributions.

To conclude, these same code/mesh NS and FaNS $O(1)$ velocity vector resolution DOF data at Re = 3,753 fully *quantify* spatial filtering annihilation of the $O(\mathrm{n}h)$ multi-scale spectral content, $\forall \mathrm{n} \leq 4$, in the NS velocity vector distributions. Also quantitatively validated is the FaNS theory significant moderation of the extrema in the NS velocity vector multi-scale distribution with scales exceeding $O(4\ h = \delta)$.

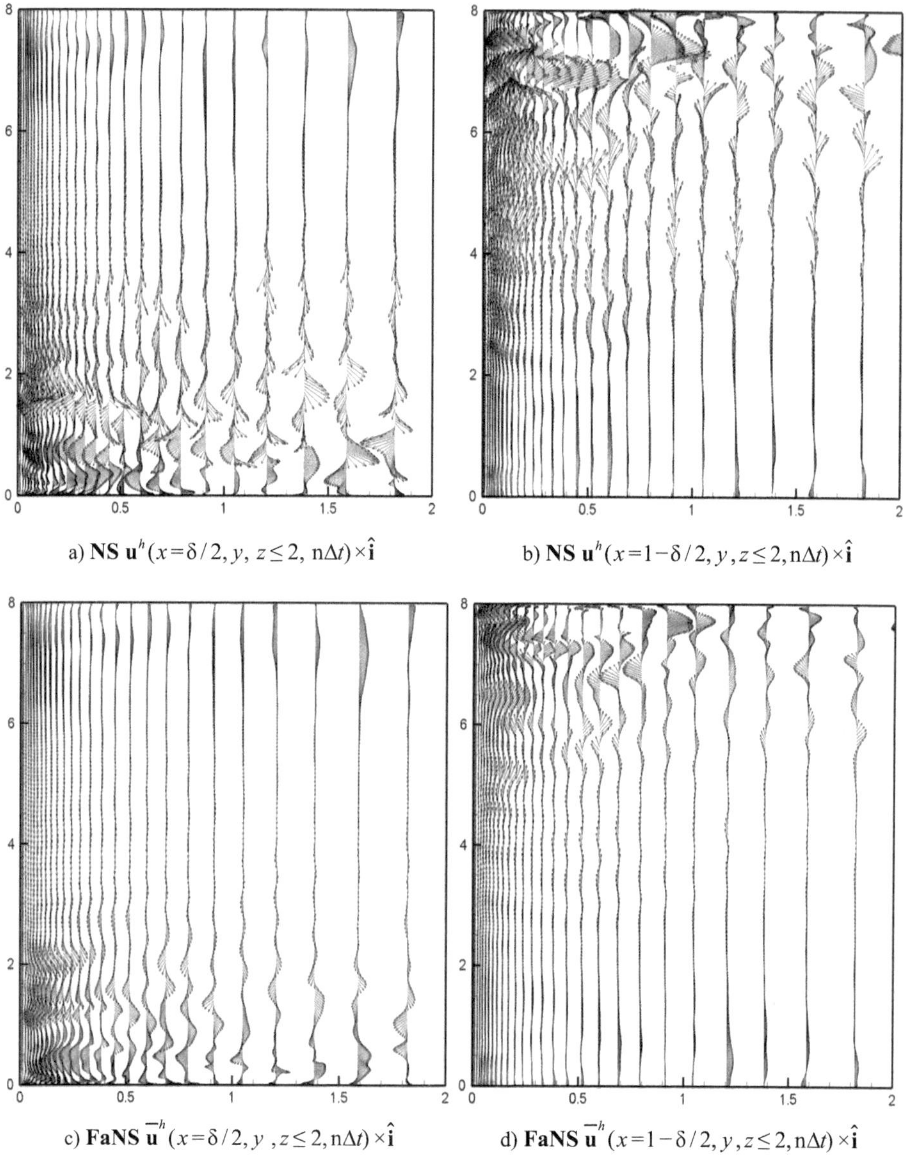

Figure 11.10: NS and FaNS theory transverse plane velocity vector resolution DOF distributions adjacent to the VFDL adiabatic endwall, periodic $\mathrm{n}\Delta t$=600 s snapshot on (*y, z*) planes displaced $|x| = \pm\,\delta/2$ from cold, hot ΔT DBC surfaces, symmetric half $n = 3$, 8 × 1 × 16 VFDL, Ra ≡ E+07, Re = 3,753, Pr = 0.71.

11.6 Legacy CFD unstagnation, FaNS theory $O(h^4,h^2,\delta^4)$ PDE system ∀Re < 4,000

NS and FaNS theory PDE system same code/mesh generated directly comparable data for Reynolds numbers ranging 100 < Re < 4,000 fully quantify FaNS theory *optimal* filter/mesh measure ratio $\delta/h = 4$ annihilation of *scale appropriate* NS thermal-velocity vector multi-scale features. Throughout, quantitatively *validated* is that both the NS $O(h^4,h^2)$PDE system and FaNS theory $O(1)\mathrm{D}^m\mathbf{P}{:}\mathrm{D}^m\mathrm{E}{:}\mathrm{DM}^h$:AD-BCE^h-NBC^h;$O(\delta^2)$PDE: AD-DBC^h PDE:$O(\delta^3)$ SFS tensor/vector system GWS^h+θTS FE $p = 1$ basis algorithm/code predicted DOF distributions are resolutely *monotone* ∀Re < 4,000. Further *validated* is that $O(\delta^3)$ bounding of the FaNS theory SFS tensor/vectors *does not compromise* $O(h^2)$ TEA functionality in FaNS $O(1)$ PDE predictions ∀Re < 4,000.

Not having to expend the compute resources required to CFD-resolve the full spectrum of NS state variable high-frequency $O(nh)$ multi-scale content via direct numerical simulation (DNS) is *precisely* the intent of CFD theory formulation on the spatially filtered NS PDE system. The current CFD code practice *unstagnation* inherent in ∀ Re FaNS theory has been fully quantified by the same code/mesh-generated, *exactly comparable* FaNS/NS data ∀Re < 4,000. The fundamental *physics of fluids* observable in these data is that the $O(\delta^2)$ resolved-unresolved scale interaction state variable is quantified to exert ***nil impact*** on prediction of the $O(1)$ state variable for laminar features dominant Re ≤ 4,000 thermal-velocity vector $n = 3$ flow fields.

Looking ahead, the Ra = E+08, hence Re ≈ 12,000 data in Chapter 12 *quantify* analytical ∀ Re FaNS theory *error freed* CFD *mathematics* prediction of wall attached laminar profile $O(1)$ velocity BL incipient transition, followed by separation then turbulent profile reattachment and finally relaminarization. FaNS theory is thereby quantitatively *validated* pertinent ∀Re ≤ 12,000 specifications, which completes the Re range necessary to substantiate the *truly transformative* CFD **unstagnation** represented by analytical FaNS theory. The fundamental *physics of fluids* observable in these data is that the $O(\delta^2)$ resolved-unresolved scale interaction state variable is quantified to exert ***profound impact*** (!) on prediction of the $O(1)$ state variable for $n = 3$ flow fields for Re ≥ $O(10^4)$.

References

[1] Baker, A.J. (2014). *Optimal Modified* CONTINUOUS *Galerkin* CFD, John Wiley, London.
[2] Christon, M.A., Gresho, P.M., Sutton, S.B. (2002), "Computational predictability of time- dependent natural convection flows in enclosures," *J Num Mtd Fluids*, 40.
[3] Gebhart, B., Jaluria, Y., Mahajan, R.L., Sammakia, B. (1988). *Buoyancy-Induced Flow and Transport*, Hemisphere, NY.
[4] Grubert, M.A. (2006) "Development of a Potentially Accurate and Efficient LES CFD Algorithm to Predict Heat and Mass Transport in Inhabited Spaces," PhD dissertation, University of Tennessee

[5] Le Quere, P., Benhia, M. (1998), “From onset of unsteadiness to chaos in a differentially heated square cavity,” *J Fluid Mech*, 359.

[6] Sekachev, M.A., (2013). “Essentially analytical theory closure for thermal-incompressible space filtered Navier-Stokes PDE system,” PhD dissertation, University of Tennessee

[7] Xin, S., Le Quere, P. (2001), “Linear stability analysis of natural convection flowfields in a differentially heated square cavity with conducting walls,” *Phys Fluids*, 13.

Chapter 12
Validation, $\forall$ Re pertinent FaNS theory prediction of laminar BL profile $O(1)$ velocity insipient transition, separation, turbulent BL profile reattachment, then relaminarization, Re ≈ 12,000

Libretto: FaNS theory *error freed* CFD *mathematics* $\forall$ Re pertinent PDE system predicts unsteady, periodic multi-scale wall-attached laminar profile $O(1)$ velocity incipient transition, separation, turbulent profile reattachment, then relaminarization at Re ≈ 12,000 on the VFDL symmetry place.

FaNS $O(1)$ state variable DOF distributions are validated *monotone* to algebraic convergence digit throughout the long-time unsteady thermal-separated velocity prediction process.

FaNS $O(\delta^2)$ interaction tensor is validated *realizable* and *anisotropic*, with the spectral content of DOF distributions ranging from regular, to $O(L \leftrightarrow h)$ multi-scale, to chaotic, to absent.

TEA-theorized $O(\delta^3)$ SFS tensor/vector *dispersive* functionality is validated *critical* to $O(1)$ velocity DOF monotonicity in the reach of wall jet separation.

Classic perturbation theory prediction that the FaNS theory state variable significance is scaled $O(1;\delta^2;\delta^3)$ is validated.

Critical examination of FaNS theory $O(\delta^2)$ state variable wall-adjacent DOF distributions validates the non-homogeneous DBC $O(\delta^4)$ AD:DBC GWS^h algorithm.

Net is a *comprehensive* validation that analytical $\forall$ Re FaNS theory $O(h^4,\delta^4)$ PDE system is the mathematically rigorous *unstagnation* of legacy LES heuristic closure modeling practice.

Keywords: CFD, validation, analytically closed FaNS theory, transition, relaminarization, tensor, realizable, anisotropic, $O(\delta^2)$ Dirichlet BCs, state variable $O(1;\delta^2;\delta^3)$ scaling

12.1 Introduction

Analytically closed space-filtered NS (FaNS theory) $O(h^4, \delta^4)$ PDE: AD-BCE:NBC:DBC: SGS system weak formulation GWS^h algorithm/code is quantified to enable long-time stable prediction of steady to unsteady, periodic multi-scale thermal-velocity vector fields possessing a full range of laminar spectral features for 100 < Re < 4,000. The key *physics of fluids* takeaway in these data is that the $O(\delta^2)$ resolved-unresolved scale interaction tensor/thermal vector exert nil impact on $O(1)$ state variable distributions

https://doi.org/10.1515/9783112206430-013

throughout. The $n = 3$ VFDL manufactured statement exercise enabling this validation is extended to Ra $\equiv$ PrRe^2 = E + 08, hence Re ≈ 12,000 for air, for which the $O(\delta^2)$ resolved-unresolved scale interaction tensor and thermal vector are validated ***the dominant*** (!) *physics of fluids* influence on $O(1)$ state variable prediction.

The FaNS theory $O(1;\delta^2;\delta^3)$ state variable *error freed* $O(1;\delta^2)$ PDE system GWS^h + θTS FE $p = 1$ basis algorithm at Ra $\equiv$ E + 08 (ΔT $\equiv$ 111°K, 200°R), for air correlates with U = 0.92 m/s (2.87 f/s), hence Re = 11,868, generates stabilization-***free*** and modeling-***free*** (!) prediction of unsteady wall-attached laminar profile $O(1)$ velocity vector insipient transition, separation, turbulent profile reattachment, then relaminarization on the VFDL symmetry plane. The unsteady, periodic *filtered* thermal-velocity vector field generated by the GWS^h long-time stable algorithm/code preserves all multi-scale content possessing measure distributions O(L) to $O(h)$ throughout thus generated $O(1)$ state variable monotone distributions.

This chapter's content thereby completes the quantitative *validation* of the ∀Re pertinence of bounded domain well-posed FaNS theory $O(1)$ state variable $O(h^4,\delta^4)$PDE:AD-BCE^h-NBC^h; $O(\delta^2)$ state variable $O(h^2,\delta^4)$PDE:AD-DBC^h; TEA $O(\delta^3)$ SFS tensor/vector GWS^h + θTS FE $p = 1$ basis algorithm/code. That generated $O(1)$ state variable DOF distributions are long-time strictly *monotone* to iterative convergence digit validates that the $O(\delta^3)$ bounding on the unresolved scale SFS tensor/vector exerts *nil* impact on $O(1)$ state variable stability and $O(h^2)$ TEA functionality.

FaNS theory $O(\delta^2)$ resolved-unresolved scale interaction state variable member DOF distributions that exhibit regularity are *validated* monotone. Regarding one absolutely fundamental VVUQ requirement, that the scaled significance of FaNS state variable distribution is validated $O(1;\delta^2;\delta^3)$ enables *rigorous quantification* that mesh resolution of the $O(1)$ state variable distribution is ***adequate***, a *premier* accomplishment (!) in the practice of CFD.

12.2 Global synopsis, FaNS theory prediction of laminar-turbulent transition, Re ≈ 12,000

The IC for FaNS theory Ra $\equiv$ E + 08 VFDL exercise is code self-generated via Raleigh number continuation from the Ra = E + 07 FaNS nΔt = 600 s solution snapshot. Following sufficient time steps nΔt to wash out the IC error, the FaNS $O(h^4,\delta^4)$ PDE:AD-BCE^h-NBC^h: $O(h^2,\delta^4)$PDE:AD-DBC^h:$O(\delta^3)$SFS tensor/vector system GWS^h + θTS FE $p = 1$ basis algorithm/code generates periodic unsteady, multi-scale resolved scale state variable distributions at Δt = 600 s past IC error washout exhibiting the full range of horizontal, vertical and longitudinal direction thermal-velocity vector features hypothesized for the $n = 2$ VFDL [1]. The fully coupled implicit GWS^h + θTS code predicted $O(\delta^2)$ resolved-unresolved scale interaction state variable cartesian resolution distribution snapshots at nΔt = 600 s past IC washout exhibit spectral content ranging organized, to multi-scale disorganized, to chaotic to completely absent.

The global characterization synopsis of the Re = 11,868 unsteady FaNS $O(1;\delta^2)$ state variable distribution nΔt = 600 s snapshots employs unitized scale isotherm, isovel, and isotensor color flood perspectives with contours with select vector overlay. To facilitate *visualization* of $O(1)$ velocity $n = 3$ distributions, the velocity vector is projected as the $\bar{\mathbf{u}}^h(x, y, z = 8, \text{n}\Delta t)$ DOF distribution on the VFDL symmetry plane or as a perspective color flood projection of $\bar{\mathbf{u}}^h(\mathbf{x}, \text{n}\Delta t)$ speed $\bar{u}^h(x, y, z, \text{n}\Delta t)$ distribution on the entire VFDL domain Ω.

State variable member coordinate notation $|x,y,z)| = (\delta/2, \pm\delta/2)$ signifies the omission of the $O(1)$ velocity DOF data from the VFDL surface node to the mesh plane containing node 6, located at a wall-normal distance of $\pm\delta/2$, recall Figure 10.4. Thereby eliminated from these graphics are the *filtered* NS velocity DOF generated by the $O(\delta^4)$ AD-BCE^h-NBC^h algorithm at the local enrichment node set within each wall-adjacent element.

FaNS theory $O(1)$ $\bar{\Theta}^h(x, y, z = 8, 4, 0 + \delta/2, \text{n}\Delta t)$ and $\bar{u}^h(x, y, z = 8, 4, 0 + \delta/2, \text{n}\Delta t)$ distribution nΔt = 600 s snapshots are unitized scale perspective visualized in Figure 12.1 as the cascade of color flood distributions on the VFDL symmetry plane $z = 8$, on the midplane $z = 4$ and on the plane displaced $z = \delta/2$ from the adiabatic endwall plane $z = 0$, view to the symmetry plane, hot ΔT DBC at the right. The ΔT DBC-induced VFDL core vertical temperature gradient exhibits only modest z-dependence (Figure 12.1a). Conversely, the maintenance of the wall jet velocity's organized distribution in the vertical traverse adjacent to the adiabatic endwall is notably z-dependent. This is quantified by the persistence of the jet core speed extremum (red) (Figure 12.1b), which quantifies that the endwall is a momentum sink. The essential overall absence of fluid motion (speed) within the VFDL core is clearly displayed.

The superposed isotherms/isovels in Figure 12.1 clearly visualize the several horizontally directed, core-penetrating thermal plume ejections engendered upon disorganization of the local wall jet velocity distribution. Wall jet terminations at each horizontal surface create adjacent recirculation regions of significant transverse (x,y) span. Away from these features, the quiescence of the VFDL central core region is quantified by the nil magnitude (dark blue) envelope of the speed distribution (Figure 12.2a), view to the symmetry plane. The vertical extent of wall jet organization, as quantified by core extremum (red) persistence, terminates at ~75% vertical surface travel on the cold ΔT DBC surface, alternatively at ~45% vertical travel adjacent to the hot ΔT DBC surface.

The numerous penetrations of $O(1)$ wall jet velocity $\bar{\mathbf{u}}^h(\mathbf{x}, \text{n}\Delta t)$ into the VFDL core are visualized in the $\bar{u}^h(|x| = \pm\delta/2, y, z, \text{n}\Delta t)$ speed distribution unitized scale color flood perspective (Figure 12.2a), viewed from the symmetry plane, hot ΔT BC at the right. The symmetry plane cold ΔT BC surface-attached wall jet "freestream" boundary remains smooth at nil magnitude (dark blue) up to ~60% of travel, as detailed on the forefront surface in Figure 12.2a. The "freestream" bounding surface of the wall jet attached to the cold ΔT DBC surface on the symmetry plane remains smooth at nil

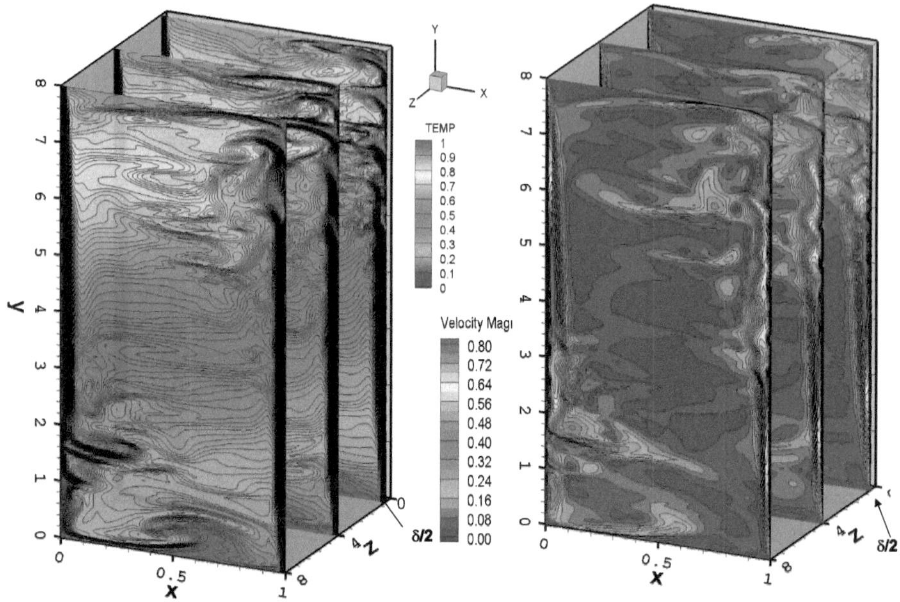

a) $\overline{\Theta}^h(x, y, z = 8, 4, 0 + \delta/2, \mathrm{n}\Delta t)$ b) $\overline{u}^h(x, y, z = 8, 4, 0 + \delta/2, \mathrm{n}\Delta t)$

Figure 12.1: Global unitized scale perspective flood cascade projection of FaNS theory $O(1)$ state variable distributions, $\mathrm{n}\Delta t = 600$ s snapshot with contours, view to the symmetry plane, symmetric half, $n = 3$, $8 \times 1 \times 16$ VFDL, Ra $\equiv$ E + 08, Re = 11,868, Pr = 0.71: (a) $\overline{\Theta}^h(x, y, z = 8, 4, 0 + \delta/2, \mathrm{n}\Delta t)$, (b) $\overline{u}^h(x, y, z = 8, 4, 0 + \delta/2, \mathrm{n}\Delta t)$.

magnitude (dark blue) up to ~60% of vertical travel on the forefront surface in Figure 12.2a.

Surface undulations of progressively increased significance (medium blue) initiate fluid penetration into the VFDL interior at vertical surface mid-elevations, followed by increased significance (light blue), strongly z-dependent fluid penetration into the entire VFDL core in the terminal ~15% of wall jet vertical travel. Conversely, adjacent to the adiabatic endwall (the graphic rear plane), the smoothness of the wall jet "freestream" surface is immediately compromised by an increase in speed (light blue), with continued persistence to nominal mid-elevations of the VFDL domain.

Viewpoint alteration of the $\overline{u}^h(|x| = \delta/2, y, z, \mathrm{n}\Delta t)$ speed flood perspective to the adiabatic endwall (Figure 12.2b), cold ΔT DBC at the right, provides clear quantification that the AD-BCE^h-NBC^h algorithm generated convolution of the NS velocity efflux limits wall jet core velocity non-D significance to ~0.10 (dark blue). This perspective also enables direct comparison of this $O(1)$ speed distribution envelope with the $O(\delta^2)$ resolved-unresolved scale interaction tensor diagonal resolution $\tau_{22}^h(\mathbf{x}, \mathrm{n}\Delta t) \equiv 2(\overline{\overline{u_2}\, u_2'})^h$ distribution color flood (Figure 12.2c), which possesses the extremum significance of all tensor resolutions.

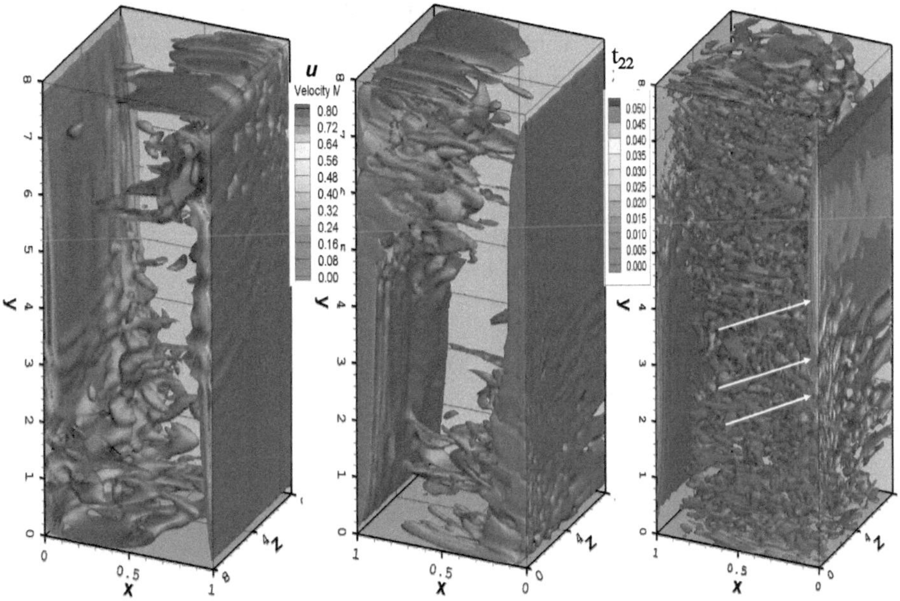

a) $\overline{u}^h(|x| = \pm\delta/2, y, z, \mathrm{n}\Delta t)$ b) $\overline{u}^h(|x| = \pm\delta/2, y, z, \mathrm{n}\Delta t)$ c) $O(\delta^2)\ \tau^h_{22}(\mathbf{x}, \mathrm{n}\Delta t)$

Figure 12.2: FaNS theory $O(1)$ speed and extremum significance $O(\delta^2)$ interaction tensor diagonal resolution $\mathrm{n}\Delta t$= 600 s distribution snapshots unitized scale flood perspectives, symmetric half $n = 3\ 8 \times 1 \times 16$ VFDL, Ra ≡ E + 08, Pr = 0.71, Re = 11,868: view to the symmetry plane, (a) $|\overline{\mathbf{u}}^h(\mathbf{x}, \mathrm{n}\Delta t)| \Rightarrow \overline{u}^h(|x| = \delta/2, y, z, \mathrm{n}\Delta t)$: view to the endwall, (b) $\overline{u}^h(|x| = \pm\delta/2, y, z, \mathrm{n}\Delta t)$, (c) $O(\delta^2)$ tensor diagonal resolution $\tau^h_{22}(\mathbf{x}, \mathrm{n}\Delta t)$.

Clearly evident therein is the transition of $O(\delta^2)$ tensor $\tau^h_{22}(\mathbf{x}, \mathrm{n}\Delta t)$ nil significance at initiation of vertical wall to uniform until approximately mid-elevation, whereupon it transitions abruptly to *isolated* $O(\delta)$ scale features with non-D magnitude ranging from 0.15 (light blue) to 0.25 (green) in a nil significance background as a weak function of z. The $\tau^h_{22}(\mathbf{x}, \mathrm{n}\Delta t)$ wall-adjacent distribution then transitions to *non-existent* (!) throughout the terminal ~15% of wall jet vertical surface travel.

The just visible $O(h{\leftrightarrow}\delta)$ scale non-D extrema at 0.50 (red) populate the reach of the VFDL adiabatic endwall vertical corner bounded by $x = 0 + O(\delta)$, $2 \le y < 4$, $0 \le z \le 2$. Elsewhere, throughout the VFDL, the appearance of $\tau^h_{22}(\mathbf{x}, \mathrm{n}\Delta t)$ is *chaotic* with exceptions being the local minor magnitude (dark to light blue) $O(\delta)$ scale features and the select $O(\mathrm{L})$ scale tubular structures.

The global synopsis of FaNS $O(1;\delta^2)$ thermal member distributions in Figure 12.3. compares horizontal plane integer elevation-sequenced color floods of the $\overline{\Theta}^h(\mathbf{x}, \mathrm{n}\Delta t) \Rightarrow \overline{\Theta}^h(x = \pm\delta/2, y = 1 + \delta/2, 2, .., 7, 8 - \delta/2,\ z, \mathrm{n}\Delta t)$ distribution with the thermal vector vertical resolution $v_2^{\Theta h}(\mathbf{x}, \mathrm{n}\Delta t) \ \Rightarrow\ (\overline{\Theta u'_2} + \overline{\Theta' \bar{u}_2})^h(x = \pm\delta/2, y, z, \mathrm{n}\Delta t)$ extremum significance color flood, viewed to the endwall, hot ΔT DBC on the left. The regularity of the $\overline{\Theta}^h(x = \pm\delta/2, y = \ldots,\ z, \mathrm{n}\Delta t)$ horizontal planar distributions quantifies that the VFDL core is "stair-stepped" isothermal. The "ripples" in the isotherm distributions adjacent to the upper left and lower right hot/cold ΔT DBC mid-elevations are due to the 3-D velocity vector $\overline{\mathbf{u}}^h(\mathbf{x}, n\Delta t)$ distributions present in the VFDL endwall-sidewall juncture vertical corners, recall Figure 11.11.

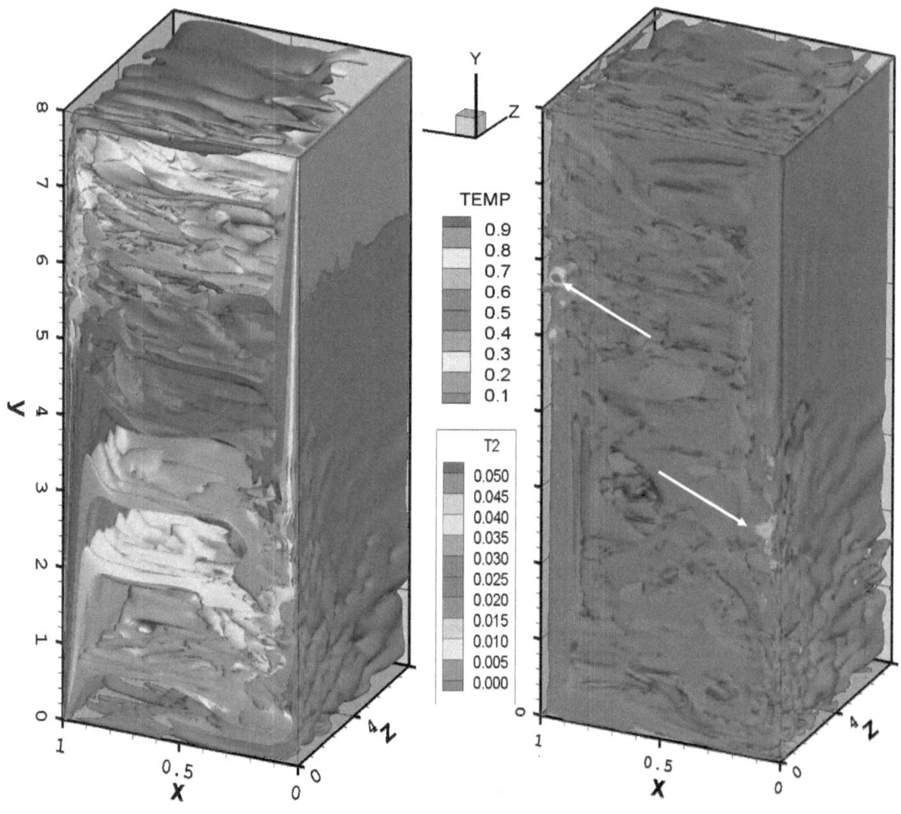

a) $\overline{\Theta}^h(x=\pm\delta/2, y=\ldots, z, n\Delta t)$ b) $\left(\overline{\overline{\Theta} u_2'} + \overline{\Theta' \overline{u}_2}\right)^h (x=\pm\delta/2, y, z, n\Delta t)$

Figure 12.3: Unitized scale global perspective of FaNS theory GWS^h algorithm $O(1)$ temperature and extremum significance $O(\delta^2)$ thermal vector nΔt= 600 s snapshots, view to adiabatic endwall, symmetric half $n = 3$, $8 \times 1 \times 16$ VFDL, Ra ≡ E + 08, Pr = 0.71, Re = 11,868: (a) $\overline{\Theta}^h(x = \pm\delta/2, y = 1 + \delta/2, ,2,.,7, 8 - \delta/2, z, n\Delta t)$, (b) $(\overline{\overline{\Theta} u'_2 + \overline{\Theta' \bar{u}_2}})^h (x = \pm\delta/2, y, z, n\Delta t)$.

The regularity of $O(\delta^2)$ thermal vector resolution $(\overline{\overline{\Theta} u'_2 + \overline{\Theta' \bar{u}_2}})^h (x = \pm\delta/2, y, z, n\Delta t)$ distribution mirrors that of $O(1)$ temperature in the reach of wall jet acceleration to VFDL nominal mid-elevation. It then transitions to z-dependent disorganized at nil significance (dark blue). Mirror-symmetric $O(\delta)$ scale 0.03 magnitude (light green) features are present in both VFDL vertical corners at ~75% wall jet vertical travel. Their 0.05 extrema (red) are $O(h)$ scale entities barely visible in the VFDL endwall vertical corners at vertical elevations of profile transition.

To summarize, the key *physics of fluids* observables in these global synoptic snapshots of ∀ Re FaNS theory $O(1{:}\delta^2)$PDE:AD-BCE^h-NBC^h-DBC^h:$O(\delta^3)$SGS system GWS^h + θTS FE $p = 1$ basis algorithm/code state variable unsteady periodic distribution perspective nΔt = 600 s snapshots at Re = 11,868 include:

- $O(1;\delta^2)$ state variable distributions quantify long-time stable persistence of a fully three-dimensional, unsteady, multi-scale space-filtered thermal-velocity vector field.
- ΔT DBC vertical surface attached accelerating wall jet distributions transition to disorganized just past mid-elevations as a weak function of distance from the endwall.
- Wall jet distribution regularity transitioning to disorganized initiates modest significance, local multi-scale $\bar{u}^h(x, y, z, \mathrm{n}\Delta t)$ features that mature to penetration of the VFDL core.
- Wall jet velocity disorganization coincides with wall-normal thermal plume ejections.
- Extremum significance $O(\delta^2)$ tensor diagonal resolution $\tau_{22}^h(\mathbf{x}, \mathrm{n}\Delta t)$ distribution possesses regularity *only* within the reach of wall jet velocity acceleration.
- Elsewhere, the significance of $\tau_{22}^h(x, \mathrm{n}\Delta t)$ distribution ranges from nil magnitude, disorganized to chaotic, except for isolated $O(\delta)$ scale bursts of nominal to extremum significance (red).
- $O(1)$ temperature distribution regularity persists throughout wall jet acceleration, which then transitions to disorganized near mid-elevations of the VFDL.
- VFDL central core temperature is essentially planar isothermal, due to the ΔT DBC-induced vertical gradient prior to z-dependent unsteady disruptions of the wall attached jet velocity.
- Extremum significance $O(\delta^2)$ thermal vector $v_2^{\Theta h}(\mathbf{x}, \mathrm{n}\Delta t)$ distribution regularity matches that of dominant significance $\tau_{22}^h(\mathbf{x}, \mathrm{n}\Delta t)$ during wall jet velocity acceleration.
- Elsewhere $v_2^{\Theta h}(\mathbf{x}, \mathrm{n}\Delta t)$ distribution appearance is disorganized, not chaotic, with isolated $O(\delta)$ scale extrema populating the vertical corners of the VFDL.
- That $O(\delta^2)$ state variable distributions possess nil magnitude remote from the extrema quantifies their influence on the unresolved scale $O(1)$ predictions is negligible.
- $O(1)$ state variable distributions are universally *monotone* to iterative convergence digit.
- $O(\delta^2)$ state variable distributions are *monotone* only in regions of regularity.
- $O(1;\delta^2)$ distribution monotonicity *validates* that the SFS tensor/vector $O(\delta^3)$ constraint does not compromise the resolved scale $O(h^2)$ TE annihilation functionality.
- ∀ Re FaNS pertinent $O(h^4,\delta^4)$PDE:AD-BCE^h-NBC^h:$O(h^2,\delta^4)$PDE:AD-DBC^h:$O(\delta^3)$SFS functional GWS^h + θTS FE $p = 1$ basis, implicitly coupled Newton Jacobian convergent algorithm/code is *validated* long time stable totally *absent* artificial stabilization.

12.3 Validation, ∀ Re FaNS theory $O(1; \delta^2)$ PDE:AD-BCE-NBC-DBC: $O(\delta^3)$ SGS algorithm prediction of laminar BL profile $O(1)$ velocity transition to turbulent

The complexity of the velocity vector distributions existent over the terminal quarter elevations of the VFDL symmetry plane is visualized in the FaNS $O(1)$ velocity $\mathrm{n}\Delta t = 600$ s snapshot true-scale perspective (Figure 12.4a). The symmetry plane distribution of $\overline{\mathbf{u}}^h(x, y, z = 8, \mathrm{n}\Delta t)$ non-D vector superposed on the corresponding $O(1)$ $\overline{\Theta}^h(x, y, z = 8, \mathrm{n}\Delta t)$ distribution color flood are quantitatively detailed in unitized scale (Figure 12.4b).

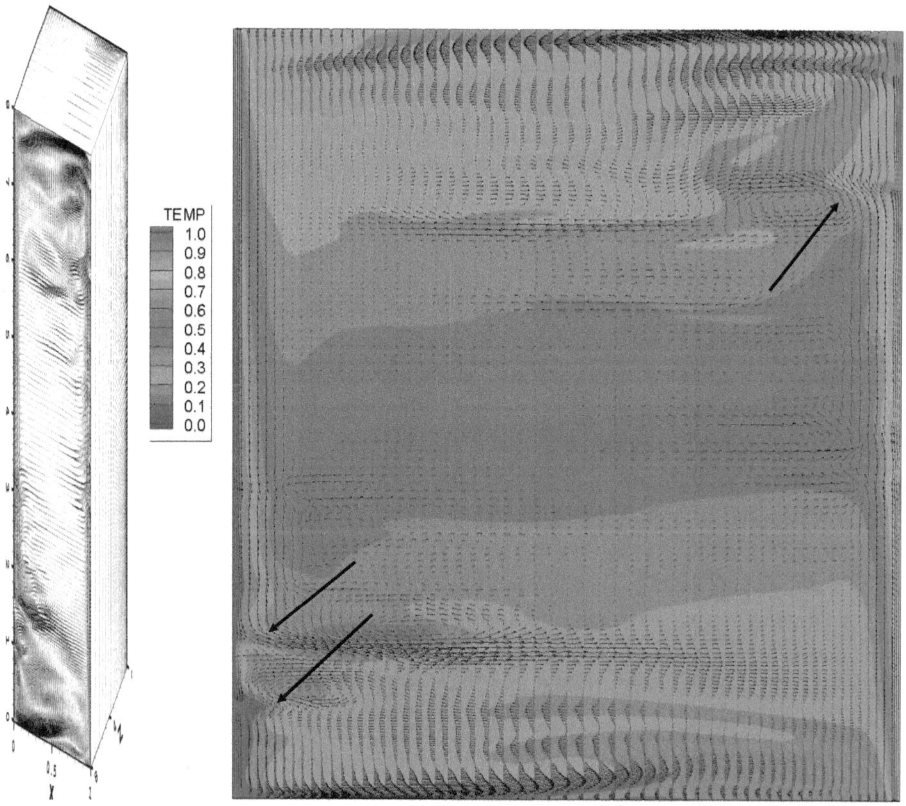

a) $\overline{\mathbf{u}}^h(x, y, z=8, \mathrm{n}\Delta t)$ true scale

b) $\overline{\Theta}^h(x, y, z=8, \mathrm{n}\Delta t)$ with $\overline{\mathbf{u}}^h(x, y, z=8, \mathrm{n}\Delta t)$

Figure 12.4: FaNS theory algorithm/code generated $O(1)$ state variable symmetry plane $\mathrm{n}\Delta t = 600$ s snapshot distributions, symmetric half $n = 3$, $8 \times 1 \times 16$ VFDL, Ra = E + 08, Pr = 0.71, Re = 11,868: (a) symmetry plane surface $\overline{\mathbf{u}}^h(x, y, z = 8, \mathrm{n}\Delta t)$ true scale 8×1 perspective, (b) $\overline{\Theta}^h(x, y, z = 8, \mathrm{n}\Delta t)$ unitized scale flood with $\overline{\mathbf{u}}^h(x, y, z = 8, \mathrm{n}\Delta t)$, superposed arrows denote thermal plume ejections.

Following VFDL horizontal surface velocity redirections to the vertical, the "freestream surfaces" of the thus-induced attached wall jet velocity distributions are smooth to 40±% vertical travel, whereafter just visible undulations appear in each of their freestream surfaces. These are the precursors to z-axis vortex embeddings into each wall jet which, when mature, generate the wall-normal major magnitude thermal plume ejections into the VFDL core, each of which is denoted by an arrow in Figure 12.4b.

Thereafter, horizontal surface encounter-induced redirection of each wall jet vertical traverse generates VFDL horizontal surface-adjacent major magnitude velocity distributions, plus a range of nearby recirculation features (Figure 12.4b). Net quantified is that FaNS theory predicted strictly *monotone* $O(1)$ state variable DOF distributions on the $n = 3$, $8 \times 1 \times 16$ symmetric half VFDL symmetry plane, exhibit the wall-dominated unsteady, multi-scale thermal-velocity vector distribution postulated for the $n = 2$, 8×1 thermal cavity, [1].

Piecewise linear interpolation of each nΔt= 600 s snapshot of wall jet velocity $\overline{\mathbf{u}}^h \bullet \widehat{\mathbf{j}} = \overline{v}^h(0.8 \leq x \leq 1, y = 1, 3, 5, 7, z = 8, \mathrm{n}\Delta t)$ DOF $\{\mathrm{V}(\ldots, z = 8, \mathrm{n}\Delta t)\}$ distribution on the symmetry plane is overlaid in Figure 12.5, hot ΔT DBC on the adjacent surface. Comparing the vertical surface encounter redirected first profile $\overline{v}^h(\ldots, y = 1, \ldots, \mathrm{n}\Delta t)$ DOF interpolation with that of the next "downstream" direction $\overline{v}^h(\ldots, y = 3, \ldots, \mathrm{n}\Delta t)$ DOF interpolation quantifies the significant acceleration registered in the wall jet approach to VFDL vertical surface mid-elevations. Note that the negative DOF dominant in the $\overline{v}^h(\ldots, y = 1, \ldots, \mathrm{n}\Delta t)$ DOF distribution are totally absent in the $\overline{v}^h(\ldots, y = 3, \ldots, \mathrm{n}\Delta t)$ DOF interpolation.

The next "downstream" $\overline{v}^h(\ldots, y = 5, \ldots, \mathrm{n}\Delta t)$ DOF interpolation quantifies significant wall jet deceleration, which is the precursor to wall jet separation at VFDL elevation $y \approx 6.5$, (the right arrow in Figure 12.4b. The DOF populating the reattached wall jet velocity $\overline{v}^h(\ldots, y = 7, \ldots, \mathrm{n}\Delta t)$ distribution interpolation are uniformly non-negative.

Rigorous *physics of fluids* characterization of wall-attached velocity vector distributions accrues to graphing of the DOF data in the boundary layer (BL) *similarity coordinate* system u^+, $\log y^+$ with definitions $u^+(y) \equiv \overline{v}^h(y)/u_\tau$, $u_\tau \equiv \sqrt{\tau_W/\rho}$, τ_W the wall shear stress for density ρ and the wall-distant Reynolds number $y^+ \equiv u_\tau y/\upsilon$ and fluid kinematic viscosity υ. The key attribute of similarity scaling is that $u^+ = y^+$ *exactly* correlates all laminar *and* turbulent profile BL distributions within the reach of the directly wall-adjacent, viscosity-dominated BL sublayer span $0 < y^+ \leq 6$ (Figure 10.3a).

The linear correlation $u^+ = y^+$ remains valid for Re-dependent laminar profile BL velocity to well past $y^+ = 6$, the abscissa at which, for sufficiently large Re, a laminar profile velocity initiates *transition* to a turbulent profile over the BL sublayer span $6 < y^+ \leq 30$. Fully turbulent incompressible BL velocity distributions are then universally *log–law* similarity coordinates correlated on $30 < y^+ \leq 800$ (Figure 10.3a).

On the symmetry plane, the BL sublayer reach $0 < y^+ \leq 30$ of FaNS theory $O(1;\delta^2)$ PDE: AD:SGS system generated $O(1)$ $\overline{\mathbf{u}}^h \bullet \widehat{\mathbf{j}} = \overline{v}^h(x, y, z = 8, \mathrm{n}\Delta t)$ hot ΔT DBC surface adjacent $\{\mathrm{V}^+(0.8 \leq x \leq 1, y = 1, 3, 5, 7, z = 8, \mathrm{n}\Delta t)\}$ DOF distribution nΔt = 600 s snapshots are

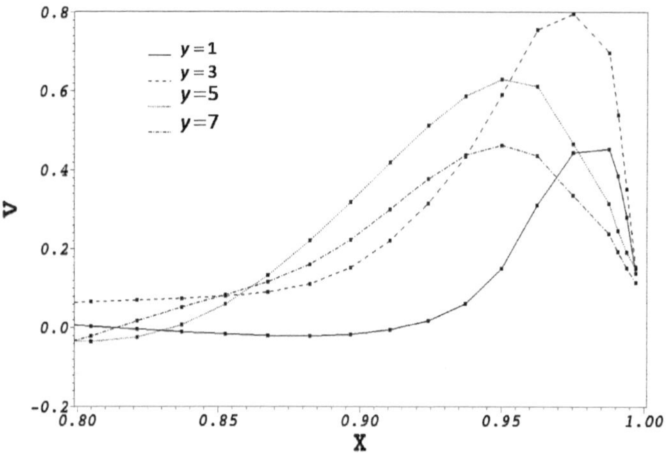

Figure 12.5: FaNS theory algorithm/code-generated VFDL symmetry plane wall jet velocity, $\bar{u}^h \bullet \hat{j} = v^h(0.8 \le x \le 1, y = 1, 3, 5, 7, z = 8, n\Delta t)n\Delta t = 600$ s snapshot DOF $\{V(\ldots, z = 8, n\Delta t)\}$ DOF piecewise linear interpolations at indicated elevations, hot ΔT DBC surface adjacent, Ra = E + 08, Pr = 0.71, Re = 11,868.

the open symbols in similarity coordinates scaled Figure 12.6a–d. As x is the VFDL wall normal coordinate, x+ now replaces traditional y+ in the following discussion.

Net of the rigorous *uniform* measure δ requirement, the similarity coordinate of each $O(1)$ velocity wall node DOF is finite at x+ < 0.2. That each $O(1)$ velocity wall node DOF abscissa is thus bounded *quantifies* that GWS^h $O(1)$PDE:AD-BCE^h-NBC^h algorithm functionality lies well within the viscous-dominated BL sublayer reach $y^+ \le 6$. Resultant is that the DOF for each $O(1)$ velocity wall node is modestly non-zero, specifically, $V^+(x = 0 + x^+, y = 1,3,5,7, z = 8, n\Delta t) \le 2 \pm 1$. The NS velocity DOF distribution wall-normal local node refinement, Figure 10.4, is thus validated for AD:BCE algorithm prediction of the $O(1)$ velocity non-homogeneous Neumann BC.

The comparative *validation* BL sublayer profiles graphed in Figure 12.6 are the laminar Blasius BL no-slip BC correlation $u^+ = y^+$ for Re = 12,000, the dashed line labeled V+ = x+, and the $Re_\tau = 180$ hard turbulent validation BL no-slip BC distribution, the solid line labeled V+(Wilcox). The DOF distributions for wall jet accelerating velocity $\{V^+(\ldots, y = 1, \ldots, n\Delta t)\}$ and $\{V^+(\ldots, y = 3, \ldots, n\Delta t)\}$ distributions, labeled V+(y = 1) and V+(y = 3) in Figure 12.6a and b, lie curvilinear parallel to the universal correlation V+ = x+. The continuation DOF in each of these distributions then *monotonically* approach V+ = x+ and each terminal DOF *exactly overlies laminar* Blasius at x+ = 10 ±, a coordinate located beyond profile correlation initial alteration to turbulent at $y^+ = 6$.

Prior to wall jet separation at VFDL elevation $y \approx 6.5$, the $\Delta t = 600$ s wall jet velocity $\{V^+(\ldots, y = 5, \ldots, n\Delta t)\}$ DOF distribution, labeled V+(y = 5) in Figure 12.6c, departs Blasius V+ = x+ at x+ ≈ 9 ±. Of key *physics of fluids* CFD prediction *significance*, the V^+ DOF continuation distribution at similarity coordinates x+=8, 11, and 15 lies *exactly*

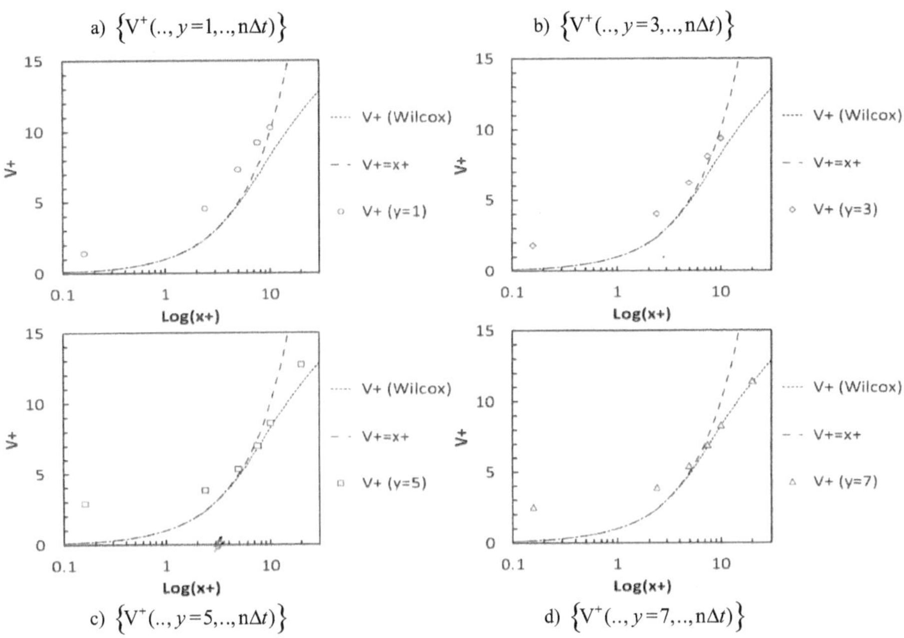

Figure 12.6: FaNS theory algorithm/code symmetry plane $O(1)$ wall-parallel jet velocity $\overline{v^+}^h$ $(x, y, z = 8, n\Delta t)$ similarity coordinate $\{V^+ (0.8 \le x \le 1, y = 1,3,5,7, z = 8, n\Delta t)\}$ DOF distributions, periodic nΔt= 600 s snapshot, hot ΔT BC surface adjacent, 8 × 1 × 16 symmetric half VFDL, Ra = E + 08, Pr = 0.71, Re = 11,868: (a) $\{V^+ (\ldots, y = 1, \ldots, n\Delta t)\}$, (b) $\{V^+ (\ldots, y = 3, \ldots, n\Delta t)\}$, (c) $\{V^+ (\ldots, y = 5, \ldots, n\Delta t)\}$, (d) $\{V^+ (\ldots, y = 7, \ldots, n\Delta t)\}$.

in between laminar V+ = x+ and transitioned-to-turbulent V+(Wilcox) on 8 ≤ x+ < 20. Thereby, the similarity coordinates characterization of $\{V^+ (\ldots, y = 5, \ldots, n\Delta t)\}$ DOF distributions *validates* that ∀ Re FaNS theory $O(1)$PDE:$O(\delta^4)$AD:BCE:NBC$O(\delta^4)$ PDEAD-DBC:$O(\delta^3)$SGS algorithm has predicted that the wall-attached laminar profile $O(1)$ velocity is undergoing *insipient transition* (!) to turbulent.

Following separation centered at VFDL wall elevation $y \approx 6.5$, the DOF distribution of the reattached wall jet velocity$\{V^+ (\ldots, y = 7, \ldots, n\Delta t)\}$, labeled V+($y$ = 7) in Figure 12.6d, departs from V+ = x+ at the laminar sublayer limit x+ = 6, and the profile continuation DOF at coordinates x+ = 8, 11, and 20 *exactly overlie* (!) the turbulent validation profile V+(Wilcox). Thereby, analytical ∀ Re FaNS theory CFD is thus *validated* to have predicted that the *reattached* $O(1)$ velocity DOF distribution $\{V^+ (\ldots, y = 7, \ldots, n\Delta t)\}$ profile following separation has transitioned to turbulent on 8 ≤ x+ ≤ 20.

To summarize, at Re = 11,868, the FaNS theory $O(1)$ resolved scale $O(h^4,\delta^4)$ PDE:AD-BCE^h-NBC^h:$O(\delta^2)$ resolved-unresolved scale interaction $O(h^2,\delta^4)$ PDE:AD-DBC^h:$O(\delta^3)$ SGS tensor/vector GWS^h + θTS FE $p = 1$ basis algorithm/Newton code is *validated* to have *rigorous mathematics* predicted wall-attached space-filtered $O(1)$ velocity laminar profile insipient transition, separation and turbulent profile reattachment, and finally relaminarization as VFDL unsteady periodicity enabled. This constitutes the truly transformational ***physics of fluids* CFD *mathematics*** resolution of the century-long persistent *stagnation* in legacy CFD practice.

> ". . . laminar–turbulent boundary layer transition, . . ."
>
> ". . . single most *critical* . . turbulent . . transition, . . separation . . . . "

This occurrence, coupled with the prior chapter's detailed thermal-velocity vector predictions ∀Re < 4,000, quantitatively ***validates*** that the FaNS theory PDE:AD:BCE:NBC:DBC:SGS system's continuous weak formulation algorithm indeed exhibits ∀Re pertinence! This *functionally practical,* well-posed, bounded domain CFD advancement is *entirely reliant* on the $O(1)$ AD-BCE^h-NBC^h and $O(\delta^2)$ AD-DBC^h GWS^h algorithm resolutions, as required by rigorous *error freed* spatial filtering convolution with a *uniform* measure δ Gaussian filter.

And finally, absolutely critical to *long time stable* $O(1)$ FaNS $O(h^4,m^4,\Delta t^3)$ PDE:AD-BCE^h-NBC^h:$O(\delta^2)$ variable $O(h^2,\delta^4)$PDE:AD-DBC^h:$O(\delta^3)$ SGS functional GWS^h algorithm/code is matrix differential calculus enabled identification of *all* nonlinear element matrix contributions to the *robustly* convergent Newton iteration algorithm Jacobian.

12.4 Validation, $O(\delta^3)$ SFS tensor/vector $O(h^2)$ TEA functionality

The second-order rational Pade' polynomial interpolation of the Gaussian filter Fourier transform inverse predicts the $O(\delta^3)$ significance constraint for SFS tensor/vectors. Validation that this scaling does not compromise $O(1)$ FaNS $O(h^4,m^4,\Delta t^3)$PDE $O(h^2)$ TEA functionality accrues to analysis of the symmetry plane separation bubble-thermal ejection event centered at elevation $y \approx 6.5$, denoted by the red arrow in Figure 12.7a. This highly dynamic periodic thermal-velocity feature is bounded "upstream" by the wall jet velocity laminar profile incipient transition (Figure 12.6c) and "downstream" by the reattached turbulent profile wall jet velocity (Figure 12.6d).

That VFDL symmetry plane $y = 6.5$ centric $\overline{\Theta}^h(1.0 \geq x > 0.87,\ 6.0 \leq y < 6.9,\ z = 8,\ n\Delta t)$ distribution unitized scale $n\Delta t = 600$ s snapshot color flood with $\overline{\mathbf{u}}^h(x, y, z = 8, n\Delta t) \times \hat{\mathrm{k}}$ vector overlay (Figure 12.7b) are both resolutely *monotone* validates that $O(\delta^3)$ scaling of the SFS tensor distribution and thermal vector exerts *nil impact* on $O(1)$ state variable $O(h^2)$ TEA functionality.

Appreciation of the genuine scale of this highly dynamic fundamental *physics of fluids* event accrues to graphing the $\overline{\Theta}^h(1.0 \geq x > 0.87,\ 6.0 \leq y < 6.9,\ z = 8,\ n\Delta t)$ distribution in true scale (Figure 12.7c). The wall-normal span of this separation/thermal burst event adjacent to the hot ΔT DBC surface is about triple that of the onset transitioning laminar profile wall jet BL thickness. The BL thickness of the reattached, transitioned-to-turbulent profile wall jet is nominally double that of the onset. The axial span of

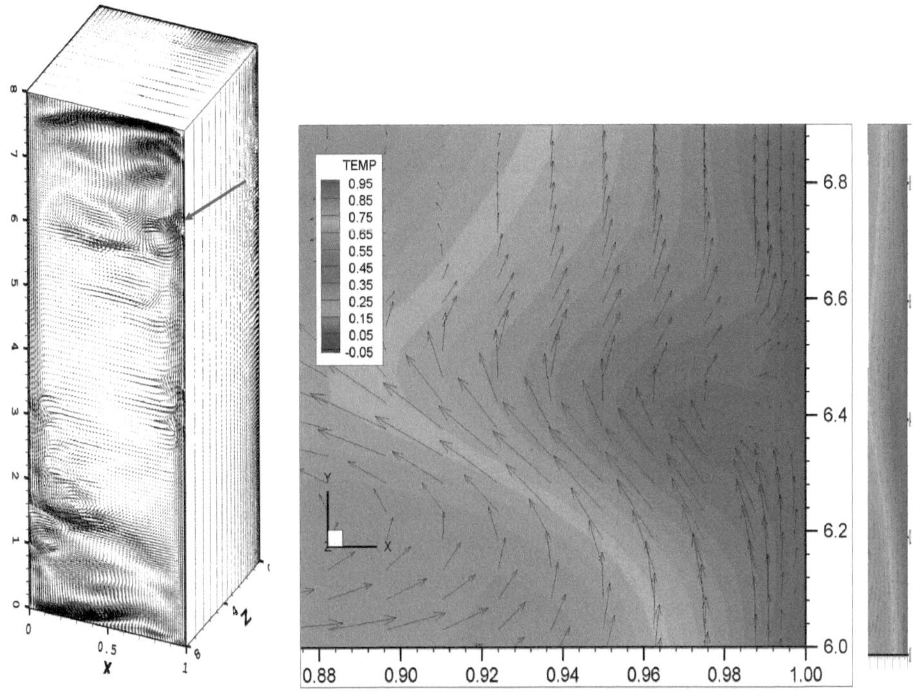

a) perspective b) $\overline{\Theta}^h(x, y, z=8, \mathrm{n}\Delta t)$ with $\overline{\mathbf{u}}^h(x, y, z=8, \mathrm{n}\Delta t)\times\hat{k}$ c) true scale

Figure 12.7: FaNS theory $O(1)$ distributions in the separation bubble-thermal ejection reach centered at $y = 6.5$, symmetry plane hot ΔT BC surface adjacent periodic $\mathrm{n}\Delta t = 600$ s snapshot, symmetric half $n = 3$, $8 \times 1 \times 16$ VFDL, Ra = E + 08, Pr = 0.71, Re = 11,868: (a) $\overline{\mathbf{u}}^h(x, y, z = 8, \mathrm{n}\Delta t) \times \hat{k}$ symmetry plane perspective, (b) $\overline{\Theta}^h(x, y, z = 8, \mathrm{n}\Delta t)$ unitized scale flood with $\overline{\mathbf{u}}^h(x, y, z = 8, \mathrm{n}\Delta t) \times \hat{k}$ overlay, (c) $\overline{\Theta}^h(1.0 \geq x > 0.87, 6.0 \leq y < 6.9, z = 8, \mathrm{n}\Delta t)$ true scale.

the separation bubble exceeds that of the onset wall jet BL thickness by an order of magnitude.

To summarize, the Re = 11,868 data presentations in Figures 12.6 and 12.7 collectively *validate every* critical aspect of the *physics of fluids* prediction fidelity resident in analytical FaNS theory *error freed* CFD *mathematics* $O(h^4,\delta^4,h^2)$ PDE system algorithm/code resolutely *monotone* wall-attached space-filtered resolved-scale velocity laminar BL profile transition to turbulent, then relaminarization. That the UT CFD Lab team failed to recognize this in your first author's third year (!) of retirement, hence thoroughly data-mine this treasure trove, is a *physics of fluids* diagnostics opportunity truly fumbled!

12.5 Validation, FaNS state variable $O(1;\delta^2;\delta^3)$ significance scaling, TEA non-positive definite SFS tensor/vector, VVUQ pertinence

The nΔt = 600 s snapshots of the Cartesian resolution of FaNS theory $O(\delta^2;\delta^3)$ state variable members possessing dominant significance in the VFDL symmetry plane separation-thermal ejection feature, each with $O(1)$ velocity vector overlay, are unitized scale compared in Figure 12.8a, b, and d, each with pertinent color bar. Also graphed is the dominant significance $D^m\overline{\mathbf{P}}$ convection term shear resolution distribution (Figure 12.8c). That each $O(1;\delta^2;\delta^3)$ significance member resolution is resolutely *monotone* completes *validation* that the FaNS theory SFS tensor/vector $O(\delta^3)$ constraint does not compromise its $O(h^2)$ TEA functionality.

The key observables in these separation region $O(1;\delta^2;\delta^3)$ state variable resolutions, with non-D and dimensional (D) significance notations include:

- $O(1)$ symmetry plane velocity vector $\overline{\mathbf{u}}^h \times \hat{\mathrm{k}}$ significance is universally $O(1)$ non-D, D.
- $O(1)$ $\overline{\Theta}^h$significance is $O(10^0)$ non-D, $O(10^2)$ D (Figure 12.7b).
- $O(\delta^2)$ interaction thermal vector vertical resolution $\left(\overline{\bar{u}_2\Theta' + u'_2\bar{\Theta}}\right)^h$ significance is $O(<10^2)$ non-D, $O(<10^{-1})$ D, *except* for the extremum (dark blue) distribution at $O(10^0)$ D significance preceding the onset of wall jet insipient transition (Figure 12.8a).
- $O(\delta^2)$ interaction tensor shear resolution $\left(\overline{\bar{u}_1u'_2 + u'_1\bar{u}_2}\right)^h$significance is $O(10^{-3})$ non-D and D, at essentially nil magnitude (green) throughout Figure 12.8b.
- $O(1)$ $D^m\overline{\mathbf{P}}$ convection tensor shear resolution $\overline{u_1^h}\ \overline{u_2^h}$ significance is $O(10^{-1})$ non-D and D, *same sign* modest extrema (yellow) bound separation initiation (Figure 12.8c.
- $O(\delta^3)$ SFS tensor shear resolution $\left(\overline{u'_1 u'_2}\right)^h$ significance is $O(10^{-2})$ both non-D and D, except for the *opposite sign* $O(\delta)$ significance extrema (red, dark blue) bounding the initiation of separation (Figure 12.8d).

Absolutely critical to FaNS theory $O(1)$ state variable distribution validated monotonicity, also code long-time algebraic stability, is the TEA-theorized *non-positive definite* $O(\delta^3)$ SFS tensor distribution (Figure 12.8d). For $C_S(\delta) \equiv \delta/2$, it is clearly quantified therein that the opposite sign extrema (red, blue) in the $\left(\overline{u'_1 u'_2}\right)^h$ distribution, at $O(\delta) \gg O(\delta^3)$ significance (!), precisely bound the $O(1)$ velocity distribution at the separation bubble onset. Recall that the functionality of the TEA-theorized SFS tensor distribution with negative magnitude is *dispersive*, not diffusive (!).

That the significance of SFS tensor resolution $\left(\overline{u'_1 u'_2}\right)^h$ surrounding the separation event is the theorized $O(\delta^3)$ (light green) quantifies FaNS code on-the-fly realization that little need exists elsewhere for $O(h^2)$ TE annihilation, an attribute of TEA theory first documented in Chapter 6.4. These features belie the legacy LES practice assumption that the action of space-filtered unresolved scale entities is compensation for the omission of the *diffusive* impact of fluid viscosity.

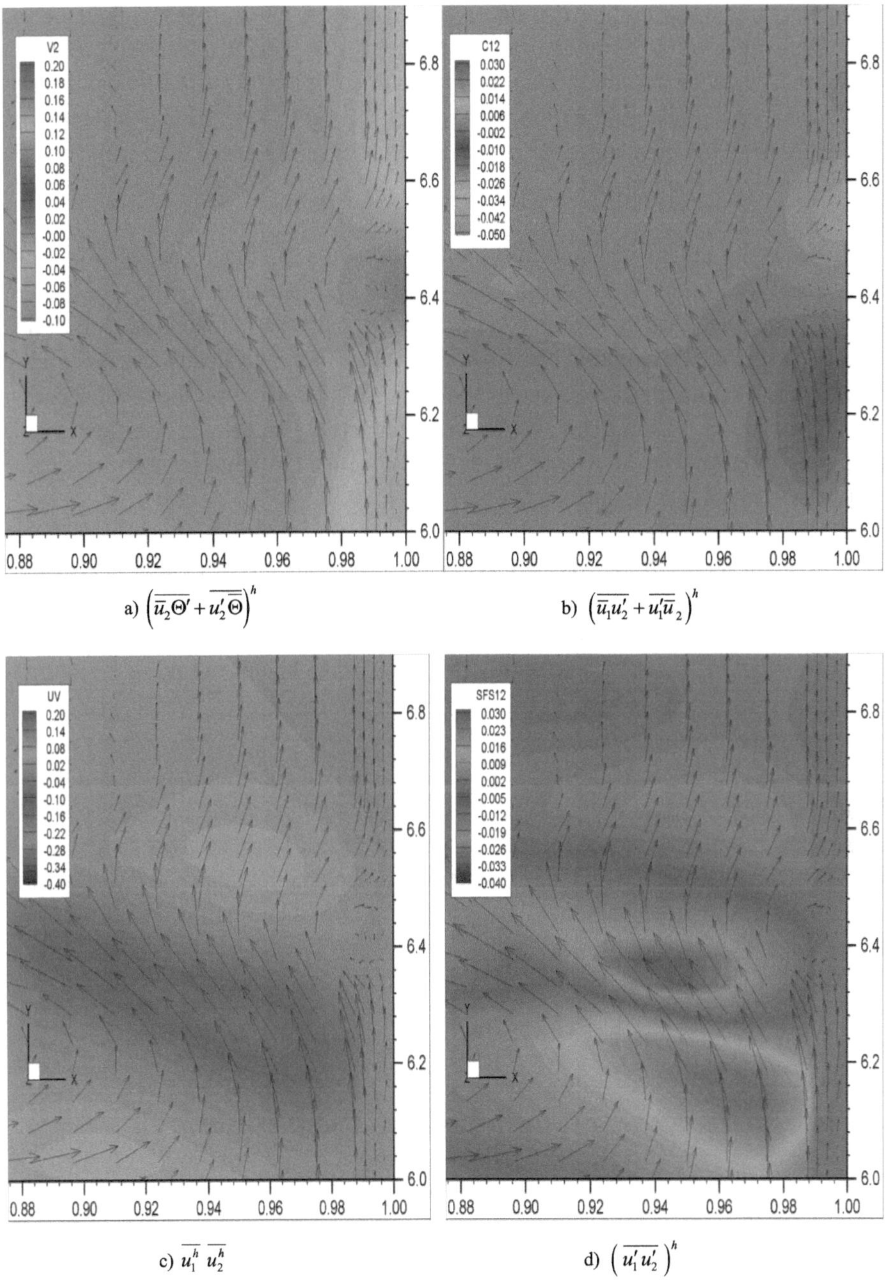

Figure 12.8: FaNS-predicted symmetry plane hot ΔT BC surface adjacent flowfield separation features, significant $O(\delta^2;\delta^3)$ state variable member distributions with $O(1)$ velocity vector overlay, $n\Delta t = 600$ s, symmetric half $n = 3$ 8 × 1 × 16 VFDL, Ra = E + 08, Pr = 0.71, Re = 11,868: (a) $O(\delta^2)$ vector resolution$\left(\overline{\bar{u}_2\Theta'} + \overline{u'_2\bar{\Theta}}\right)^h$, (b) $O(\delta^2)$ tensor resolution $\left(\overline{\bar{u}_1 u'_2} + \overline{u'_1\bar{u}_2}\right)^h$, (c) $O(1)$ velocity tensor product resolution $\overline{u_1^h}\ \overline{u_2^h}$, (d) $O(\delta^3)$ SFS tensor resolution $\left(\overline{u'_1 u'_2}\right)^h$.

The Gaussian uniform filter measure is $\delta \equiv 0.05$ non-D (~0.01 m, 0.0333 ft D); hence, the D significance of FaNS theory orderings are $O(\delta^2) \cong$ E–03 ft and $O(\delta^3) \cong$ 4E-05 ft. Within the wall jet separation-thermal plume symmetry plane feature, the significance of the $O(1)$ vector/scalar resolution $\{\overline{q}^h(\mathbf{x}, n\Delta t)\} \Rightarrow \{|\overline{u}_j^h|, \overline{T}^h\}^T$ global measures is $O\{\text{E}+00, \text{E}+02\}^T$ (D). The dominant significance $O(\delta^2)$ resolved-unresolved scale interaction state variable resolution $\{q^h_{\delta 2}(\mathbf{x}, n\Delta t)\} \Rightarrow \{\tau^h_{12}, v^h_2\}^T$ global measures is $O\{\text{E–03}, \text{~E–01}\}^T$ (D).

Thereby, the global measure ratios for FaNS $O(:\delta^2;1)$ state variable dominant significance resolution members in the thermal plume event at ~E–01/E + 02 and E–03/E + 00 are *each* precisely O(E–03), i.e., *identical* with FaNS-theorized $O(\delta^2)$. Thus validated in this highly dynamic *physics-of-fluids* transitioning velocity thermal plume feature is classic fluid mechanics perturbation theory identification that the FaNS theory state variable is scaled $O(1;\delta^2;\delta^3)$. This fundamental attribute of FaNS theory thus enables the sought VVUQ topic *uncertainty quantification* (UQ) that mesh resolution of the $O(1)$ state variable is *adequate* (!).

To close, legacy RaNS/LES standard practice for turbulent Prandtl number modeling is alteration of the **DP** eddy viscosity closure model by a scalar coefficient. Dominant FaNS $O(\delta^2)$ thermal vector resolution $v_2^{\Theta h} \equiv \left(\overline{\overline{u}_2\Theta' + u'_2\overline{\Theta}}\right)^h$ significance at $O(<10^{-1})$ D is one order larger than that for the dominant $O(\delta^2)$ tensor resolution $\tau^h_{12} \equiv (\overline{\overline{u}_1 u'_2 + u'_1 \overline{u}_2})^h$ at $O(10^{-2})$ D. Within the reach of the laminar profile wall jet transitioning onset to separation (Figure 12.8a), the significance of $v_2^{\Theta h} \equiv (\overline{\overline{u}_2\Theta' + u'_2\overline{\Theta}})^h$ is even one order larger at $O(10^0)$ D.

These FaNS theory execution identified significance disparities predict that legacy practice turbulent Prandtl number models will lack the fidelity required to accurately resolve the actions in dynamic thermal-fluid transport at large Re. This conjecture is equally pertinent to legacy turbulent Schmidt number modeling practices regarding the prediction of multi-species mass transport (compressible mixing), which is of specific pertinence to hypersonic aerothermodynamics or flowfields with chemical reactions.

12.6 Validation, ∀ Re FaNS theory $O(\delta^2)$ resolved-unresolved scale interaction tensor is both realizable and anisotropic

The diagonal Cartesian resolution of FaNS theory predicted $O(\delta^2)$ resolved-unresolved scale interaction tensor nΔt = 600 s snapshots are unitized scale perspective compared in Figure 12.9, view to the adiabatic endwall, hot ΔT DBC on the right. The range of D significance $(0.0 \le \tau^h_{22}(\mathbf{x}, \text{n}\Delta t) \le 0.50)$, $(0.0 \le \tau^h_{11}(\mathbf{x}, \text{n}\Delta t) \le 0.08)$, $(0.0 \le \tau^h_{33}(\mathbf{x}, \text{n}\Delta t) \le 0.01)$ quantifies that the tensor is indeed highly *anisotropic*. Furthermore, the fact that the DOF in these tensor resolutions are uniformly non-negative *validates* that the FaNS theory $O(\delta^2)$ tensor $\tau_{ij}^h \equiv (\overline{\bar{u}_i u'_j} + \overline{u'_i \bar{u}_j})^h$ is also *realizable*, a fundamental requirement of NS tensor field theory, Pope [3].

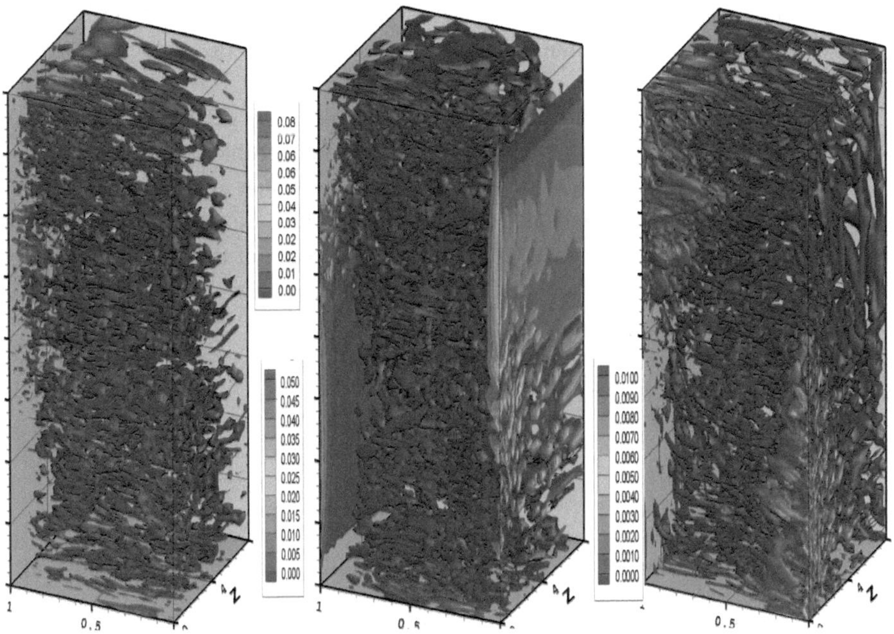

a) $(0.0 \le \tau_{11}^h(\mathbf{x},n\Delta t) \le 0.08)$ b) $(0.0 \le \tau_{22}^h(\mathbf{x},n\Delta t) \le 0.50)$ c) $(0.0 \le \tau_{33}^h(\mathbf{x},n\Delta t) \le 0.01)$

Figure 12.9: FaNS theory $O(\delta^2)$ resolved-unresolved scale interaction tensor diagonal Cartesian resolution $\tau_{jj}^h(\mathbf{x}, n\Delta t)$ distribution nΔt= 600 s snapshots unitized scale perspectives, view to adiabatic endwall, symmetric half n = 3, 8 × 1 × 16 VFDL, Ra = E + 08, Pr = 0.71, Re = 11,868: (a) $(0.0 \le \tau_{11}^h(\mathbf{x}, n\Delta t) \le 0.08)$, (b) $(0.0 \le \tau_{22}^h(\mathbf{x}, n\Delta t) \le 0.50)$, (c) $(0.0 \le \tau_{33}^h(\mathbf{x}, n\Delta t) \le 0.01)$.

12.7 Validation, AD algorithm non-homogeneous DBC for FaNS $O(\delta^2)$ state variable resolution PDE system

Prediction of the Cartesian resolution of FaNS theory space filtered $O(\delta^2)$ state variable involves a decoupled set of linear harmonic EBV PDEs sourced by the product of $O(1)$ velocity gradient with the pertinent $O(1)$ state variable member gradient. The fundamental attribute of linear EBV PDE solutions is dependence on *a priori* defined BCs on the entire bounding surface of the PDE domain Ω. FaNS theory resolution is the GWS^h:AD-DBC^h algorithm for prediction of the non-homogeneous Dirichlet BC for each DOF located at a bounding surface node $\mathbf{x}_b \in \partial\Omega$.

The dominant significance $O(\delta^2)$ tensor Cartesian diagonal resolution $\tau_{22}^h(\mathbf{x}, n\Delta t)$ distribution nΔt= 600 s snapshot is unitized scale perspective graphed in Figure 12.9b, view to the adiabatic endwall, cold ΔT DBC on the right. Its appearance is nil magnitude (dark blue), chaotic throughout the entire core of the VFDL, as well as on the horizontal, adiabatic endwall, and symmetry plane bounding surfaces. Its nil significance at the vertical surface initiation of the wall jet rapidly increases to a transversely uniform distribution at 0.30 ± D significance (green) throughout laminar profile wall jet acceleration to nominal mid-elevation of the VFDL surface as a weak function of z.

Thereafter, throughout the laminar profile insipient transition, and separation, the $\tau_{22}^{h}(\mathbf{x}, \mathrm{n}\Delta t)$ wall-adjacent distribution morphs to denumerable 0.25 ± D significance $O(\delta)$ scale entities, plus a half dozen isolated $O(h)$ scale entities at 0.5 D extrema (red) just visible in the VFDL vertical corner reach spanning $4\pm \geq y \geq 2.5\pm$. This resolution then transitions to *nil* significance (!) throughout the reach of the wall jet turbulent profile reattachment to its termination at the VFDL horizontal surface. This absence is best visualized at the lower right corner of the symmetry plane in Figure 12.9b.

The AD-DBC^h algorithm statement for prediction of $\tau_{22}^{h}(\mathbf{x}, \mathrm{n}\Delta t)$ DOF at nodes $\mathbf{x}_b \in \partial\Omega_{\hat{n}_k}$ on a vertical surface

$$\tau_{22}^{h}(\mathbf{x}_b \in \partial\Omega_{\hat{n}_k}) \equiv \left(g_\delta {}^* \tau_{22}^{\mathrm{NS}}\right) \Rightarrow \mathrm{T22}(t)\Big|_{\mathbf{x}_b \in \partial\Omega_{\hat{n}_k}} \equiv \frac{\delta^2}{2\gamma}\int_\Omega g_\delta(x_m - x_l)^* \left(\frac{\partial \overline{u_2^h}}{\partial x_1}\right)^2_{x_l, t} d\tau$$

confirms wall node T22(t) DOF magnitude is sourced by the square of the $O(1)$ velocity vertical resolution strain rate. The specific PDE source term $2\dfrac{\partial \bar{u}_2}{\partial x_k}\dfrac{\partial \bar{u}_2}{\partial x_k}$ is also the square of the velocity vertical resolution strain rate. The evolution of these $O(1)$ distributions energizes the transversely uniform 0.25 ± D significance of $\tau_{22}^{h}(\mathbf{x}, \mathrm{n}\Delta t)$ throughout the reach of the wall jet laminar profile velocity acceleration. Although 0.25 ± D significance exceeds $O(\delta^2)$ by *two orders* (!), the spatially *uniform* distribution of $\tau_{22}^{h}(\mathbf{x}, \mathrm{n}\Delta t)$ renders it of *nil* impact on prediction of laminar profile $O(1)$ velocity as the action of all $O(\delta^2)$ tensor resolutions in the $O(1)$ $\mathrm{D}^m\overline{\mathbf{P}}$ kinetic flux vector is *modulo* divergence.

The AD-DBC^h algorithm statement for prediction of the mid-significance $O(\delta^2)$ tensor diagonal resolution $\tau_{11}^{h}(\mathbf{x}, \mathrm{n}\Delta t)$ on a horizontal surface

$$\mathrm{T11}(t)\big|_{\mathbf{x}_b \in \partial\Omega_{\hat{n}_k}} \equiv \frac{\delta^2}{2\gamma}\int_\Omega g_\delta(x_m - x_l)^* \left(\frac{\partial \overline{u_1}^h}{\partial x_2}\right)^2_{x_l, t} d\tau$$

confirms the DOF at nodes $\mathbf{x}_b \in \partial\Omega_{\hat{n}_k}$ are similarly sourced by the wall jet streamwise strain rate squared, plus source term contribution totals $2\dfrac{\partial \bar{u}_1}{\partial x_k}\dfrac{\partial \bar{u}_1}{\partial x_k}$. That this distribution nΔt= 600 s snapshot exhibits *zero* (!) presence throughout the wall jet laminar profile velocity insipient transition is unitized scale perspective validated in Figure 12.9a, view to the adiabatic endwall. Its homogeneous chaotic nil magnitude distribution in the VFDL core matches that of $\tau_{22}^{h}(\mathbf{x}, \mathrm{n}\Delta t)$.

Its distribution distinction resides along the horizontal surface adjacent to the z-dependent progression of mid-magnitude 0.04 ± D (green) $O(\delta \leftrightarrow \mathrm{L})$ scale entities at the vertical wall jet termination at VFDL horizontal surfaces, best visualized on the upper surface of Figure 12.9a. Thereby, as with $\tau_{22}^h(\mathbf{x}, \mathrm{n}\Delta t)$, the uniformity of the $\tau_{11}^h(\mathbf{x}, \mathrm{n}\Delta t)$ distribution nΔt= 600 s snapshot assures that it exerts nil impact on $O(1)$ velocity wall jet acceleration on a horizontal surface. The exception to this statement can occur at the "edges" of these uniformly distributed $O(\delta \leftrightarrow \mathrm{L})$ scale horizontal surface features.

The nΔt= 600 s snapshot of the least significant tensor diagonal resolution $\tau_{33}^h(\mathbf{x}, \mathrm{n}\Delta t)$ distribution confirms it is nil magnitude, disorganized but not chaotic (Figure 12.9c). The exceptions are a range of $O(\mathrm{L})$ scale mid-magnitude 0.04 ± D (green) tubular "slivers" adjacent to horizontal surfaces, and an isolated 0.08 ± D (red) extremum $O(\delta)$ scale sliver. Similar magnitude isolated $O(\delta)$ scale features exist in the VFDL vertical corners at elevations coincident with wall jet separation on the symmetry plane.

Continuing, the uniformly less significant shear resolutions of the $O(\delta^2)$ $\tau_{ij}^h(\mathbf{x}, \mathrm{n}\Delta t)$ tensor distribution nΔt=600 s snapshots are unitized scale perspective compared in Figure 12.10, view to the adiabatic endwall. Each of these possesses the essentially identical *nil* significance, opposite sign (yellow to green) uniformly chaotic appearance except for the regions exhibiting total absence. Clearly visible distinctions include the $\tau_{12}^h(\mathbf{x}, \mathrm{n}\Delta t)$ regular distribution in the VFDL upper wall reach, also the elongated tubular structures in the distributions of $\tau_{13}^h(\mathbf{x}, \mathrm{n}\Delta t)$ and $\tau_{23}^h(\mathbf{x}, \mathrm{n}\Delta t)$ in the first half of laminar profile wall jet acceleration.

The Cartesian resolution of FaNS theory $O(\delta^2)$ thermal vector $v_j^{\Theta h}(\mathbf{x}, \mathrm{n}\Delta t)$ distribution nΔt= 600 s snapshots is unitized scale perspective compared in Figure 12.11, view to the adiabatic endwall, cold ΔT DBC at the right. Their essential nil significance matches that of the $O(\delta^2)$ tensor shear resolutions except for isolated $O(h \leftrightarrow \delta)$ scale local embeddings of extremum (red, yellow) significance. Also notable is that the dominant significance $v_2^h(\mathbf{x}, \mathrm{n}\Delta t)$ resolution distribution (Figure 12.11b), mimics the regularity of $\tau_{22}^h(\mathbf{x}, \mathrm{n}\Delta t)$ within the reach of the wall jet laminar profile velocity acceleration to VFDL mid-elevations, albeit at genuinely nil significance.

The AD-DBC^h algorithm statement for convolution of the $O(\delta^2)$ thermal vector resolution $v_2^{\Theta h}(\mathbf{x}, \mathrm{n}\Delta t)$ reduced to a vertical surface is

$$\mathrm{V2}^{\Theta}(t)|_{\mathbf{x}_b \in \partial\Omega_{\hat{n}_k}} \approx \frac{\delta^2}{2\gamma} \int_{\Omega^+} g_\delta(x_m - x_l)^* \left(\frac{\partial \overline{\Theta}^h}{\partial x_1} \frac{\partial \overline{u}_2^h}{\partial x_1} \right) \mathrm{d}\tau = 0$$

Thus predicted is that the magnitude of $v_2^{\Theta h}(\mathbf{x}, \mathrm{n}\Delta t)$ DOF at nodes $\mathbf{x}_b \in \partial\Omega$ on a vertical surface is zero to $O(\mathrm{Re}^{-1/2})$ which is clearly validated in Figure 12.11b. Therein, $O(h)$ scale features with $O(<\delta)$ D significance (green) exist at the VFDL vertical corner mid-elevation. The local $O(\delta^2)$ significance extremum (red) is the barely visible $O(h)$ scale "dot" embedded in the mid-magnitude (green) $O(\delta)$ scale symmetry

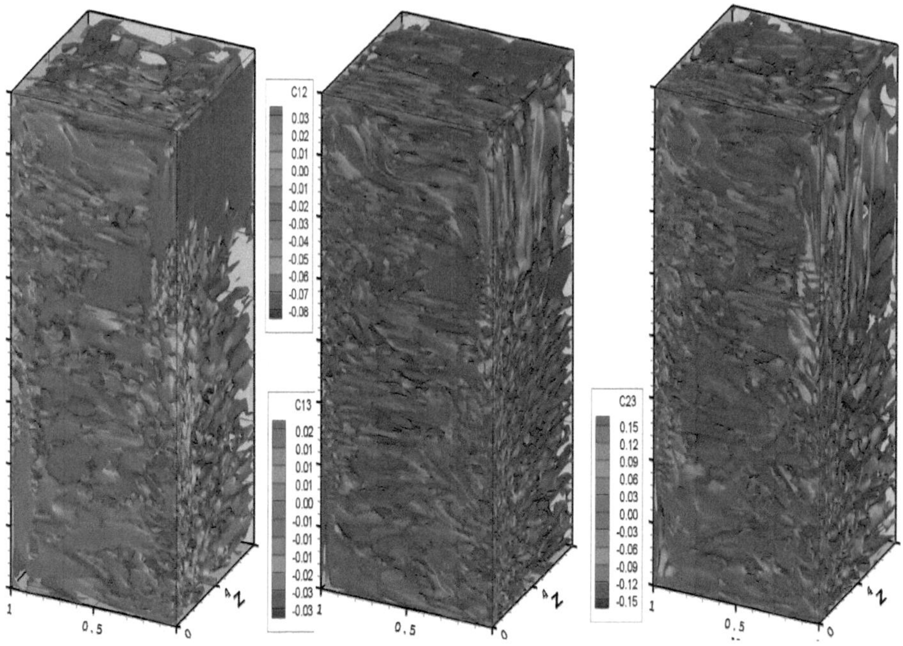

a) $\left(-0.08 \le \tau^h_{12}(\mathbf{x}, \mathrm{n}\Delta t) \le 0.03\right)$ b) $\left(-0.03 \le \tau^h_{13}(\mathbf{x}, \mathrm{n}\Delta t) \le 0.02\right)$ c) $\left(-0.15 \le \tau^h_{23}(\mathbf{x}, \mathrm{n}\Delta t) \le 0.15\right)$

Figure 12.10: FaNS theory algorithm/code $O(\delta^2)$ resolved-unresolved scale interaction tensor shear resolutions $\tau^h_{ij}(\mathbf{x}, \mathrm{n}\Delta t)$, unitized scale perspective, view to adiabatic endwall, nΔt= 600 s snapshot, symmetric half $n = 3$, 8 × 1 × 16 VFDL, Ra = E + 08, Re = 11,868, Pr = 0.71: (a) $\left(-0.08 \le \tau^h_{12}(\mathbf{x}, \mathrm{n}\Delta t) \le 0.03\right)$, (b) $\left(-0.03 \le \tau^h_{13}(\mathbf{x}, \mathrm{n}\Delta t) \le 0.02\right)$, and (c) $\left(-0.15 \le \tau^h_{23}(\mathbf{x}, \mathrm{n}\Delta t) \le 0.15\right)$.

plane feature at the VFDL elevation of the wall jet turbulent profile velocity reattachment.

The AD-DBC^h algorithm statement for convolution of the $O(\delta^2)$ thermal vector $\{v_1^{\Theta h}(\mathbf{x}, \mathrm{n}\Delta t)\}$ on a vertical surface

$$\mathrm{V1}^{\Theta}(t)|_{\mathbf{x}_b \in \partial\Omega_{\hat{n}_k}} \approx \frac{\delta^2}{2\gamma}\int_{\Omega^+} g_\delta(x_m - x_l)^* \left(\frac{\partial \overline{\Theta}^{\,h}}{\partial x_1}\;\frac{\partial \overline{u}_1^{\,h}}{\partial x_1}\right) \mathrm{d}\tau$$

predicts the magnitude of DOF at nodes $\mathbf{x}_b \in \partial\Omega$ is nil significance throughout the entire wall jet laminar profile velocity reach, as clearly *validated* in Figure 12.11a. Isolated $O(\delta)$ significance $O(h)$ scale extrema (red) occur at wall jet separation VFDL elevations, as well as $O(\delta)$ scale streaks at the top surface initiation of the horizontal wall jet.

That the overall distribution significance of the $O(\delta^2)$ thermal vector resolution $v_3^h(\mathbf{x}, \mathrm{n}\Delta t)$ is mid-range (green) confirms it exerts *nil* impact on the prediction of $\overline{\Theta}^{\,h}(\mathbf{x}, t)$ (Figure 12.11c). The exceptions are several $O(\delta)$ significance (light blue) $O(h)$

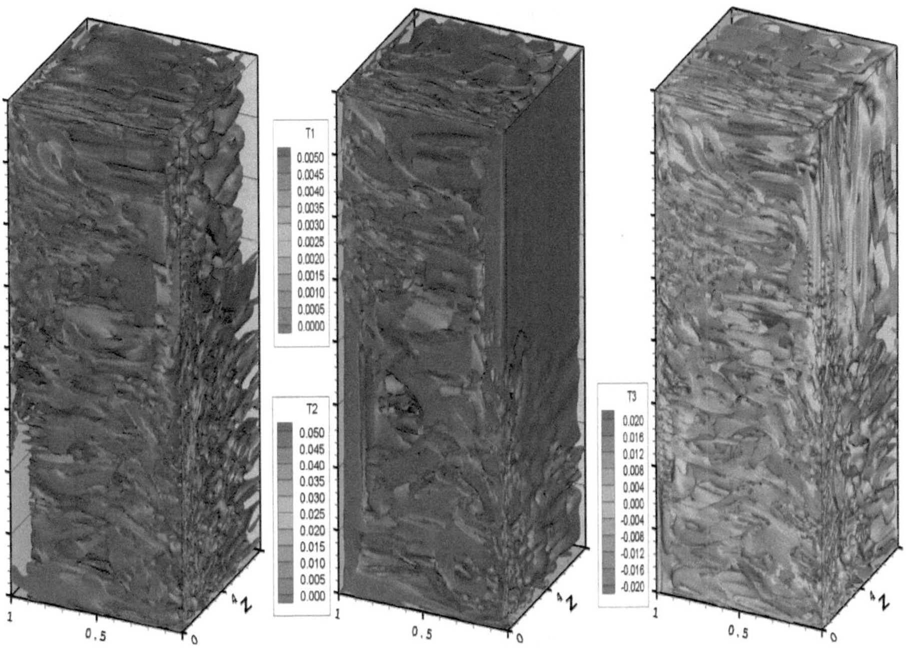

a) $\left(0.00 \leq v_1^h(\mathbf{x}, n\Delta t) \leq 0.005\right)$ b) $\left(0.00 \leq v_2^h(\mathbf{x}, n\Delta t) \leq 0.05\right)$ c) $\left(-0.02 \leq v_3^h(\mathbf{x}, n\Delta t) \leq 0.02\right)$

Figure 12.11: FaNS theory $O(\delta^2)$ resolved-unresolved scale interaction thermal vector resolutions $v_j^{\Theta h}(\mathbf{x}, n\Delta t)$ distribution nΔt= 600 s snapshots unitized scale perspective, view to adiabatic endwall, symmetric half $n = 3\ 8 \times 1 \times 16$ VFDL, Ra = E + 08, Pr = 0.71, Re = 11,868: (a) $\left(0.0\,0 \leq v_1^h(\mathbf{x}, n\Delta t) \leq 0.005\right)$, (b) $\left(0.00 \leq v_2^h(\mathbf{x}, n\Delta t) \leq 0.05\right)$, (c)$\left(-0.02 \leq v_3^h(\mathbf{x}, n\Delta t) \leq 0.02\right)$.

width streaks at the inception of vertical wall jet turning, as well as $O(h)$ scale "dots" embedded in the vertical corner within the insipient separation reach of the wall jet.

To conclude, FaNS theory $O(\delta^2)$ resolved-unresolved scale interaction tensor/vector Cartesian resolution distribution nΔt= 600 s snapshot global perspectives exhibit their theorized $O(\delta^2)$ significance, except locally in the VFDL geometry-induced regions experiencing intense thermal-velocity vector dynamics. Of key fundamental CFD *physics of fluids* impact is the fact that the *anisotropic* $O(\delta^2)$ tensor diagonal resolution *dominates* laminar profile transition to turbulent. Further, the lesser significant shear resolutions bear *no resemblance* to that/those generated by legacy RaNS/LES "physics based" $O(\delta^2)$ typically linear *isotropic* closure models. This predicts their assured inadequacy in characterizing velocity vector distributions that are not essentially uni-directional, that is, aerodynamic.

Finally, the $O(\delta)$ significance isolated $O(h \leftrightarrow \delta)$ scale features in $\tau_{ij}^h(\mathbf{x}, n\Delta t)$ and $v_j^{\Theta h}(\mathbf{x}, n\Delta t)$ within the reach of wall jet laminar insipient transition and separation, in an otherwise chaotic-to-disorganized nil background, quantitatively *validates* that the $O(\delta^2)$ resolved-unresolved scale state variable exerts ***dominant impact*** on the prediction of the $O(1)$ state variable *locally* in flowfield reaches exhibiting dynamically vigorous multi-scale thermal-fluid evolution for Re specifications exceeding those associated with strictly laminar phenomena.

12.8 Validation, ∀ Re FaNS theory $O(1;\delta^2;\delta^3)$ scaled state variable, VVUQ pertinence

The National Academy of Sciences *Verification, Validation, Uncertainly Quantification* (VVUQ) document, NAS (2012), cites topics *required* to be addressed to validate the ***reliability*** of complex computational methodologies. In addition to those detailed in Chapter 6, each of which has been weak formulation continuous Galerkin TEA-altered *error freed* CFD *mathematics* validated, and VVUQ compliant FaNS theory PDE system attributes that correlate precisely with previously unstated topics include:

> . . . Development of reference solutions, including *"manufactured statements,"* for the kinds of complex mathematical models described.
>
> . . . Development of methods to estimate *error bounds* when meshes cannot resolve important scales. An example is turbulent fluid flow.

The $n = 3$ virtual fluid dynamics laboratory (VFDL) *manufactured statement* has enabled ***thorough*** quantitative validation of the spatial filter uniform measure δ, bounded domain well-posed analytically closed ∀ Re FaNS theory $O(1,\delta^2)$ PDE:AD-BCE^h-NBC^h-DBC^h:SGS system GWS^h + θTS FE $p = 1$ basis algorithm for Re specifications generating single-scale laminar to that inducing insipient transition, separation, turbulent profile reattachment then relaminarization. Specific to the "turbulent fluid flow" VVUQ topic, *validation* that the FaNS $O(1;\delta^2)$ state variable global significance ratio was indeed $O(\delta^3)$ enables ***rigorous uncertainty quantification*** (UQ) that mesh resolution for the $O(1)$ state variable distribution for the VFDL unsteady, dynamic multi-scale transitioning symmetry plane event is ***adequate***.

To conclude, the weak formulation continuous GWS^h + θTS FE $p = 1$ basis algorithm written on *error-freed* CFD *mathematics* PDE statement for FaNS theory $O(1;\delta^2;\delta^3)$ scaled incompressible space-filtered NS state variable

$$\{\overline{q}\,(\mathbf{x},t)\} = \left\{\overline{\varphi},\ \overline{u}_i,\ \overline{\Theta},\ \overline{Y_a},\ \overline{P};\ \varphi,\ u_i,\ \Theta,\ Y_a,\ P;\ \tau_{ij}, v_j^{\Theta}, v_j^{Ya};\ \overline{u'_j u'_i},\ \overline{u'_j\Theta'},\ \overline{u'_j Y'_a}\right\}^T$$

is the truly transformative, long-time stable, Newton-convergent, phake physics stabilization ***free*** and closure modeling ***free*** statement validated *complex* mathematical models ***fully VVUQ compliant***! And finally, the word *"models"* in these VVUQ statements must be replaced with "*theories*" as analytical FaNS theory is ***totally devoid*** of any notion (!) of a model.

12.9 Extraction, NS velocity vector from FaNS spatially filtered $O(1)$ velocity vector

The coding complexity for AD-BCE^h-NBC^h GWS^h algorithm prediction of the NS wall-normal velocity resolution $\mathbf{u}^h_{\mathrm{AD}}(\mathbf{x}, \mathrm{n}\Delta t)$distribution was moderated by extending prediction to the $n = 3$ NS velocity vector $u_j(x_i,t)$ on the entirety of the FaNS domain Ω^+. For AD-BCE^h statement alteration to the NS $n = 3$ velocity

$$\text{AD: } u_i(x_k,t) \approx \overline{u}_i(x_k,t) - \frac{\delta^2}{4\gamma}\nabla^2\overline{u}_i(x_k,t) + O(\delta^4)$$

restated as the *differential statement* (DS),

$$\text{DS: } \mathcal{D}(u_i(x_k,t)) \equiv u_i - \overline{u}_i + \frac{\delta^2}{4\gamma}\nabla^2\overline{u}_i = 0 + O(\delta^4)$$

Green-Gauss divergence theorem enforcement generates the GWS^h element matrix for NS velocity $u_i(x_j, t)$ for $1 \le I \le 3$

$$\mathrm{GWS}^h(\mathrm{t})\text{: } [\mathrm{M200}]_\mathrm{e}\{UI\}_\mathrm{e} = [\mathrm{M200}]_\mathrm{e}\left\{\overline{U}I\right\}_\mathrm{e} + \frac{\delta^2}{4\gamma}[\mathrm{M2}LL]_\mathrm{e}\left\{\overline{U}I\right\}_\mathrm{e} - \frac{\delta^2}{4\gamma}[\mathrm{m20}L]_\mathrm{e}\hat{\mathrm{n}}_\mathrm{l}\left\{\overline{U}I\right\}_\mathrm{e}$$

Thus identified is the potential for same code-resident ***dual*** NS-FaNS theory GWS^h + θTS FE basis algorithms written on *error freed* $O(h^4, m^4, \delta^4, \Delta t^4)$ CFD *mathematics*. This *diagnostic insight* potential for unsteady NS velocity mechanisms that energize unsteady FaNS theory wall-adjacent $O(1)$ distributions was realized well after the defense of the Sekachev [4] dissertation, hence is a totally unexplored topic.

Unitized scale perspective projections of the nΔt=600 s snapshots of FaNS theory $O(1)$ $|\overline{\mathbf{u}}^h(\mathbf{x}, \mathrm{n}\Delta t)| \Rightarrow \overline{u}^h(x = \pm\,\delta/2, y = \pm\,\delta/2, z, \mathrm{n}\Delta t)$ and the AD algorithm-generated NS $|\mathbf{u}^h_{\mathrm{AD}}(\mathbf{x}, \mathrm{n}\Delta t)| \Rightarrow u^h_{\mathrm{AD}}(x = \pm\,\delta/16, y = \pm\,\delta/16, z, \mathrm{n}\Delta t)$ speed distributions are compared in Figure 12.12 view to the symmetry plane, hot ΔT DBC surface on the right. That the nil-speed enveloping surface distributions within the VFDL core are essentially identical is not unexpected. Conversely, the clearly visible distinctions in the NS u^h $O(\delta{\leftrightarrow}L)$ scale multiple *extremum* (red) significance features directly adjacent to the upper half of the forefront VFDL surface are due to the NS velocity DOF data located at the local mesh refinement nodes sequentially displaced δ/16 distant from the hot ΔT DBC surface within each wall-adjacent element, that is, nodes 1 through 4 in Figure 10.4.

It is *precisely* these AD-BCE^h algorithm-generated NS u^h_{AD} local refinement nodes 1–5 DOF data that energize NS velocity efflux convolution for the $O(1)$ velocity non-

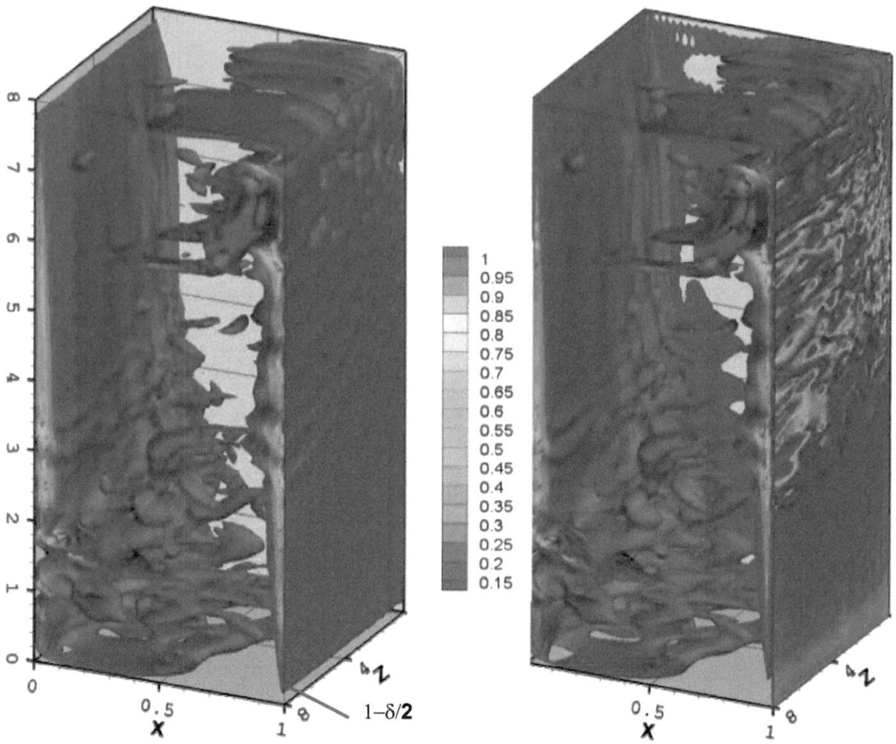

a) $\bar{u}^h\left(x=\pm\delta/2, y=\pm\delta/2, z, \text{n}\Delta t\right)$ b) $u^h_{\text{AD}}\left(x=\pm\delta/16, y=\pm\delta/16, z, \text{n}\Delta t\right)$

Figure 12.12: Extraction of unfiltered NS velocity $\mathbf{u}^h_{\text{AD}}(\mathbf{x}, \text{n}\Delta t)$ distribution from FaNS theory $\text{D}^m\bar{\mathbf{P}}$:AD GWS^h algorithm generated $O(1)$ space-filtered velocity, $n\Delta t$=600 s snapshot unitized scale perspectives, view to symmetry plane, symmetric half n = 3, 8 × 1 × 16 VFDL, Ra ≡ E + 08, Pr = 0.71, Re = 11,868: (a) $\bar{u}^h(x = \pm\,\delta/2, y = \pm\,\delta/2, z, \text{n}\Delta t)$, (b) $u^h_{\text{AD}}(x = \pm\,\delta/16, y = \pm\,\delta/16, z, \text{n}\Delta t)$.

homogeneous Neumann NBC, hence exert *fundamental* impact (!) on FaNS theory-predicted vertical wall jet $O(1)$ velocity distribution destabilization leading to prediction of wall jet laminar profile velocity insipient transition, separation and turbulent profile reattachment.

Recalling the USAF BOLT hypersonic missile experiment, Chapter 8, the alternative to conducting hypersonic missile atmospheric re-entry experiments is now the thoroughly validated FaNS theory CFD prediction of the hypersonic shock layer submerged, non-equilibrium real gas aerothermodynamics dominated boundary layer laminar-turbulent transition on missile surfaces involved in radiation-convection heat transfer. The CFD formulation sequence to accomplish this objective involves transition from textbook RANS unsteady DY PDE

$$\text{RaNS DY:} \frac{\partial Y}{\partial t} + u_j \frac{\partial Y}{\partial x_j} - \frac{\partial}{\partial x_j}\left(\frac{1 + \mathrm{Re}^T}{\mathrm{ReSc}}\right)\frac{\partial Y}{\partial x_j} = 0$$

to *error freed* CFD *mathematics,* equilibrium variable γ air RaNS $O(h^4, m^4, \Delta t^3)$ $\mathrm{D}^m\mathrm{Y}$

$$\text{RaNS TEA altered } \mathrm{D}^m Y\text{: } \left[1 - \frac{\gamma \Delta t^2}{6}\frac{\partial}{\partial x_j}\left(u_j u_k \frac{\partial}{\partial x_k}\right)\right]\frac{\partial Y}{\partial t} + u_j \frac{\partial Y}{\partial x_j}$$

$$- \frac{\partial}{\partial x_j}\left[\left(\frac{1 + \mathrm{Re}^T}{\mathrm{ReSc}}\right)\frac{\partial Y}{\partial x_j} + \frac{h^2 ReSc}{12\left(1 + \mathrm{Re}^T\right)} u_j u_k \frac{\partial Y}{\partial x_k}\right]$$

$$= 0 + O(h^4, m^4, \Delta t^3)$$

to RaNS $O(h^4, m^4, \Delta t^4)$ $\mathrm{D}^m \rho \mathrm{Y}^\alpha$ for $1 \le \alpha \le n$ mass/mole fraction multi-species.$\rho\, Y^\alpha$

$$\text{RaNS } \mathrm{D}^m \rho\, Y^\alpha\text{: } \left[1 - \frac{\gamma \Delta t^2}{6}\frac{\partial}{\partial x_j}\left(u_j u_k \frac{\partial}{\partial x_k}\right)\right]\frac{\partial \rho\, Y^\alpha}{\partial t} + \left(1 + \frac{h^2 \mathrm{Re}}{6\Delta t}\right) u_j \frac{\partial \rho\, Y^\alpha}{\partial x_j}$$

$$- \frac{\partial}{\partial x_j}\left[\left(\frac{h^2}{6\Delta t} + \frac{1 + \mathrm{Re}^T}{\mathrm{ReSc}}\right)\frac{\partial \rho\, Y^\alpha}{\partial x_j} + \frac{h^2 \mathrm{ReSc}}{12\left(1 + \mathrm{Re}^T\right)} u_j u_k \frac{\partial \rho\, Y^\alpha}{\partial x_k}\right]$$

$$= 0 + O(h^4, m^4, \Delta t^4)$$

to ∀ Re FaNS theory non-equlibrium multi-species $O(h^4, m^4, \delta^4, \Delta t^4)$ $\mathrm{D}^m \overline{\rho\, Y}_\alpha$ PDE:AD-BCE-NBC:$O(h^2, \delta^4)$ PDE:AD-BCE-DBC: $O(\delta^3)$ SFS tensor/vector non-equilibrium algorithm statement

$$\mathrm{D}^m \overline{\rho\, Y}_\alpha\text{: } \left(1 - \frac{\gamma \Delta t^2}{6}\right)\frac{\partial}{\partial x_j}\left(\overline{u_j u_k}\frac{\partial}{\partial x_k}\right)\frac{\partial \overline{\rho\, Y_\alpha}}{\partial t} + \left(1 + \frac{\mathrm{Re} h^2}{6\Delta t}\right)\frac{\partial}{\partial x_j}\left(\bar{u}_j \overline{\rho\, Y_\alpha} + v_j^{\overline{\rho Y_\alpha}}\right)$$

$$- \int_{\partial\Omega} g_\delta(x_k - s_k) * \left[\frac{\mathrm{D}^*}{\mathrm{ReSc}}\frac{\partial \rho Y_\alpha}{\partial x_j}\right]\hat{n}_j\, \mathrm{d}s - s(\rho Y_{\alpha|\beta})$$

$$- \frac{\partial}{\partial x_j}\left[\left(\frac{\mathrm{D}^*}{\mathrm{ReSc}} + \frac{h^2}{6\Delta t}\right)\frac{\partial \overline{\rho\, Y_\alpha}}{\partial x_j} + \frac{\mathrm{C_S}(\delta) h^2 \mathrm{ReSc}}{12\,\mathrm{D}^*}\bar{u}_j \bar{u}_k \frac{\partial \overline{\rho\, Y_\alpha}}{\partial x_k}\right] = 0 + O(h^4, m^4, \Delta t^4)$$

coupled with $O(\delta^2)$ resolved-unresolved scale interaction mass fraction *vector* harmonic PDE:AD-BCE^h-DBC^h algorithm prediction of $v_j^{\overline{\rho Y_\alpha}} \equiv \overline{\bar{u}_j \rho Y'_\alpha} + \overline{u'_j \overline{\rho Y_\alpha}}$

$$\mathrm{D}v_j^{\overline{\rho Y_\alpha}}\text{: } -\delta^2 \nabla^2 v_j^{\overline{\rho Y_\alpha}} + 4\gamma v_j^{\overline{\rho Y_\alpha}} - 2\delta^2 \frac{\partial \bar{u}_j}{\partial x_k}\frac{\partial \overline{\rho Y_\alpha}}{\partial x_k} = 0 + O(\delta^4)$$

Should this dual NS/FaNS capability prove viable in generating CFD *mathematics of physics of fluids* fundamental diagnostics of hypersonic missile non-equilibrium aerothermodynamics-radiation coupled laminar-turbulent BL transition, the $O(h^4)$ asymptotic convergence rate validated for the $O(1)$ FaNS theory $O(h^4)$ PDE system

GWS^h + θTS FE $p = 1$ basis distributions will accrue to the FE $p = 2$ basis implementation of the $O(\delta^4)$ harmonic PDE AD:BCE^h GWS^h algorithm.

12.10 Summary, ∀ Re FaNS theory *optimal* $O(h^4, m^4, \delta^4, \Delta t^4)$ CFD algorithm/code

This chapter completes documentation of the culmination of three-plus decades of UT CFD Laboratory graduate student dissertation research to identify, derive and validate mathematically rigorous resolutions to your first author's 1970s observation in joining the UTK engineering faculty,

> ". . . how to we correct what is *wrong* with the way CFD is practiced?"

The content in Part A identifies and quantitatively validates *error freed* CFD *mathematics* resolutions that completely annihilate *each* of the longstanding liabilities persistent in legacy NS/RaNS CFD theory/code practice.

The Part B content builds thereon in identification and quantitative validation of the *physics of fluids* ∀ Re FaNS theory CFD unstagnation that could (should!) in the near term replace legacy LES theory/code practice.

The full realization of these *unstagnation* advances simply awaits sufficient interest within the global CFD community to expend the effort to build the first- and higher-order effects validated *physics of fluids* fidelity ∀Re FaNS theory *error freed* CFD *mathematics* production code.

References

[1] Le Quere, P., Benhia, M. (1998). "From onset of unsteadiness to chaos in a differentially heated square cavity," *J Fluid Mech*, 359.

[2] NAP (2012). Assessing the Reliability of Complex Models: Mathematical and Statistical Foundations of Verification, Validation, and Uncertainty Quantification, National Research Council, National Academies Press, Washington DC.

[3] Pope, S.B. (2000). *Turbulent Flows*, Cambridge University Press, NY.

[4] Sekachev, M.A., (2013). "Essentially analytical theory closure for thermal-incompressible space filtered Navier-Stokes PDE system," PhD dissertation, University of Tennessee

Chapter 13
Epilogue, *error freed* CFD *mathematics* continuing impact

The last chapter rigorously derived and validated ∀Re pertinent, analytically closed space filtered NS *error freed* CFD *mathematics* PDE systems are *precisely* responsive to the absolutely ***key*** citation in the NASA forward-thinking contractor report, “CFD Vision 2030 Study, . . .”, [1].

> “. . . *single* most critical issue . . . adequately predict viscous turbulent flows . . . . . . boundary layer transition . . . flows with separation present.”

The Mujeeb [2] update to the “CFD Vision 2030 . . . .” preamble states

> “. . . Integration Committee . . . advocates for and inspires . . activities recommended by the vision study for *revolutionary advances* in state-of-the-art computational technologies . . ., ,”

implores that the Vision focus be on ***revolutionary*** advances, not legacy practice ***evolution***.

Certainly, and without exception, weak formulation continuous Galerkin FE $p = 1,2,3$ completeness degree basis algorithms written on *error freed* CFD *mathematics* are precisely responsive to the sum total of CFD Vision 2030 documentation admonitions. And of equal mathematical rigor impact, the thoroughly validated theory/practice attributes for *optimal* continuous Galerkin CFD practice are totally responsive to the US National Academy of Sciences detailed ***critical*** VVUQ topics, NAP [3], specifically:

> goal-oriented *a posteriori* error-estimation applicable . . . mathematical *statements* [***replaced*** their word *models*] more complex than linear elliptic PDEs;
>
> algorithms for goal-oriented error estimates that scale well on massively parallel architectures, ... complex grids including adaptive-mesh . . .;
>
> reference solutions, including “manufactured statements,” for . . . complex mathematical theories [*not* models];
>
> methods to estimate error bounds . . . meshes cannot resolve important scales . . . turbulent fluid flow.

Finally, and of truly *inestimable* impact on CFD *code practice*, but rarely (if ever?!) referenced in the literature, is continuous Galerkin FE basis algorithm enabled matrix differential calculus identification of ***every*** (!) nonlinear error freed contribution to the *quadratic* convergent Newton iteration algorithm jacobian. This is the ***critical enabling technology*** requisite to achieving the NASA “CFD Vision 2030 objective update, Gomez and Carey [4].

https://doi.org/10.1515/9783112206430-014

> ". . . . . seeking *convergence* demonstration for unsteady flowfield prediction in complex geometries . . ."

which is fundamentally critical to

> ". . . performing *on-demand* visualization and analysis of a multi-billion point unsteady CFD simulation . . .",

a truly illusive objective absent *robustly convergent* numerical linear algebra to eliminate generation, hence promulgation of ***non-converged*** CFD data into the literature.

References

[1] Slotnick, J., Khodadoust, A., Alonso, J., Darmofal, D., Gropp, W., Lurie, E., Mavriplis, D., (2014). "CFD Vision 2030 Study: A Path to Revolutionary Computational Aerosciences," NASA Tech Report. CR-2014-21817

[2] Mujeeb, R.M. (2018). "Marching toward the 2030 vision of CFD," Aerospace America, 56(1)

[3] NAP (2012). "Assessing the Reliability of Complex Models: Mathematical and Statistical Foundations of Verification, Validation, and Uncertainty Quantification," National Research Council, National Academies Press, Washington DC

[4] Gomez, R.J., Carey, A.W., (2020). "Progress toward the 2030 vision for CFD," Aerospace America, 58,11

Nomenclature

a_i	Coefficient set for FD/FE tri-diagonal stencils
AD	Approximate deconvolution
AD-BCE	FaNS BCE integral $GWS^h(t)$ convolution algorithm
AD-BCE-DBC	FaNS Dirichlet BC $GWS^h(t)$ algorithm
AD-BCE-NBC	FaNS Neumann BC $GWS^h(t)$ algorithm
AF	Approximate factorization
$B(\cdot)$	Bilinear form
BCE	Boundary commutation error integrals
BL	Boundary layer
c	Speed of sound, $c \equiv \sqrt{\gamma RT}$
CFL	Courant number (non-D time step), CFL = $u\Delta t/h$
C_S	Smagorinsky constant, FaNS SGS tensor coefficient
CFD	Computational fluid dynamics
CNFD	Crank-Nicolson finite difference algorithm
D	Binary diffusion coefficient
DD	Difference differential equation
$\det_e$	Jacobian determinant, global-to-element coordinate transformation
DE	Energy conservation principle, (Oh^4) PDE
D^mE	TEA theory-altered DE
DG	Discontinuous Galerkin
DM	Mass conservation principle PDE
D^mM	TEA theory-altered DM
DM^h	Incompressible NS DM enforcement (Oh^4) PDE
DOF	Degree of freedom
D**P**	Momentum vector conservation, (Oh^4) PDE
$D^m\mathbf{P}$	TEA theory altered D**P**,
$D^m\mathbf{P}^h$	Spatially discretized $D^m\mathbf{P}$
D^mQ	TEA-altered RaNS scalar member (Oh^4) PDE
DS	Differential statement
DY	Mass/mole fraction PDE
D^mY	TEA-altered DY $O(h^4,m^4)$ PDE
$e^N(x_j,t)$	Error in continuum $q^N(x_j,t)$
$e^h(q)$	PDE state variable member $O(h^2)$ truncation error TE
$e^h(m^P)$	Phase speed error, $e^h \equiv g_{\text{exact}} - g^h$
$e^h(x_j,t)$	GWS^h+θTS FE basis discrete approximation error
E	NS total internal energy, $E=\rho e + \rho\mathbf{u}^2/2$
$\lVert\{e^h(n\Delta t)\}\rVert_E$	PDE energy norm set, state variable member-specific
EBV	Elliptic boundary value, PDE system
Eu	Euler number, $\text{Eu} \equiv \rho_\infty U^2/p_\infty$,
Euler	Inviscid, non-heat-conducting NS PDE system
f	NS kinetic flux vector, $f \equiv uq$
FaNS	Analytically closed spatially filtered NS PDE system
FE	Finite element
FD	Finite difference
FV	Finite volume
FVWS	Finite volume weak statement
FV^h	Difference algebra discretized FVWS

https://doi.org/10.1515/9783112206430-015

$\{F(\{Q_{n+1}\}^p)\}$	GWS^h+θTS global algebraic statement
g	Gravity vector magnitude
g^h	CFD algorithm amplification factor
Gr	Grashof number, $Gr \equiv g\beta\Delta TL^3/\nu^2$, $\Delta T \equiv (T_{max}-T_{min})$, $\beta \equiv 1/T_{abs}$
GWS	Continuous Galerkin weak formulation
GWS^h	Galerkin spatially discrete steady weak statement
GWS^h+θTS	θ-Implicit time TS unsteady GWS^h
h	Measure of the mesh, spatially discretized variable (superscript)
H^m	Hilbert space of order m
H_0	Stagnation enthalpy
IC	Initial condition
I,J,K	FE matrix, hypermatrix tensor summation indices, $1 \leq I, J, K \leq n$
$[JAC(\{Q\}^p_{n+1})]_e$	Newton Jacobian element matrix
k	Thermal conductivity
L	Reference length scale
LES	Large Eddy Simulation space-filtered NS PDE
m	Non-D wavenumber, $m \equiv \kappa h$
m	NS momentum vector
Ma	Mach number, $Ma \equiv \lvert u_j \rvert / c$, $c \equiv \sqrt{\gamma RT}$
Ma_∞	Transonic-hypersonic onset Mach number
$[M\ldots]_e$	FE basis element matrix/hypermatrix on Ω_e
$[m\ldots]_e$	FE basis element matrix/hypermatrix on $\partial\Omega_e$
[m1010]	Radiosity theory FE p = 1 element matrix
MLT	Mixing length theory
MWR	Method of weighted residuals
NS	Navier-Stokes PDE system
n-D	n-Dimensional
non-D	Non-dimensional
$\{N(x_j)\}$	Finite element (FE) trial space basis
ODE	Ordinary differential equation
$O(\cdot)$	Exponential order of the argument
$O(h^4,m^4,\Delta t^4)$	Asymptotic convergence, TEA-altered PDE GWS^h FE p=1 algorithm
p	Pressure, FE trial space basis completeness degree
P	Kinematic pressure, incompressible NS
PDE	Partial differential equation
Pr	Prandtl number, $Pr \equiv c_p\mu_\infty/k_\infty$
q	NS generic scalar variable
$q^N(x_j,t)$ NS $q(x_j,t)$	continuum approximation
$\{q(\mathbf{x}, t)\}$	PDE system state variable
Q	Discretized NS state variable nodal DOF
RaNS	Time-averaged NS PDE
Ra	Rayleigh number, $Ra \equiv PrGr$
Re	Reynolds number, $Re \equiv \rho_\infty UL/\mu_\infty$
Re^T	Turbulent Reynolds number, $Re^T \equiv \mu^T/\mu_\infty$ or ν^T/ν_∞
Sc	Schmidt number, $Sc \equiv D/\nu$
SFS	Subfilter-scale tensor/vector
$S_{i,j}$	Stokes strain rate *isotropic* tensor, $\bar{S}_{ij} \equiv (\partial\bar{u}_i/\partial x_j + \partial\bar{u}_j/\partial x_i)/2$
S	Stokes tensor dyadic
T	Temperature

TE	Truncation error
TEA	Truncation error annihilation (theory)
TKE	Turbulent kinetic energy
TS	Taylor series
u,v,w	Velocity Cartesian resolution
(u_W^+, y^+)	BL similarity coordinate system
$\mathbf{u}$	NS velocity vector
$\mathbf{u}_{AD}^h(\mathbf{x}, n\Delta t)$	NS velocity generated by AD BCE GWS^h algorithm
$u_{AD}^h(\mathbf{x}, n\Delta t)$	velocity speed distribution
U	Non-D reference velocity
VBV	Verification, Benchmark, Validation CFD protocol
VVUQ	Verification, Validation, Uncertainly Quantification protocol
VFDL	Virtual fluid dynamics laboratory
WF	Continuum weak formulation
WS	WF extremum weak statement
Y	Mass fraction
α,β,γ,μ	Time TS coefficients
δ	Uniform measure of Gaussian filter
$\delta_{i,j}$	Kronecker delta
δ/h	Spatial filter/mesh measure ratio
δ^+	FaNS theory filter δ similarity variable, $\delta^+ \equiv u_\tau \delta/\nu$
θ	Time integration implicitness factor
Θ	Potential temperature, $\Theta \equiv (T - T_{min})/(T_{max} - T_{min})$
$\lVert \Theta_e^h \rVert_E$	DE-discretized PDE element energy norm
λ	Perfect gas law constant
κ	Wavenumber vector, Wave number of Fourier mode wavelength y, $\kappa \equiv 2\pi/\lambda$
ν, μ	Kinematic, dynamic viscosity
ν^T	Turbulent (eddy) viscosity
$v_j^\Theta \equiv \overline{\bar{u}_j \Theta'} + \overline{u'_j \bar{\Theta}}$	$O(\delta^2)$ resolved/unresolved scale interaction thermal vector
$\mathbf{v_g}(\kappa)$	Fourier group velocity, $\mathbf{v}_g(\boldsymbol{\kappa}) \equiv \nabla_\kappa \omega$
$\tau_{ij}(\mathbf{x}, t) \equiv \overline{\bar{u}_j u'_i} + \overline{u'_j \bar{u}_i}$	$O(\delta^2)$ resolved/unresolved scale interaction tensor
ρ	Fluid density
$\tau_{i,j}$	stress tensor, $\tau_{ij} = \overline{u'_i u'_j}$
$\lVert \varphi^h(t_{n+1}) \rVert_E$	DM^h energy norm evolution
ω	cyclic frequency, vorticity
Ω	PDE domain
Ω^h	Spatially discretized PDE domain, $\Omega^h \equiv \cup_e \Omega_e$
Ω_e	FE domain
$\partial\Omega$	PDE domain boundary segment
{ }	Column matrix
[ ]	Square matrix, hypermatrix
diag	Diagonal matrix
‖ ‖	Mathematical norm
∪	Union
∀	Denotes "for all"

Appendix A: The finite element toolbox

A.1 Finite element trial space basis

The continuous Galerkin weak statement GWS^N terminal decision is the trial space $\Psi_\alpha(\mathbf{x})$ mathematical support for the approximation. With roots in interpolation theory, the efficiency of computational fluid dynamics (CFD) algorithms is determined by the Lagrange linear piecewise continuous polynomial, where accuracy is controlled by the number and location of the interpolation *knots*. As extensively illustrated, interpolation *knots* transform into nodes of the finite element (FE) discretization where the DOF of the FE discretization are co-located.

The FE linear trial space *basis* consists of linear, piecewise-continuous interpolation polynomials spanning full dimensionality. For example, a two-element discretization of a $n = 1$ domain segment generates three geometric *nodes*, X1, X2, X3 (Figure A.1a). Consequently, this results in a three-coefficient continuum approximation:

$$q^N(x) \equiv \sum_{\alpha=1}^{N=3} \Psi_\alpha(x)Q_\alpha = \Psi_1(x)Q_1 + \Psi_2(x)Q_2 + \Psi_3(x)Q_3$$

The linear Lagrange polynomial definition for the generic Ψ_α centered at node X_j

$$\Psi_\alpha(x) = \begin{cases} \frac{x - X_{j-l}}{X_j - X_{j-l}}, \text{ for } X_{j-l} \le x \le X_j \\ \frac{X_{j+l} - x}{X_{j+l} - X_j}, \text{ for } X_j \le x \le X_{j+l} \\ 0, \text{ for } x < X_{j-l},\ x > X_{j+l} \end{cases}$$

emphasizes that each $\Psi_\alpha(x)$ is of *global* span, as heavy lines denoted in (Figure A.1b).

The FE trial space basis $\{N(x)\} = \{n_1(x), n_2(x)\}^T$ eliminates the global expression burden by coalescence of only the non-zero components of Ψ_α that span *each* element domain Ω_e (Figure A.1c). The resultant linear FE basis $\{N(x)\}$ matrix statement

$$\{N(x)\} \equiv \begin{Bmatrix} n_1(x) \\ n_2(x) \end{Bmatrix} = \frac{1}{(\mathrm{XR} - \mathrm{XL})_e} \begin{Bmatrix} \mathrm{XR} - x \\ x - \mathrm{XL} \end{Bmatrix}_e$$

possesses identical entries $\forall \Omega_e$ (!), wherein XR and XL denote the right and left node coordinates of a generic Ω_e. Thus, the interpolation polynomial is defined as both local *and* universal $\forall \Omega_e$, which converts GWS^N global matrix statement integral evaluations into a “DO loop” on $\Omega^h = \cup \Omega_e$, for $1 \le e \le M$.

https://doi.org/10.1515/9783112206430-016

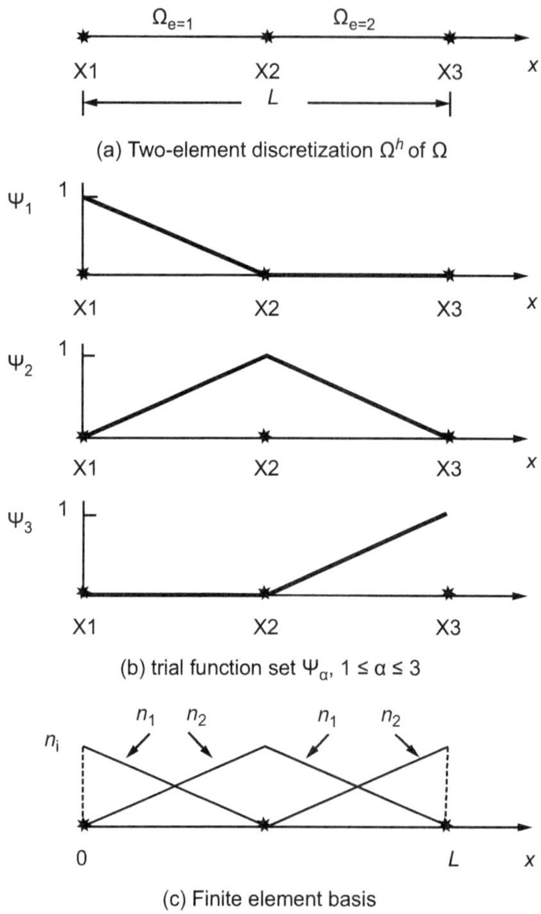

Figure A.1: Global trial space $\Psi_\alpha(x)$ coalesced to the Lagrange linear trial space basis $\{N(x)\}$.

A.2 GWS^h elementlevel matrix assembly

Consider DE reduced to steady conduction with source

$$\text{DE:} -\frac{\mathrm{d}}{\mathrm{d}x}\left(k\frac{\mathrm{d}T}{\mathrm{d}x}\right) - s = 0$$

Enforcing the Green-Gauss divergence theorem GWS$^N \Rightarrow$ GWS^h generates

$$\text{GWS}^h\text{:} \sum_{e=1}^{\mathrm{M}}\left[\sum_{j=1}^{2}\int_{\Omega_e}\frac{\mathrm{d}n_i}{\mathrm{d}x}k\frac{\mathrm{d}n_j}{\mathrm{d}x}\mathrm{d}x\,\mathrm{Q}_j - \int_{\Omega_e} n_i s\,\mathrm{d}x\right] - k\frac{\mathrm{d}T}{\mathrm{d}x}n_i\Bigg|_a^b = 0,\ 1 \le i \le 2$$

The summation index i is then replaced with FE trial space basis matrix products

$$\sum_{j=1}^{2}\int_{\Omega_e}\frac{dn_i}{dx}k\frac{dn_j}{dx}dxQ_j^e\Bigg|_{i=1,2}\Rightarrow\int_{\Omega_e}\frac{d\{N\}}{dx}k\frac{d\{N\}^T}{dx}dx\{Q\}_e$$

a certainly familiar form. As the linear FE basis derivative is $d\{N\}^T/dx = \{-1,\ 1\}/h_e$, the generic Ω_e matrix is

$$\int_{\Omega_e}\frac{d\{N\}}{dx}k\frac{d\{N\}^T}{dx}dx=\frac{1}{h_e^2}\begin{bmatrix}1 & -1\\ -1 & 1\end{bmatrix}\int_{\Omega_e}kdx=\frac{\overline{k}_e}{h_e}\begin{bmatrix}1 & -1\\ -1 & 1\end{bmatrix}$$

wherein $\overline{k}_e$ is analytically predicted to be the average conductivity on Ω_e.

Selecting XL, the origin of the local coordinate $\overline{x}$ spanning each element, the same operation on the source term nets

$$\int_{\Omega_e}n_i s dx\Bigg|_{i=1,2}\Rightarrow\int_{\Omega_e}\{N\}sdx=s\int_{\Omega_e}\begin{Bmatrix}1-\overline{x}/h_e\\ \overline{x}/h_e\end{Bmatrix}dx=\frac{sh_e}{2}\begin{Bmatrix}1\\ 1\end{Bmatrix}$$

and for endpoint efflux terms

$$-k\frac{dT}{dx}n_i\Bigg|_{i=1,2}\Rightarrow -k\frac{dT}{dx}\{N\}\Bigg|_{X_L}^{X_R}=-k\frac{dT}{dx}\begin{Bmatrix}0\\ 1\end{Bmatrix}_{e=M}+k\frac{dT}{dx}\begin{Bmatrix}1\\ 0\end{Bmatrix}_{e=1}$$

Thereby, the GWS^h linear FE trial space basis element matrix statement is

$$\text{GWS}^h:\ \frac{\overline{k_e}}{h_e}\begin{bmatrix}1 & -1\\ -1 & 1\end{bmatrix}\{Q\}_e-\frac{sh_e}{2}\begin{pmatrix}1\\ 1\end{pmatrix}-k\frac{dT}{dx}\begin{Bmatrix}-\delta_{e1}\\ \delta_{eM}\end{Bmatrix}$$

The subscript e denotes element-dependence, δ_{e1} and δ_{eM} are Kronecker delta (0,1) switches for *a priori* given flux boundary conditions (BCs).

Recalling Figure A.1(a), for an M = 2 discretization of uniform measure $h_e = h$, the GWS^h matrices for the element pair Ω_e sharing the common (middle) node are

$$\text{for } e=1:\ \frac{k}{h}\begin{bmatrix}1 & -1\\ -1 & 1\end{bmatrix}\begin{Bmatrix}QL\\ QM\end{Bmatrix}_1-\frac{sh}{2}\begin{Bmatrix}1\\ 1\end{Bmatrix}-\begin{Bmatrix}f_n\\ 0\end{Bmatrix}$$

$$\text{for } e=2:\ \frac{k}{h}\begin{bmatrix}1 & -1\\ -1 & 1\end{bmatrix}\begin{Bmatrix}QM\\ QR\end{Bmatrix}_2-\frac{sh}{2}\begin{Bmatrix}1\\ 1\end{Bmatrix}+\begin{Bmatrix}0\\ f_R\end{Bmatrix}$$

for QL, QR the respective left/right DOF with DOF QM the shared common node. GWS^h was originally written as a summation over M and it remains so:

$$\text{GWS}^h\text{:}\ \sum_{e=1}^{M}\left(\frac{k_e}{h_e}\begin{bmatrix}1 & -1\\ -1 & 1\end{bmatrix}\{Q\}_e-\frac{sh_e}{2}\begin{pmatrix}1\\1\end{pmatrix}-k\frac{\mathrm{d}T}{\mathrm{d}x}\begin{Bmatrix}-\delta_{e=1}\\ \delta_{e=M}\end{Bmatrix}\right)$$

$$=[\text{Matrix}]\begin{Bmatrix}QL\\ QM\\ QR\end{Bmatrix}-\{b\}=\{0\}$$

The answer to the query, "how to enter 2 × 2 matrices into the 3 × 3 [Matrix]?," is to add row-column zero pairs to each element matrix at locations where global degrees of freedom (DOF) are missing. Thus

$$[\text{Matrix}]\begin{Bmatrix}QL\\ QM\\ QR\end{Bmatrix}=\sum_{e=1}^{M}[\text{Matrix}]_e\{Q\}_e=\frac{k}{h}\begin{bmatrix}1 & -1 & 0\\ -1 & 1 & 0\\ 0 & 0 & 0\end{bmatrix}\begin{Bmatrix}QL\\ QM\\ 0\end{Bmatrix}$$

$$+\frac{k}{h}\begin{bmatrix}0 & 0 & 0\\ 0 & 1 & -1\\ 0 & -1 & 1\end{bmatrix}\begin{Bmatrix}0\\ QM\\ QR\end{Bmatrix}=\frac{k}{h}\begin{bmatrix}1 & -1 & 0\\ -1 & 2 & -1\\ 0 & -1 & 1\end{bmatrix}\begin{Bmatrix}QL\\ QM\\ QR\end{Bmatrix}$$

Thus identified is the generic element matrix *assembled* GWS^h

$$\text{GWS}^{\text{h}}\text{:}\quad \frac{k}{h}\begin{bmatrix}1 & -1 & 0\\ -1 & 2 & -1\\ 0 & -1 & 1\end{bmatrix}\begin{Bmatrix}QL\\ QM\\ QR\end{Bmatrix}-\frac{sh}{2}\begin{Bmatrix}1\\ 2\\ 1\end{Bmatrix}-\begin{Bmatrix}f_{e=1}\\ 0\\ -f_{e=\text{M}}\end{Bmatrix}=\{0\}$$

wherein the first-term middle row-column matrix product is *immediately* recognized as the sign-reversed Green-Gauss divergence theorem-generated second-order accurate finite difference (FD) tridiagonal stencil for the $n = 1$ Laplacian integrated over Ω_e.

A.3 Finite element $p = 1, 2, 3$ completeness degree trial space bases

The geometrical shapes for $1 \le n \le 3$ FE trial space bases are (Figure A.2):

$n = 1$: line segments

$n = 2$: triangles, quadrilaterals

$n = 3$: tetrahedra, hexahedra

Linear ($p = 1$) FE basis DOF are always vertex node coincident. The FE $p = 2, 3$ basis non-vertex disposed DOF augmentations are denoted by an "x."

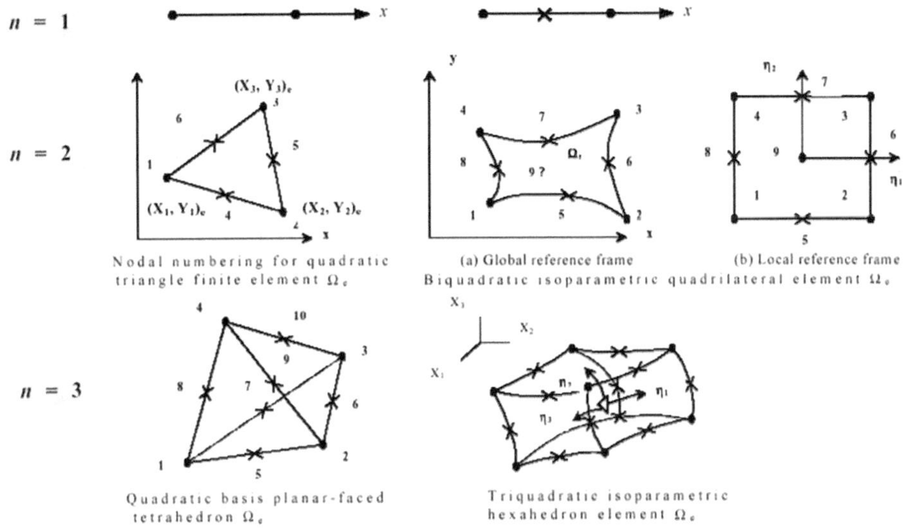

Figure A.2: FE trial space basis domains, Ω_e, $1 \leq n \leq 3$, linear/quadratic DOF locations.

FE trial space bases are always spanned by a local normalized coordinate system:

- *natural coordinate* (NC) system ζ_a, $1 \leq {}_a \leq n + 1$, lines, triangles, and tetrahedra
- *tensor product* (TP) system η_i, $1 \leq i \leq n$, lines, quadrilaterals, and hexahedra

The linear FE basis also defines these local systems as well as the global/local *coordinate transformation* $x_i = f(\zeta_a, \eta_j)$. Normalized NC FE linear basis coordinate systems

$$n = 1: \{N\} \equiv \{\zeta_1, \zeta_2\}^T$$
$$n = 2: \{N\} \equiv \{\zeta_1, \zeta_2, \zeta_3\}^T$$
$$n = 3: \{N\} \equiv \{\zeta_1, \zeta_2, \zeta_3, \zeta_4\}^T$$

are *linearly dependent* via the constraint $\sum_{a=1}^{n+1} \zeta_a \equiv 1$. Each coordinate ζ_a is zero at *every* node of Ω_e, except the node with index "a" whereat $\zeta_a = 1$. An NC p-basis discretization readily supports solution-adapted, nonuniform meshing of highly non-regular-shaped domains and boundaries.

The TP bilinear FE basis family is geometrically versatile, especially when using boundary-fitted coordinate transformations. As shown in Figure A.2, TP bilinear FE bases are spanned by *linearly independent* local η_j coordinate systems, which transition to rectangular Cartesian in the η-space unit geometry. TP bilinear FE basis for $n = 2$ and $n = 3$ element geometries are

$$\{N^{+}(\eta)\} = \frac{1}{4}\begin{Bmatrix} (1-\eta_1)(1-\eta_2) \\ (1+\eta_1)(1-\eta_2) \\ (1+\eta_1)(1+\eta_2) \\ (1-\eta_1)(1+\eta_2) \end{Bmatrix}, \quad \{N^{+}(\eta)\} = \frac{1}{8}\begin{Bmatrix} (1-\eta_1)(1-\eta_2)(1-\eta_3) \\ (1+\eta_1)(1-\eta_2)(1-\eta_3) \\ (1+\eta_1)(1+\eta_2)(1-\eta_3) \\ (1-\eta_1)(1+\eta_2)(1-\eta_3) \\ (1-\eta_1)(1-\eta_2)(1+\eta_3) \\ (1+\eta_1)(1-\eta_2)(1+\eta_3) \\ (1+\eta_1)(1+\eta_2)(1+\eta_3) \\ (1-\eta_1)(1+\eta_2)(1+\eta_3) \end{Bmatrix}$$

A.4 Finite element NC basis coordinate transformation

Invoking the chain rule, the NC linear FE basis GWS^h DE algorithm for the n-D Laplacian generic Ω_e matrix statement

$$\begin{aligned} \mathrm{GWS}^h\text{: } [\mathrm{M2}KK]_e\{Q\}_e &\equiv \int_{\Omega_e} \nabla\{N\}\cdot\nabla\{N\}^T \mathrm{d}\tau\{Q\}_e \\ &= \int_{\Omega_e} \frac{\partial\{N\}}{\partial\zeta_\alpha}\left(\frac{\partial\zeta_\alpha}{\partial x_k}\right)_e \frac{\partial\{N\}^T}{\partial\zeta_\beta}\left(\frac{\partial\zeta_\beta}{\partial x_k}\right)_e \mathrm{det}_e\, \mathrm{d}\zeta\{Q\}_e \end{aligned}$$

contains det_e, the global-local coordinate transformation determinant, as well as the element local coordinate differential element $\mathrm{d}\zeta$. Counterclockwise (CCW) node numbering is the convention (see Figure A.2) for which an elementary linear algebraic process readily predicts the NC basis $n = 2$ coordinate system

$$\begin{aligned} \zeta_1 &= [\{\mathrm{X}_2\mathrm{Y}_3 - \mathrm{X}_3\mathrm{Y}_2)_e + (\mathrm{Y}_2 - \mathrm{Y}_3)_e x + (\mathrm{X}_3 - \mathrm{X}_2)_e y]/\mathrm{det}_e \\ \zeta_2 &= [\{\mathrm{X}_3\mathrm{Y}_1 - \mathrm{X}_1\mathrm{Y}_3)_e + (\mathrm{Y}_3 - \mathrm{Y}_1)_e x + (\mathrm{X}_1 - \mathrm{X}_3)_e y]/\mathrm{det}_e \\ \zeta_3 &= [\{\mathrm{X}_1\mathrm{Y}_2 - \mathrm{X}_2\mathrm{Y}_1)_e + (\mathrm{Y}_1 - \mathrm{Y}_2)_e x + (\mathrm{X}_2 - \mathrm{X}_1)_e y]/\mathrm{det}_e \end{aligned}$$

also the determinant (det) which equates to the area of the triangle Ω_e

$$2A_e \equiv \det \begin{bmatrix} 1 & \mathrm{X}_1 & Y_1 \\ 1 & \mathrm{X}_2 & Y_2 \\ 1 & \mathrm{X}_2 & Y_2 \end{bmatrix}_e$$

The $n = 3$ NC system ζ_α definition involves a 4×4 matrix determinant solution

$$\{N_1(\zeta)\} = \begin{Bmatrix} \zeta_1 \\ \zeta_2 \\ \zeta_3 \\ \zeta_4 \end{Bmatrix} \equiv \frac{1}{6\mathrm{V}_e}\begin{Bmatrix} a_1 + b_1x + c_1y + d_1z \\ -(a_2 + b_2x + c_2y + d_2z) \\ a_3 + b_3x + c_3y + d_3z \\ -(a_4 + b_4x + c_4y + d_4z) \end{Bmatrix}$$

$$a_i \equiv \det \begin{bmatrix} X_j & Y_j & Z_j \\ X_m & Y_m & Z_m \\ X_p & Y_p & Z_p \end{bmatrix}, \; b_i \equiv -\det \begin{bmatrix} 1 & Y_j & Z_j \\ 1 & Y_m & Z_m \\ 1 & Y_p & Z_p \end{bmatrix}_e$$

$$c_i \equiv \det \begin{bmatrix} X_j & 1 & Z_j \\ X_m & 1 & Z_m \\ X_p & 1 & Z_p \end{bmatrix}_e, \; d_i \equiv -\det \begin{bmatrix} X_j & Y_j & 1 \\ X_m & Y_m & 1 \\ X_p & Y_p & 1 \end{bmatrix}_e$$

$$6V_e \equiv \det \begin{bmatrix} 1 & X_1 & Y_1 & Z_1 \\ 1 & X_2 & Y_2 & Z_2 \\ 1 & X_3 & Y_3 & Z_3 \\ 1 & X_4 & Y_4 & Z_4 \end{bmatrix}$$

wherein indices (i, j, m, p) permute (1,2,3,4) over the CCW Ω_e vertex node numbering (Figure A.2).

As NC bases are linear, spatial differentiation generates a string of {0,1} column matrices:

$$\frac{\partial\{N\}}{\partial\zeta_\alpha} = \{1,0,0,0,\ \}^T, \{0,1,0,0,\ \}^T, \{0,0,1,0,\ \}^T, \{0,0,0,0,1\}^T$$

Hence, GWS^h DE coordinate-transformed square matrix entries are each the Kronecker delta, $\delta_{\alpha\beta}$.

$$\frac{\partial\{N\}}{\partial\zeta_\alpha}\frac{\partial\{N\}^T}{\partial\zeta_\beta} \Rightarrow \left[\mathrm{M2}\delta_{\alpha\beta}\right]_{n+1,n+1}$$

which are unity at matrix column-row location α,β, and elsewhere are zero. Therefore,

$$\begin{aligned} \mathrm{GWS}^h\!: [\mathrm{M2}KK]_e\{Q\}_e &\equiv \int_{\Omega_e} \nabla\{N\}\cdot\nabla\{N\}^T \mathrm{d}\tau\{Q\}_e \\ &= \frac{\det_e}{n!}\left(\frac{\partial\zeta_\alpha}{\partial x_k}\frac{\partial\zeta_\beta}{\partial x_k}\right)_e \left[\mathrm{M2}\delta_{\alpha\beta}\right]\{Q\}_e \end{aligned}$$

Matrix statement completion requires the inverse coordinate transformation. Formation is elementary for $n = 2$:

$$\left[\frac{\partial\zeta_\alpha}{\partial x_i}\right]_e = \frac{1}{\det_e}\begin{bmatrix} Y_2 - Y_3, & X_3 - X_2 \\ Y_3 - Y_1, & X_1 - X_3 \\ Y_1 - Y_2, & X_2 - X_1 \end{bmatrix}_e$$

and determinantal for $n = 3$

$$\left[\frac{\partial \zeta_a}{\partial x_i}\right]_e = \left[\frac{\partial x_i}{\partial \zeta_a}\right]_e^{-1} = \frac{1}{\det_e}\begin{bmatrix}\text{matrix of}\\ i\ \text{cofactors}\\ \text{transposed}\end{bmatrix}_e$$

Letting subscripted $\zeta_{a\,k}$ denote Ω_e vertex node differences ΔX, ΔY, the linear NC basis GWS^h algorithm $n = 2$ element matrix for the Laplacisn is

$$\mathrm{GWS}^h\text{: } [\mathrm{M2}KK]_e\{Q\}_e \equiv \int_{\Omega_e} \nabla\{N\}\cdot\nabla\{N\}^T \mathrm{d}\tau\{Q\}_e$$

$$= \frac{1}{\det_e n!}\begin{bmatrix} \zeta_{11}^2+\zeta_{12}^2, & \zeta_{11}\zeta_{21}+\zeta_{12}\zeta_{22}, & \zeta_{11}\zeta_{31}+\zeta_{12}\zeta_{32} \\ \zeta_{21}\zeta_{11}+\zeta_{22}\zeta_{12}, & \zeta_{21}^2+\zeta_{22}^2, & \zeta_{21}\zeta_{31}+\zeta_{22}\zeta_{32} \\ \zeta_{31}\zeta_{11}+\zeta_{32}\zeta_{12}, & \zeta_{31}\zeta_{21}+\zeta_{32}\zeta_{22}, & \zeta_{31}^2+\zeta_{32}^2 \end{bmatrix}\{Q\}_e$$

For the linear NC basis, the n = 3 element matrix subscripted $\zeta_{a\,k}$ entries denote Ω_e vertex node difference products ΔXΔY, ΔYΔZ, ΔXΔZ:

$$\mathrm{GWS}^h\text{: } [\mathrm{M2}KK]_e\{Q\}_e \equiv \int_{\Omega_e} \nabla\{N\}\cdot\nabla\{N\}^T \mathrm{d}\tau\{Q\}_e$$

$$= \frac{1}{\det_e n!}\begin{bmatrix} \zeta_{11}^2+\zeta_{12}^2+\zeta_{13}^2, & \zeta_{11}\zeta_{21}+\zeta_{12}\zeta_{22}+\zeta_{13}\zeta_{23}, & \cdots \\ \zeta_{21}\zeta_{11}+\zeta_{22}\zeta_{12}+\zeta_{23}\zeta_{13}, & \zeta_{21}^2+\zeta_{22}^2+\zeta_{23}^2, & \cdots \\ \zeta_{31}\zeta_{11}+\zeta_{32}\zeta_{12}+\zeta_{33}\zeta_{13}, & \zeta_{31}\zeta_{21}+\zeta_{32}\zeta_{22}+\zeta_{33}\zeta_{23}, & \cdots \\ \zeta_{41}\zeta_{11}+\zeta_{42}\zeta_{12}+\zeta_{43}\zeta_{13}, & \zeta_{41}\zeta_{21}+\zeta_{42}\zeta_{22}+\zeta_{43}\zeta_{23}, & \cdots \end{bmatrix}\{Q\}_e$$

As a final point, the integration of undifferentiated linear NC bases $\forall n$ is analytical

$$\int_{\hat{\Omega}_e} \zeta_1^p\,\zeta_2^q\,\zeta_3^r\,\zeta_4^s\,d\tau = \det_e \frac{p!q!r!s!}{(n+p+q+r+s)!}$$

A.5 Finite element TP basis coordinate transformation

Invoking the chain rule, the bilinear TP basis GWS^h DE algorithm matrix for n-D Laplacian on a generic Ω_e

$$\mathrm{GWS}^h\text{: } [\mathrm{M2}KK]_e\{Q\}_e \equiv \int_{\Omega_e} \nabla\{N^+\}\cdot\nabla\{N^+\}^T \mathrm{d}\tau\{Q\}_e$$

$$= \int_{\Omega_e} \frac{\partial\{N^+\}}{\partial\eta_j}\left(\frac{\partial\eta_j}{\partial x_k}\right)_e \frac{\partial\{N^+\}^T}{\partial\eta_i}\left(\frac{\partial\eta_i}{\partial x_k}\right)_e \det_e \mathrm{d}\eta\{Q\}_e$$

likewise inserts $\det_e$, the global-local coordinate transformation determinant. In contrast to the NC basis completion, the TP basis inverse coordinate transformation $\eta_j = \eta_j(x_i)$ is not *a priori* known. The resolution is x_i coordinate system interpolation on Ω_e in terms of the element nodal coordinate set $\{X_i\}_e$ and the TP basis $\{N^+(\eta_j)\}$ for $1 \le i \le n$

$$x_i = x_i(\eta_j) \equiv \left\{N^+(\eta_j)\right\}^T \{X_i\}_e$$

The resultant forward coordinate transformation on Ω_e is

$$\left[\frac{\partial x_i}{\partial \eta_j}\right]_e = \frac{\partial}{\partial \eta_j}\{N^+(\eta)\}^T\{X_i\}_e$$

with inverse transformation determinantal operation

$$\left[\frac{\partial \eta_j}{\partial x_i}\right]_e = \left[\frac{\partial x_i}{\partial \eta_j}\right]_e^{-1} = \frac{1}{\det_e}\begin{bmatrix}\text{matrix of}\\ i\ \text{cofactors}\\ \text{transposed}\end{bmatrix}_e$$

The $n = 2$ matrix is readily stated explicitly

$$\left[\frac{\partial \eta_j}{\partial x_i}\right]_e = \left[\frac{\partial x_i}{\partial \eta_j}\right]_e^{-1} = \frac{1}{\det_e}\begin{bmatrix}\dfrac{\partial x_2}{\partial \eta_2}, & -\dfrac{\partial x_1}{\partial \eta_2}\\ -\dfrac{\partial x_2}{\partial \eta_1}, & \dfrac{\partial x_1}{\partial \eta_1}\end{bmatrix}_e$$

while the $n = 3$ inverse transformation remains i cofactor matrix expressed.

All n elements of the inverse coordinate transformation matrix are η-dependent, in contrast to the NC basis transformation being constants, as is the $\det_e$. To $n = 2$ illustrate, the elements of the TP basis transformation $n = 2$ matrix are:

$$\left(\frac{\partial \eta_2}{\partial x_2}\right)_e = \frac{1}{\det_e}\left(\frac{\partial x_1}{\partial \eta_1}\right)_e = \frac{1}{\det_e}\frac{\partial}{\partial \eta_1}\{N_1^+(\eta)\}^T\{X_1\}_e$$

$$= \frac{1}{4\det_e}\left[(1-\eta_2)(X_2 - X_1)_e + (1+\eta_2)(X_3 - X_4)_e\right]$$

$$\left(\frac{\partial \eta_1}{\partial x_1}\right)_e = \frac{1}{4\det_e}\left[(1-\eta_1)(Y_4 - Y_1)_e + (1+\eta_1)(Y_3 - Y_2)_e\right]$$

$$\left(\frac{\partial \eta_1}{\partial x_2}\right)_e = \frac{-1}{4\det_e}\left[(1-\eta_1)(X_4 - X_1)_e + (1+\eta_1)(X_3 - X_2)_e\right]$$

$$\left(\frac{\partial \eta_2}{\partial x_1}\right)_e = \frac{-1}{4\det_e}\left[(1-\eta_2)(Y_2 - Y_1)_e + (1+\eta_2)(Y_3 - Y_4)_e\right]$$

It is thus obvious that TP basis η-derivatives are also η-dependent, e.g., for $n = 3$

$$\frac{\partial}{\partial \eta_j}\{N_k^+(\eta)\} = \frac{1}{8}\left\{\begin{matrix} -(1-\eta_2)(1-\eta_3) \\ (1-\eta_2)(1-\eta_3) \\ (1+\eta_2)(1-\eta_3) \\ -(1+\eta_2)(1-\eta_3) \\ -(1-\eta_2)(1+\eta_3) \\ (1-\eta_2)(1+\eta_3) \\ (1+\eta_2)(1+\eta_3) \\ -(1+\eta_2)(1+\eta_3) \end{matrix}\right\}_{j=1}, \text{ etc}$$

with the other two matrices formed via obvious j index alterations on the $(1 \pm \eta_j)$ term therein. The $n = 2$ dual matrix set is

$$\frac{\partial}{\partial \eta_j}\{N_k^+(\eta)\} = \frac{1}{4}\left\{\begin{matrix} -(1-\eta_2) \\ (1-\eta_2) \\ (1+\eta_2) \\ -(1+\eta_2) \end{matrix}\right\}_{j=1}, \frac{1}{4}\left\{\begin{matrix} -(1-\eta_1) \\ -(1+\eta_1) \\ (1+\eta_1) \\ (1-\eta_1) \end{matrix}\right\}_{j=2}$$

Therefore, TP linear basis derivatives are distributions that preclude TP basis derivative extractions, also inverse coordinate transformation products, from GWS^h element matrices. The resolution is numerical quadrature evaluation of GWS^h TP basis element matrices. Since Ω_e is rectangular Cartesian in transform space, Gauss quadrature readily generates integral approximations with *a priori* given accuracy.

For η_{ji} denoting TP basis coordinate transformations, the generic Ω_e $[\text{M2}KK]_e$ matrix element, "to be integrated" is compactly expressed as

$$\begin{aligned}(\text{m2}kk_{\alpha\beta})_e &\equiv \iiint_{\hat{\Omega}_e} \frac{\partial N_\alpha}{\partial \eta_j}\frac{\partial N_\beta}{\partial \eta_k}\left(\eta_{ji}\eta_{ki}\right)_e \det_e^{-1} d\eta \\ &\equiv \int_{\hat{\Omega}} \left(\text{integrand}_{j,k,i}^{\alpha,\beta}(\boldsymbol{\eta})\right) d\eta,\ 1 \le j,\ k, i \le n\end{aligned}$$

The Gauss quadrature approximation of any matrix element is the summation cascade

$$(\text{m2}kk_{\alpha\beta})_e \cong \sum_{i=1}^{n}\sum_{j=1}^{n}\sum_{k=1}^{n}\sum_{p=1}^{P}\sum_{q=1}^{Q}\sum_{r=1}^{R} H_p H_q H_r\left(\text{integrand}_{j,k,i}^{\alpha,\beta}(\boldsymbol{\eta} \Rightarrow \eta_{j,k,i}^{p,q,r})\right)$$

wherein p,q,r denote specific η_j coordinates in Ω_e. Quadrature summation limits $(P,Q,R) \le 3$ with associated $\eta_{j\ k,\ i}$ coordinates and weights $H_{p,q,r}$ are defined in Table A.1.

Theory predicts that a Pth order (P,Q,R in $n = 3$) Gauss quadrature rule integrates a degree $2P$–1 polynomial exactly. Therefore, (P,Q,R) is the highest degree polynomial in $(m2kk_{\alpha\beta})_e$ dependent. That degree $2P$–1 integration is dependent on exact is moderated by $\det_e^{-1}$ as each matrix element is thus a *rational* polynomial.

Table A.1: Gauss numerical quadrature rule coefficients, $n = 2, 3$.

P, Q, R	Coordinates, $\eta_{j,k,i}^{p,q,r}$	Weights, $H_{p,q,r}$
1	0.0	1.0
2	±1/√3	1.0
3	0.0	8/9
	±√0.6	5/9

A.6 Nonlinearity, FE p-basis hypermatrix identifications

Completing the quadrature loop for all elements of $[\mathrm{M2}KK]_e$ extends the summation cascade to

$$[\mathrm{M2}KK]_e \cong \sum_{\alpha,\beta}^{n} \sum_{i}^{n} \sum_{j}^{n} \sum_{k}^{n} \sum_{p}^{P} \sum_{q}^{Q} \sum_{r}^{R} H_p H_q H_r \left(\mathrm{integrand}_{j,k,i}^{\alpha,\beta}(\boldsymbol{\eta} \Rightarrow \eta_{j,k,i}^{p,q,r})\right)$$

Once code preprocessed, resultant is the linear FE basis GWS^h NS/RaNS/FaNS PDE system algorithm *matrix library* for $[\mathrm{M2}KK]_e$ for domain discretization $\Omega \approx \Omega^h \equiv \cup\, \Omega_e$. This naming is distinct from historically referenced FE matrix libraries, which is classically the reservoir of various completeness degree trial space bases used in structural mechanics analyses.

Matrix library preprocessing is required for all algorithmic element matrices. The key to code execution efficiency is in the extraction of element-independent data prior to matrix library generation, in particular ∀ DOF element matrices. This enables matrix library generation external from the CFD code's deeply embedded element "DO loop," which is a *key* contribution to execution efficiency. This is critical for the PDE system TEA theory alteration which generates significantly nonlinear calculus terms.

The resolution to this is the generic element first-order hypermatrix library, as thoroughly introduced in the monograph chapters. Consider $O(h^2)$ TEA-altered incompressible NS/FaNS $\mathrm{D}^m\mathbf{P}$

$$\mathrm{D}^m\mathbf{P}\colon \frac{\partial u_i}{\partial t} + \frac{\partial}{\partial x_j}\left[\begin{matrix} u_j u_i + \mathbf{P}^*\big|_{n+1}^{p}\,\delta_{ij} - \frac{1}{\mathrm{Re}}\left(\frac{\partial u_i}{\partial x_j} + \frac{\partial u_j}{\partial x_i}\right) \\ -\frac{h^2 Re}{12}\left(u_j u_k \frac{\partial u_i}{\partial x_k} + u_i u_k \frac{\partial u_j}{\partial x_k}\right) \end{matrix}\right] - \frac{\mathrm{Gr}}{\mathrm{Re}^2}\,\Theta \widehat{g}_i = 0 + O(h^4)$$

The nonlinear convection term therein permeates the PDE system. Recalling u_j is divergence-free, repositioning the embedded scalar matrix product, removing element-independent data from within the integrand, and imposing the Green-Gauss divergence theorem generate the definition of this term "first-order hypermatrix"

$$\int_{\Omega_e} \{N\}\frac{\partial}{\partial x_j}\left(\{UJ\}_e^T\{N\}\{N\}^T\{UI\}_e\right)d\tau = \{UJ\}_e^T\int_{\Omega_e}\{N\}\{N\}\frac{\partial}{\partial x_j}\{N\}^T d\tau\{UI\}_e$$
$$\equiv \{UJ\}_e^T[M300J]_e\{UI\}_e$$

Selecting the NC linear basis, the $n = 3$ transformed integral generates the hypermatrix set $[M300J]_e$

$$\int_{\Omega_e}\{N\}\{N\}\frac{\partial\{N\}^T}{\partial\zeta_\alpha}\frac{\partial\zeta_\alpha}{\partial x_j}\det_e d\zeta = \int_0^1\begin{Bmatrix}\zeta_1\\ \zeta_2\\ \zeta_3\\ \zeta_4\end{Bmatrix}\begin{Bmatrix}\zeta_1\\ \zeta_2\\ \zeta_3\\ \zeta_4\end{Bmatrix}\frac{\partial\{N\}^T}{\partial\zeta_\alpha}\frac{1}{\det_e}\begin{bmatrix}\text{matrix of}\\ j\,\text{cofactors}\\ \text{transposed}\end{bmatrix}_e \det_e d\zeta$$

The pair of $\det_e$ cancel, $\dfrac{\partial\{N\}^T}{\partial\zeta_\alpha} = \{1,0,0,0,\},\{0,1,0,0,\},\{0,0,1,0,\},\{0,0,0,0,1\}$, and the integration limits are (0, 1).

Alteration to the TP bilinear basis is the hypermatrix set $[M300J]_e$.

$$\int_{\Omega_e}\{N^+\}\{N^+\}\frac{\partial\{N^+\}^T}{\partial\eta_k}\frac{\partial\eta_k}{\partial x_j}det_e d\eta$$

$$= \frac{1}{8^3}\int_{-1}^1\begin{Bmatrix}(1-\eta_1)(1-\eta_2)(1-\eta_3)\\ (1+\eta_1)(1-\eta_2)(1-\eta_3)\\ (1+\eta_1)(1+\eta_2)(1-\eta_3)\\ (1-\eta_1)(1+\eta_2)(1-\eta_3)\\ (1-\eta_1)(1-\eta_2)(1+\eta_3)\\ (1+\eta_1)(1-\eta_2)(1+\eta_3)\\ (1+\eta_1)(1+\eta_2)(1+\eta_3)\\ (1-\eta_1)(1+\eta_2)(1+\eta_3)\end{Bmatrix}\begin{Bmatrix}(1-\eta_1)(1-\eta_2)(1-\eta_3)\\ (1+\eta_1)(1-\eta_2)(1-\eta_3)\\ (1+\eta_1)(1+\eta_2)(1-\eta_3)\\ (1-\eta_1)(1+\eta_2)(1-\eta_3)\\ (1-\eta_1)(1-\eta_2)(1+\eta_3)\\ (1+\eta_1)(1-\eta_2)(1+\eta_3)\\ (1+\eta_1)(1+\eta_2)(1+\eta_3)\\ (1-\eta_1)(1+\eta_2)(1+\eta_3)\end{Bmatrix}$$
$$\frac{\partial\{N^+\}^T}{\partial\eta_k}\frac{1}{\det_e}\begin{bmatrix}\text{matrix of}\\ j\,\text{cofactors}\\ \text{transposed}\end{bmatrix}_e \det_e d\eta$$

$$\cong \sum_{\alpha,\beta}^{n}\sum_{i}^{n}\sum_{j}^{n}\sum_{k}^{n}\sum_{p}^{3}\sum_{q}^{3}\sum_{r}^{3}H_pH_qH_r\left(\text{integrand}_{j,k,i}^{\alpha,\beta}(\boldsymbol{\eta}\Rightarrow\eta_{j,k,i}^{p,q,r})\right) \equiv [M300J]_e$$

The $\det_e$ again cancel, integration limits are ±1, $\partial\{N^+\}^T/\partial\eta_k$ quantification was previously detailed, and the $P \geq 3$ Gauss quadrature rule is required for accuracy.

GWS^h Newton Jacobian contributions $[\text{JAC\_}\{UI\},\{UI\}]_e[\text{JAC}(\{UI,UJ\})]_e$ contain $[M300J]_e$, which is altered to $[M3J00]_e$. For the TP linear basis, the integration process generates the $[M3J00]_e$ hypermatrix set

$$\int_{\Omega_e} \frac{\partial \{N^+\}^T}{\partial \eta_k} \frac{\partial \eta_k}{\partial x_j} \{N^+\}\{N^+\} \det_e \mathrm{d}\eta$$

$$= \frac{1}{8^3} \int_{-1}^{1} \frac{\partial \{N^+\}^T}{\partial \eta_k} \frac{1}{\det_e} \begin{bmatrix} \text{matrix of} \\ j \text{ cofactors} \\ \text{transposed} \end{bmatrix}_e \begin{Bmatrix} (1-\eta_1)(1-\eta_2)(1-\eta_3) \\ (1+\eta_1)(1-\eta_2)(1-\eta_3) \\ (1+\eta_1)(1+\eta_2)(1-\eta_3) \\ (1-\eta_1)(1+\eta_2)(1-\eta_3) \\ (1-\eta_1)(1-\eta_2)(1+\eta_3) \\ (1+\eta_1)(1-\eta_2)(1+\eta_3) \\ (1+\eta_1)(1+\eta_2)(1+\eta_3) \\ (1-\eta_1)(1+\eta_2)(1+\eta_3) \end{Bmatrix}$$

$$\begin{Bmatrix} (1-\eta_1)(1-\eta_2)(1-\eta_3) \\ (1+\eta_1)(1-\eta_2)(1-\eta_3) \\ (1+\eta_1)(1+\eta_2)(1-\eta_3) \\ (1-\eta_1)(1+\eta_2)(1-\eta_3) \\ (1-\eta_1)(1-\eta_2)(1+\eta_3) \\ (1+\eta_1)(1-\eta_2)(1+\eta_3) \\ (1+\eta_1)(1+\eta_2)(1+\eta_3) \\ (1-\eta_1)(1+\eta_2)(1+\eta_3) \end{Bmatrix} \det_e \mathrm{d}\eta \, \mathrm{diag}[\delta_{IJ}]$$

$$\cong \sum_{\alpha,\beta}^{n} \sum_{i}^{n} \sum_{j}^{n} \sum_{k}^{n} \sum_{p}^{3} \sum_{q}^{3} \sum_{r}^{3} H_p H_q H_r \left(\text{integrand}_{j,k,i}^{\alpha,\beta} (\boldsymbol{\eta} \Rightarrow \eta_{j,k,i}^{p,q,r}) \right) \equiv [\mathrm{M3}J00]_e$$

and the $\det_e$ again cancel.

The representative dual-differentiated nonlinear hypermatrix set is the NS/FaNS PDE system $O(h^2)$ TEA augmentation matrix $[\mathrm{M30}JK]_e$

$$- \int_{\Omega_e} \{N\} \frac{\partial}{\partial x_j} \left(\frac{h^2 \mathrm{Re}}{12} u_j u_k \frac{\partial u_i}{\partial x_k} \right) \mathrm{d}\tau = \frac{h_e^2 \mathrm{Re}}{12} \{\mathrm{U}J * \mathrm{U}K\}_e^T \int_{\Omega_e} \{N\} \frac{\partial \{N\}}{\partial x_j} \frac{\partial \{N\}^T}{\partial x_k} d\tau \{\mathrm{U}I\}_e$$

$$\approx \frac{h_e^2 \mathrm{Re}}{12} \{\mathrm{U}J * \mathrm{U}K\}_e^T [\mathrm{M30}JK]_e \{\mathrm{U}I\}_e$$

The TP linear basis completion statement is

$$\int_{\Omega_e} \{N^+\} \frac{\partial \{N^+\}}{\partial \eta_m} \frac{\partial \eta_m}{\partial x_j} \frac{\partial \{N^+\}^T}{\partial \eta_m} \frac{\partial \eta_m}{\partial x_k} \det{}_e \, \mathrm{d}\eta = \frac{1}{8^3} \int_{-1}^{1} \left\{ \begin{matrix} (1-\eta_1)(1-\eta_2)(1-\eta_3) \\ (1+\eta_1)(1-\eta_2)(1-\eta_3) \\ (1+\eta_1)(1+\eta_2)(1-\eta_3) \\ (1-\eta_1)(1+\eta_2)(1-\eta_3) \\ (1-\eta_1)(1-\eta_2)(1+\eta_3) \\ (1+\eta_1)(1-\eta_2)(1+\eta_3) \\ (1+\eta_1)(1+\eta_2)(1+\eta_3) \\ (1-\eta_1)\,(1+\eta_2)(1+\eta_3) \end{matrix} \right\}$$

$$\frac{\partial \{N^+\}}{\partial \eta_m} \frac{1}{\det_e} \begin{bmatrix} \text{matrix of} \\ j \text{ cofactors} \\ \text{transposed} \end{bmatrix}_e \frac{\partial \{N^+\}^T}{\partial \eta_m} \frac{1}{\det_e} \begin{bmatrix} \text{matrix of} \\ k \text{ cofactors} \\ \text{transposed} \end{bmatrix}_e \det{}_e d\eta$$

$$\cong \sum_{\alpha,\beta}^{n} \sum_{i}^{n} \sum_{j}^{n} \sum_{k}^{n} \sum_{p}^{3} \sum_{q}^{3} \sum_{r}^{3} H_p H_q H_r \left(\text{integrand}_{\mathrm{j,k,i}}^{\alpha,\beta} \, (\boldsymbol{\eta} \Rightarrow \eta_{j,k,i}^{p,q,r}) \right)$$

$$\equiv [\mathrm{M30}JK]_e$$

and one inverse $\det_e$ remains.

Alteration of illustrated hypermatrix completions for compressible fluid NS/FaNS algorithms involves symbolic manipulations at best, and while detailed integral statements at first appearance look mind-boggling, the Gauss quadrature preprocessor package makes truly short work of these details.

Appendix B: Theory, weak formulation optimal continuous Galerkin FE $p = 1,2,3$ basis CFD algorithms

B.1 Derivation, mathematically rigorous weak formulation theory

The mathematical characterization of fluid mechanics' continuum conservation principles is fundamentally challenged by their ubiquitous nonlinearity. Notwithstanding, the nonlinear NS transonic potential DM PDE system does uniquely enable rigorous derivation of weak formulation continuous Galerkin CFD FE $p = 1,2,3$ completeness degree algorithms. The transonic potential DM EBV PDE, the statements of density nonlinearity and intrinsic symmetric quadratic form are

$$\text{DM:} \quad -\frac{\partial}{\partial x_j}\left(\rho\frac{\partial\varphi}{\partial x_j}\right) = 0,$$

$$\rho = \rho_\infty\left[\frac{T_{0\infty}}{T_\infty} - \frac{\gamma-1}{2\gamma T_\infty}\nabla\varphi\cdot\nabla\varphi\right]^{1/(\gamma-1)},$$

$$\left\|\varphi,\varphi\right\|_E \equiv \frac{1}{2}\int_\Omega [\rho\nabla\varphi\cdot\nabla\varphi]d\tau$$

Rigorous derivation of accuracy, hence asymptotic convergence, involves manipulations on an abstract *bilinear form* $B(u,v)$ populating a region Ω of Euclidean space for arbitrary functions $u(\mathbf{x})$ and $v(\mathbf{x})$:

$$B(u,v) \equiv \int_\Omega \nabla u\cdot k_1\nabla v\,d\tau + \int_{\partial\Omega n} k_2 uv\,d\sigma = b(v)$$

Parameter k_1 denotes the solution-generated *distributed* data on Ω with $b(v)$ the remaining *a priori* given data on $\partial\Omega$ independent of u,v. For reference, in the bilinear form specific to the weak formulation of the transonic full potential DM PDE:

$$B(u\rightarrow\varphi^N, v\rightarrow\Psi_\beta) = \int_\Omega \nabla\Psi_\beta\cdot\rho\,\nabla\varphi^N\,d\tau$$

$\varphi^N(\mathbf{x})$ as DM weak form approximates the solution and Ψ_β represents the space of test functions $\forall\beta$.

Abstract weak form theory replaces "find $\varphi^N(\mathbf{x})$ to EBV PDE + BCs" with find $u \in H^1$ such that $B(u,v) = b(v)$ $\forall v \in H^1$ and $v(\partial\Omega_D) \equiv 0$ and $\forall v \in H^1$ and $v(\partial\Omega_D) \equiv 0$. Therein, $\partial\Omega_D$ denotes domain boundary segments on which *a priori* given Dirichlet data exist. This restriction is symbolized $v \in H_0^1$ as fluxes on such $\partial\Omega$ segments are *a priori* unknown.

https://doi.org/10.1515/9783112206430-017

The process of "converting" find $u \in H^1$ such that $B(u, v) = b(v) \ldots$ to computable form, i.e., a discretely implemented weak statement WS^h, first, replaces v by its interpolation v^h. Then, u is replaced by its spatially discrete approximation $u^h(\mathbf{x})$, a process based on interpolation theory as well. These *much smaller* v^h and u^h function spaces are termed *finite-dimensional subspaces,* symbolized as H^h. Note that *all* such functions in H^h also reside in H^1.

Implementing v interpolation v^h first generates the abstract weak form:

$$\text{find } u \in H^1 \text{ such that } B(u, v^h) = b(v^h), \ \forall v^h \subset H^h \in H_0^1$$

Then for u interpolation altered to $u^h \subset H^h \in H^1$ generates the terminal abstract statement:

$$\text{find } u^h \subset H^h \text{ such that } B(u^h, v^h) = b(v^h) \ \forall v^h \subset H^h \in H_0^1 \text{ and } u^h \subset H^h \in H^1$$

This completes the mathematical concepts and tools necessary to *measure accuracy.*

Subtracting the latter expression from the former yields

$$B(u - u^h, v^h) = 0, \ \forall \ u \in H^1, \ v^h \subset H^h \in H_0^1, u^h \subset H^h \in H^1$$

as $u - u^h \equiv e^h$ is the weak form solution approximation error. This bilinear form states that the error is *orthogonal* in H^h, the finite-dimensional sub-space of H^1 from which *all* eligible functions are extracted. Specifically, this bilinear form is valid for functions supporting non-Galerkin WS^h predictions, such as artificially stabilized non-Galerkin and/or finite volume weak form algorithms, and *all* test functions $v^h \subset H^h$ remain absolutely arbitrary.

The fact that GWS^h error e^h is orthogonal in H^h for *any* v^h leads to the query:

> "Is there an *optimal* choice for v^h such that the *best approximation* u^h is identified?"

The Hilbert space H^h is represented geometrically in Figure B.1 as a plane containing *all* eligible functions u^h and v^h, which are interpreted as vectors lying therein. Since H^h is not large enough to contain the exact solution to $B(u, v)$, it is represented as the out-of-plane vector u. Consequently, the exact solution u is resolvable into its projection $u^h \subset H^h$ lying in the H^h plane, and the plane-orthogonal complement $u - Pu^h$ where P denotes "perpendicular" to the plane H^h.

Vector addition of u^h and $u - Pu^h$ is equal to u for all u^h in H^h. Any v^h candidate also lies in H^h and the magnitude of its projection $u - P{*}v^h$ is larger than $u - Pu^h$ since $u - P{*}v^h$ is not perpendicular to the plane H^h. Therefore, u^h is the *optimal* choice for v^h, which indeed defines the *optimal* continuous Galerkin GWS^h.

This geometric interpretation does yield to mathematical precision. The bilinear form

$$\text{find } u \in H^1 \text{ such that } B(u, v^h) = b(v^h), \ \forall v^h \subset H^h \in H_0^1$$

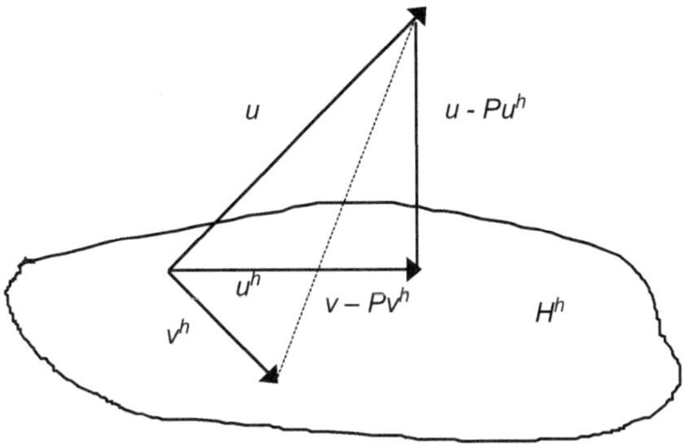

Figure B.1: Geometric illustration of the best approximation.

can be written for any argument pair. Replacing both u and v^h with $u - v^h$ is interpreted as the error associated with v^h being the interpolation of u. These alterations

$$\mathrm{B}(u - u^h, v^h) = 0,\ \forall u \in H^1,\ v^h \subset H^h \in H_0^1,\ u^h \subset H^h \in H^1$$

then transition to $\mathrm{B}(u - v^h,\ u - v^h) = 2\|u - v^h\|_E$ which is precisely twice the energy norm of the interpolation error. Adding and subtracting u^h from each argument

$$\begin{aligned} 2\|u - v^h\|_E &= \mathrm{B}(u - u^h - v^h + u^h, u - u^h - v^h + u^h) = \mathrm{B}(u - u^h, u - u^h, u^h - v^h,\ u^h - v^h) \\ &= \mathrm{B}(e^h, e^h)\ + B(u^h - v^h,\ u^h - v^h) \end{aligned}$$

remains the *error* in the weak form solution u^h for *any* v^h. As the right-side terms are non-negative symmetric quadratic forms, the energy norm measurement of the distance between u and v^h is a *minimum* when u^h and v^h are identical, which causes the second term to vanish.

To summarize, the error in the approximate solution u^h to $\mathrm{B}(u^h, v^h)$ is the *smallest possible* $\forall v^h$ when v^h and u^h are *identical.* This abstract weak form theory has precisely predicted that the continuous WF *Galerkin criterion* is *optimal,* as previously asserted, since u^h is the *perpendicular projection* of u onto H^h in the abstract sense. Thereby, continuous Galerkin weak statement solution u^h is the *best possible* (!) among all candidates in H^h as measured in DM energy norm $\|u^h\|_E$ and/or Hilbert H^1 norm $\|u^h\|_{H^1}$.

It remains to establish a *bound* on GWS^h error $e^h \equiv u - u^h$ in terms of *data* specifications. These manipulations on abstract bilinear form $B(u - v^h,\ u - v^h) = 2\|u - v^h\|_E$ quantify the continuous Galerkin solution error for any function $u^h \subset H^h$ is exactly twice the energy norm $B(u - u^h, u - u^h)\ = B(e^h, e^h)\ =\ 2\|e^h\|_E$. Recall $B(u,v)$ for transonic DM contains solution-generated nonlinear *data* distribution $k_1(\mathbf{x}) \subset \Omega$ while $b(v)$ denotes *a priori* given data on Ω.

Therefore, for $B(u-u^h, u-u^h) = B(e^h, e^h) = 2\|e^h\|_E$ to be mathematically *well-posed* requires certain data properties exist, specifically *continuity* and *coerciveness*. In the continuum, where $(u, v) \in H^1$ with measures $\|u\|_{H^1}$ and $\|v\|_{H^1 0}$, *continuity* requires $B(u-u^h, u-u^h) = B(e^h, e^h) = 2\|e^h\|_E$ to be bounded. The NS potential DM bilinear form is $B(\rho\nabla\varphi, \nabla\Psi) \leq max\{\rho(\varphi(\mathbf{x}))\}\ B(\nabla\varphi, \nabla\Psi)$. Reverting to abstract notation, the notion of *continuity* for the well-posed EBV PDE + BC statement requires

$$\mathrm{B}(u, v, \kappa, h, k) \leq \beta \mathrm{B}(u, v) \leq \beta\|u, v\|_{H^1} \leq \beta\|u\|_{H^1}\|v\|_{H^1}$$

for β denoting the maximum search result with the final expression established via the triangle inequality [2].

The principle of *coerciveness* is the counterpart to *continuity*. For potential flow, the DM symmetric bilinear form, which coincides with the energy norm, must have a data minimum bound of the form $\mathrm{B}(\rho\nabla\varphi, \nabla\varphi) \leq min\{\rho(\varphi(\mathbf{x}))\}\ \mathrm{B}(\nabla\varphi, \nabla\varphi)$, as detailed in Oden (1994). With *continuity* and *coerciveness* quantified, the bound on the error e^h can be established. Combining $\mathrm{B}(u-u^h, u-u^h) = \mathrm{B}(e^h, e^h) = 2\|e^h\|_E$ with $\mathrm{B}(u, u) \geq \alpha\|u\|^2_{H^1}$, $\forall\ u \in H^1$, along with the argument exchange $u \to e^h$, coerciveness requires $\|e^h\|_E \geq \alpha\|e^h\|^2_{H^1}$. The bilinear form on the right side admits the useful manipulation:

$$\begin{aligned}\|e^h\|^2_{H^1} = \mathrm{B}(e^h, e^h) &= \mathrm{B}(u-u^h,\ u-u^h \pm v^h) = \mathrm{B}(e^h, (u-v^h) + (v^h-u^h)) \\ &= \mathrm{B}(e^h, u-v^h) + \mathrm{B}(e^h, v^h-u^h)\end{aligned}$$

and the last line in the second term vanishes for the weak form Galerkin criterion. Combining and rearranging yields the result for arbitrary v^h

$$\|e^h\|^2_{H^1} = \mathrm{B}(e^h, e^h) \leq \frac{1}{\alpha}\mathrm{B}(e^h, u-v^h) \leq \frac{\beta}{\alpha}\|e^h\|_{H^1}\|u-v^h\|_{H^1}$$

via the constraint of continuity. Completing the obvious cancellation, the terminal statement on error bounding in $u^h \subset H^h \in H^1$ is

$$\|e^h\|_{H^1} \leq \frac{\beta}{\alpha}\|u-v^h\|_{H^1}, \quad \forall\ v^h$$

Thereby, the weak form continuous GWS^h error e^h is bounded above by the material *data* constant β/α times the distance between the exact solution u and *any* function $v^h \subset H^h \in H^1$. The smaller is α the larger is the bound and as $\alpha \to 0$ the problem likely becomes unstable. Finally, since v^h is absolutely arbitrary, selecting it as the *interpolant* u^I of u bounds e^h by the error associated with the interpolation of the exact solution for reasonable β/α.

GWS^h approximation error $e^h = u - u^h$ *bounding* by interpolation error has been elegantly abstract analysis derived. Interpreting the right side of $\|e^h\|_{H^1} \leq \frac{\beta}{\alpha}\|u-v^h\|_{H^1}$, $\forall\ v^h$ as interpolation error, the *asymptotic convergence* rate estimate at the level of the individual Ω_e where h is the measure of the element Ω_e and ρ relates to the particular geometry of Ω_e [3],

$$\left\| u - v^h \right\|_{H^m, \Omega_e} \leq \ C_k \frac{h_e^{\,p+1}}{\rho^m} \left\| u \right\|_{r, \Omega^e}$$

The two *independent* measures that characterize the geometry of *any* $n = 3$ discretization cell Ω_e are the diameter of the largest inscribed sphere which contacts at least three surfaces of Ω_e, and the diameter of the smallest circumscribed sphere which intersects at least three vertex nodes of Ω_e. Thereby, the error generated by exact solution interpolation is bounded by the H^r norm of the analytical solution u multiplied by a FE basis degree p-dependent constant C_p times the Ω_e measure h_e raised to $p+1$ power divided by the Ω_e geometry inscribed-circumscribed sphere *pair* ratio ρ raised to m which for NS potential second-order EBV DM is $m = 1$. The H^r norm is the *smoothness* constraint of the exact solution, i.e., u must possess at least r derivatives that are square-integrable to exist therein.

Thus emphasized is the key property of weak formulation Galerkin GWS^h. As $m = 1$ $\forall$ second-order EBV PDE + BC systems, the exact solution u must possess two derivatives in all combinations. Conversely, the GWS^h weak form approximation u^h need possess only a single derivative. Hence, r is *never smaller* than m, and r is required to grow as order $2p$ in the progression to higher completeness degree p FE bases. The precise statement of this key issue is

$$u \ \in \ H^r, \ \textit{while} \ \ u^h \subset H^h \ \in \ H^m$$

The two Ω_e geometry scales ρ_e and h_e, in the asymptotic convergence statement can be reduced to the mesh measure h_e alone by requiring that any mesh refinement sequence $\Omega^h \Rightarrow \Omega^{h/2} \Rightarrow \Omega^{h/4}$ ... be *regular*, i.e., maintain $h_e/\rho_e \leq \sigma$ a nominal constant. Substituting $\rho_e^{\;-m} \leq \sigma h_e^{\;-m}$ then generates a *regular* refinement *local* interpolation error bound.

$$\left\| u - v^h \right\|_{m, \Omega_e} \leq \ C_{p\sigma} \ h_e^{\,p+1-m} \left\| u \right\|_{r, \Omega_e}$$

with a new constant $C_{p\sigma}$ dependent at most on the FE basis degree p.

The *global* asymptotic error estimate results upon summation over all $\Omega_e \subset \Omega^h$. Then, replacing the interpolation v^h by the GWS^h solution u^h, which is itself bounded by the interpolation error, then defines the GWS^h solution approximation error *asymptotic convergence rate* under *regular* mesh refinement to be

$$\left\| e^h \right\|_{m, \Omega} \equiv \sum_{e=1}^{M} \left\| u - u^h \right\|_{m, \Omega_e} \ \leq \ C \ h^{p+1-m} \left\| u \right\|_{r, \Omega}$$

which requires the exact solution $^*u^*$ be *sufficiently smooth* such that it resides in H^r for $r > m$.

As exact solutions are never known, $B(u,v) \equiv \int_\Omega \nabla u \cdot k_1 \nabla v d\tau + \int_{\partial\Omega_n} k_2 uv d\sigma = b(v)$ provides the resolution that equates the abstract bilinear form $B(u,v)$ to $b(v)$, the placeholder $\forall$ *a priori* given *data*. Strictly algebraic $b(v)$ functions reside in L2, and correspondingly in H^0. For DM $b(v)$ definition, as these data can exist on both Ω and $\partial\Omega$, the asymptotic convergence estimate transitions to

$$\|e^h\|_{m,\Omega} \equiv \sum_{e=1}^{M} \|u - u^h\|_{m,\Omega_e} \leq C\, h^{p+1-m}\left[\|\text{data}\|_{0,\Omega} + \|\text{data}\|_{0,\partial\Omega}\right]$$

This determination must acknowledge $r - m$ dominance for non-smooth data generating u absent the required smoothness $u \in H^r$. The resolution of this key theoretical issue involves replacing the exponent $p+1-m$ with $\gamma \equiv min(p+1-m,\ r-m)$. Thereby, the *terminal* asymptotic error estimate in H^m is

$$\|e^h\|_{m,\Omega} \leq C\, h^{\gamma}\left[\|\text{data}\|_{0,\Omega} + \|\text{data}\|_{0,\partial\Omega}\right],\ \gamma \equiv \min(p+1-m,\ r-m)$$

Finally, the CFD preferred practice is to replace error estimation in H^m with the NS/RaNS PDE system state variable member-specific energy norm set, which explicitly embeds solution-generated nonlinear data distributions. In terms of the NS transonic potential DM intrinsic energy norm $\|\varphi,\varphi\|_E \equiv \frac{1}{2}\int_\Omega [\rho\nabla\varphi \cdot \nabla\varphi]d\tau$, and recalling that $m = 1$ $\forall$second-order EBV PDE systems:

$$\text{GWS}^h\!:\ \|e^h\|_E \leq C\, h^{2\alpha}\left(\|\text{data}\|_{L2,\Omega}, \|\text{data}\|_{L2,\partial\Omega}\right),\quad \alpha \equiv \min(p,\ r-1)$$

This is valid for FE $p = 1,2,3$ completeness degree Lagrange polynomial trial space bases. By extension, the weak formulation of NS/RaNS PDE continuous GWS^h FE $p = 1$, 2, 3 completeness degree basis algorithm asymptotic convergence rates in the PDE state variable member-specific energy norm $\|e^h\|_E$ are predicted to be equal to 2, 4, and 6, respectively, for *data* non-smoothness L2 norm bounded.

Acknowledgment: This material is extracted from lecture notes prepared by Prof. J. Tinsley Oden for presentation at our annual summer short course, *Finite Element Computational Fluid Dynamics*, hosted in alternating years at the University of Tennessee/Knoxville and the University of Texas/Austin.

B.2 Validation, NS PDE GWS^h asymptotic convergence, solution optimality

Validation that the weak form continuous GWS^h CFD approximation error $\|e^h(t)\|_E$ is the *extremum* proved to be precisely tractable for the laminar incompressible $n = 2$ NS PDE system Reynolds ordered to the BL PDE system for state variable $\{q(x_j)\} = \{u, v\}^T$:

$$\text{DM: } \frac{\partial v}{\partial y} + \frac{\partial u}{\partial x} = 0, \ \mathbf{DP} \bullet \hat{i}\text{: } u\frac{\partial u}{\partial x} + v\frac{\partial u}{\partial y} - \frac{1}{\text{Re}}\frac{\partial^2 u}{\partial y^2} + \frac{\text{d}P}{\text{d}x} = 0$$

This PDE system is dual initial value, $\mathbf{DP} \bullet \hat{i}$ in the streamwise coordinate direction x and DM in the wall-normal coordinate direction y. As the pressure gradient dP/dx is imposed uniformly across the BL thickness δ, $\|\text{d}P/\text{d}x\|_{\text{L2},\Omega}$ is the smooth data domain bounding norm. The NS laminar BL $O(h^2)$ $\mathbf{DP} \bullet \hat{i}$ PDE GWS^h + θTS FE $p = 1$ basis algorithm asymptotic error estimate now extended to initial value for θ = 0.5 is:

$$\|e^h(\text{n}\Delta x)\|_E \le Ch^2\|\text{d}P/\text{d}x\|_{\text{L2},\Omega} + C_2\Delta x^3\|u^h(x_0)\|^2_{H^1(\Omega)}$$

Precisely designed $u(y, x = 0)$ DOF initial condition (IC) distribution for wall-normal uniform mesh refinement sequence $10 \le \text{M}_n \le 160$ guaranteed that the IC norm $\|u^h(x_0)\|^2_{H^1(\Omega)}$ was independent of any specific mesh M_n $\Omega^{h/\text{n}}$ as of course is $\|\text{d}P/\text{d}x\|_{\text{L2},\Omega}$.

A solution *verity* for any GWS^h + θTS algorithm is

$$\| q^h(\Delta x) \|_E + \|e^h(\Delta x)\|_E \equiv \|q(\Delta x)\|_E$$

The factor of two uniform mesh refinement sequence $\Omega^{h/\text{n}}$, for $n = 2, 4, 8, 16, \ldots$.

$\| q^h(\Delta x) \|_E + \| e^h \|(\Delta x)_E \equiv \| q(\Delta x) \|_E \equiv \| q^{h/2}(\Delta x) \|_E + \|e^{h/2}(\Delta x)\|_E = \ldots = \|q^{h/\text{n}}(\Delta x) \|_E + \|e^{h/\text{n}}(\Delta x)\|_E$ enables $\|e^h(\Delta x)\|_E$ error quantification, provided bounding norms $\|\text{data}\|_{\text{L2},\Omega}$, $\|\text{data}\|_{\text{L2},\partial\Omega}$, and $\|\{q^h(t_0)\}\|^2_{H^1(\Omega)}$ are independent of any specific $\Omega^{h/\text{n}}$.

Substituting GWS$^\text{h}$ + θTS measure h and $h/2$ solutions into the defined uniform refinement string admits clearing all unknown constants, hence

$$\|e^h(\Delta x)\|_E = 2^2 \|e^{h/2}(\Delta x)\|_E$$

Inserting the left-side term definition, the measure $h/2$ mesh GWS^h + θTS solution $u^{h/2}(y, \text{n}\Delta x)$ energy norm error is *quantified* to be:

$$\|e^{h/2}(n\Delta x)\|_E = \left(\| q^{h/2}(n\Delta x)\|_E - \| q^h(n\Delta x) \|_E\right) / (2^2 - 1)$$

For uniform wall-normal mesh refinement sequence $10 \le \text{M}_\text{n} \le 160$, and for Re = E05, GWS^h + θTS FE $p = 1$ basis-predicted convergence rate is *validated* to be quadratic via the open symbols in Figure B.2a. Throughout the mesh M_n sequence, the magnitude of GWS^h energy norm error $\|e^h(\text{n}\Delta x)\|_E$ is essentially insensitive to the domain bounding *data* $\|\text{d}P/\text{d}x\|_{\text{L2},\Omega}$ ranging adverse, neutral to favorable. This behavior is entirely distinct from the same code/mesh second-order accurate Crank-Nicolson FD (CNFD) algorithm $\|e^h(\text{n}\Delta x)\|_E$ sensitivity to data $\|\text{d}P/\text{d}x\|_{\text{L2},\Omega}$ [1, fig. 6.7].

With GWS^h asymptotic convergence theory quantitatively validated, a *regular* mesh solution-adapted *nonuniform de-refinement* sequence was executed for IC the uni-

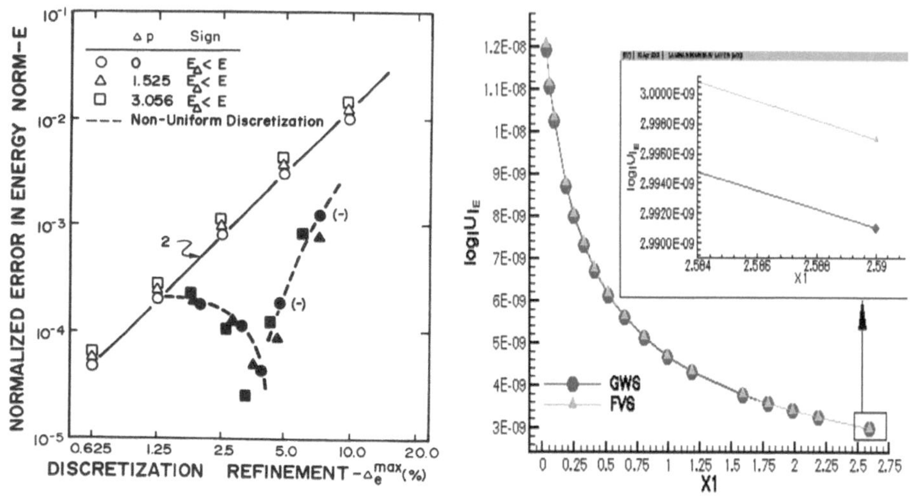

Figure B.2: Validation, $u^h(y, n\Delta x)$ asymptotic convergence in $\|e^h(n\Delta x)\|_E$, error extremization for the NS $n = 2$ BL PDE GWS^h + θTS FE $p = 1$ basis algorithm: (a) Open symbols represent the validation of quadratic convergence theory; solid symbols indicate a nonuniform mesh de-refinement sequence identifying the *optimal* mesh $M_n = 23$ solution. (b) GWS^h + θTS $\|u^h(n\Delta x \le 2.5)\|_E$ extremization in comparison with $O(h^2)$ CNFD.

form mesh $M_n = 80$ solution. Wall-normal mesh non-uniformity was progression ratio $h_{e+1} = \text{pr}\, h_e$, pr > 1, generated while maintaining the measure h of the $M_n = 80$ mesh-wall adjacent Ω_e constant. For identical Re and $\|dP/dx\|_{L2,\Omega}$ range, the mesh M_n *de-refinement* sequence $u^h(y, n\Delta x)$ energy norm error $\|e^h(n\Delta x)\|_E$, evolution, the filled symbols in Figure B.2a, experienced passage through zero. Thereafter, the (–) flagged error norm magnitude increased with continued mesh M_n de-refinement. (Recall you cannot plot zero on a semi-log graph!)

Thus quantitatively validated is the *existence* of a solution-adapted nonuniform minimal DOF *optimal* mesh $M_n = 23$ supported $u^h(y, n\Delta x)$ solution for a *genuinely* nonlinear NS PDE system. Repeating this sequence for $10 \le M_n \le 160$ for GWS^h $p = 2{,}3$ FE bases algorithms proved unsuccessful, which was later assessed to be due to state variable segregated Jacobian admitted convergence stall. The exception was the zero-pressure-gradient FE $p = 2$ basis algorithm, which indeed validated $O(h^4)$ asymptotic convergence [4].

That sole nonlinearity in **DP** • $\hat{i}$ is the convection term, which enabled quantification that the GWS^h + θTS FE $p = 1$ basis solution $u^h(y, n\Delta x)$ is the *extremum,* as the generated error is *orthogonal* to the FE $p = 1$ basis trial space union (refer to Figure B.1). For Re = E + 05 and $\|dP/dx\|_{L2,\Omega} = 0$, the same code/same M = 32 uniform mesh generated *exactly* comparable GWS^h + θTS and CNFD $\|u^h(n\Delta x)\|_E$ *evolutions* on $0 \le x \le 2.5$ (Figure B.2b) *validates* GWS^h + θTS $\|u^h(n\Delta x = 2.5)\|_E$ displacement from the IC $\|u^h(n\Delta x = 0)\|_E$ is the *extremum,* inset zoom clearly quantified.

B.3 Error quantification, non-smooth data, asymptotic convergence failure

Parameter r is a measure of *a priori* given *data* smoothness. The occurrence of $r-1<p$ renders the asymptotic convergence theory invalid, as the GWS^h solution does not reside in H^1. This degradation was encountered for a steady conduction DE: convection-radiation BC tapered fin-tube heat exchanger geometry (think Harley hog air-cooled engine), Figure B.3a.

The step change in element thermal capacity at the cylinder-fin juncture, which constitutes domain-distributed *data*, induced the GWS^h FE $p=1$ and $p=2$ basis $T^h(r)$ distributions slope discontinuity clearly displayed in Figure B.3b. That this rendered GWS^h prediction $T^h(r)$ *not* resident in H^1 is *validated* by the asymptotic convergence rate of the FE $p=2$ basis solution being identical with the $p=1$ basis prediction under uniform mesh refinement, Figure B.3, as predicted by the theory.

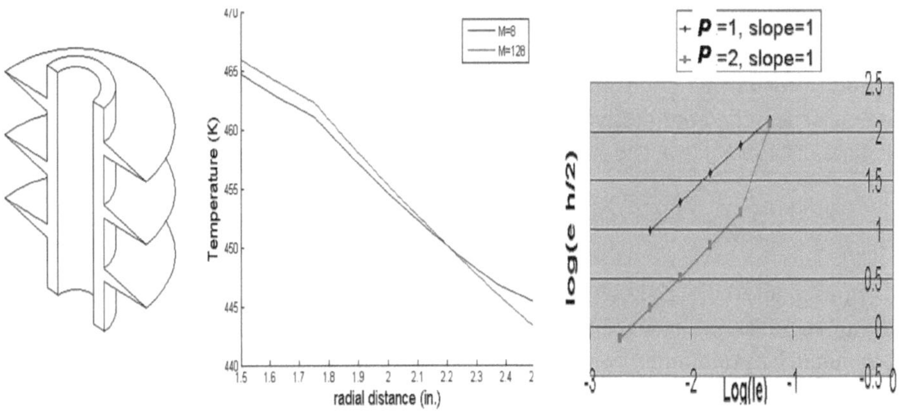

Figure B.3: Validation, GWS^h FE $p=1{,}2$ basis $T^h(r)$ asymptotic convergence in H^1, steady conduction DE: Robin BC finned tube heat exchanger: (a) air-cooled finned cylinder geometry; (b) tapered/straight fin radial $T^h(r)$, temperature distributions, uniform M = 8-128 meshes; and (c) uniform mesh refinement quantified FE $p=1{,}2$ bases $O(h^1)$ asymptotic convergence in H^1.

B.4 Theory alterations for *error freed* CFD mathematics, NS/RaNS $O(h^4)$ PDE GWS^h FE $p = 1,2,3$ algorithms, asymptotic convergence, error quantification

NS/RaNS GWS^h FE $p = 1,2,3$ completeness degree trial space basis algorithm asymptotic convergence rates in NS/RaNS $O(h^2)$ PDE energy norm $\|e^h\|_E$ are predicted 2, 4, 6 for *data* smoothness L2 norm bounded. Generalized to *unsteady* NS/RaNS TEA-altered $O(h^4)$ PDE *systems*, GWS^h + θTS FE $p = 1,2,3$ basis asymptotic convergence exponent $\alpha \equiv \min(p, r-1)$ transitions to $\alpha \equiv \min(2,2,3, r-1)$, which accounts for FE $p = 1$ basis *asymptotic convergence* rate elevation to 4. The definitions for NS/RaNS **DP** energy norms are

$$\text{NS: } \|u_i(t)\|_E \equiv \frac{1}{2\text{Re}} \int_\Omega \frac{\partial u_i(t)}{\partial x_j} \frac{\partial u_i(t)}{\partial x_j} d\tau, \quad 1 \le i \le n$$

$$\text{RaNS: } \|u_i(t)\|_E \equiv \frac{1}{2\text{Re}} \int_\Omega \frac{\partial u_i(t)}{\partial x_j} \left[\left(1 + \text{Re}^T\right) \frac{\partial u_i(t)}{\partial x_j} \right] d\tau, \quad 1 \le i \le n$$

As solution-adapted nonuniform mesh M_n refinement is becoming a requisite CFD code practice, the quantitatively validated factor of two *uniform* mesh refinement for NS/RaNS TEA $O(h^4)$ PDE GWS^h FE $p = 1,2,3$ degree basis predictions

$$\|e^{h/2}(t)\|_E = \left(\| q^{h/2}(t) \|_E - \| q^h(t) \|_E \right) / \left(2^{f(p)} - 1 \right), \ f(p = 1,2,3) = 4, 4, 6$$

is required altered to the *regular* solution adapted protocol. Recall that the caveat *regularity* requires the ratio of the Ω_e geometry independent scales ρ_e in the diameter parameter be maintained constant throughout any mesh refinement sequence to enable asymptotic convergence quantification in terms of the single mesh measure h.

The maintenance of *regularity* implies that the solution-adapted nonlinear mesh refinement domain discretization sequence with measure distributions $h_n(x_i)$ for n, $n+1$ meshes M_n correlate in essence with the number of elements Ω_e populating Ω^h for each mesh M_n. Based on this premise, *regular* solution-adapted nonuniform mesh refinement *error estimation* is predictable during on-the-fly code execution in terms of the NS/RaNS state variable member PDE energy norm set $\|\{e^h(t)\}\|_E$

$$\|\{e^h(t)\}\|_{E, M_{n+1}} = \frac{\|\{q^h(x_j, t)\}\|_{E, M_{n+1}} - \|\{q^h(x_j, t)\}\|_{E, M_n}}{(M_{n+1}/M_n)^{f(p)} - 1}, \ f(p = 1,2,3) = 4, 4, 6$$

for M_{n+1}/M_n, the ratio of element numbers populating $\Omega^h \equiv \cup \Omega_e$ for each n, $n+1$ in the sequence.

References

[1] Baker, A.J. (1983). *Finite Element Computational Fluid Mechanics*. Hemisphere, NY.

[2] Oden, J.T., Demkowicz, L.F. (1996). *Applied Functional Analysis*, CRC Press, FL.

[3] Oden, J.T., Reddy, J.N. (1976). *An Introduction to the Mathematical Theory of Finite Elements*, Wiley Interscience, NY.

[4] Soliman, M.O., Baker, A.J. (1981). "Accuracy and convergence of a finite element algorithm for laminar boundary layer flow," *J Computers Fluids*, 9.

Appendix C: *Error freed* CFD *mathematics* altered continuous Galerkin FE basis algorithm and Newton jacobian matrix statements

Libretto: Weak formulation continuous Galerkin $GWS^h + \theta TS$ FE $p = 1,2,3$ basis algorithms written on *error freed* CFD *mathematics* altered NS/RaNS/FaNS $O(h^4,m^4,\Delta t^4)$ PDE systems utilize *hypermatrix* resolution for all intrinsic *nonlinearities*. Quadratic convergent Newton algorithm Jacobian contributions are then precisely identified using matrix differential calculus operations on $GWS^h + \theta TS$ algorithm element matrices.

C.1 NS/RaNS $O(h^4,m^4,\Delta t^4)$ PDE $GWS^h + \theta TS$ FE p basis algorithm *error freed* CFD *mathematics* altered global matrix statements

The $GWS^h + \theta TS$ FE $p = 1,2,3$ basis algorithm universal flux vector statement for incompressible/compressible NS/RaNS/FaNS TEA altered $O(h^4,m^4,\Delta t^4)$ PDE systems is

$$[\mathrm{m}]\frac{\partial\{q\}}{\partial t} + \frac{\partial}{\partial x_j}\left(\{f_j\} - \{f_j^d\}\right) - \{s\} = \{0\} + O(h^4, m^4, \Delta t^4)$$

The order of entries in time derivative matrix [m] for incompressible/compressible state variables $\{q(x_j, t)\} = \{\varphi, u_i, \Theta, Y; P\}^T, \{q(x_j, t)\} = \{\rho, m_i, E, \rho Y; p\}^T$

$$[m] = \mathrm{diag}\left[(0,1), 1, 1, 1, 1, 1 - \frac{\gamma\Delta t^2}{6}\frac{\partial}{\partial x_j}\left(u_j u_k \frac{\partial}{\partial x_k}\right), 0\right]$$

accounts for the GWS algorithm intrinsic "mass" matrix, also the $O(m^4)$ scalar transport *optimal* $\gamma \equiv -0.5$ TEA functional augmentation for mass fractions Y and ρY. The first (0,1) entry in [m] distinguishes algorithm handling of density for incompressible and compressible fluids.

For incompressible NS/RaNS, the definitions for kinetic and dissipative flux vectors and source matrix for $1 \le (j, k) \le n$ are

$$\{f_j(\{q\})\} = \begin{Bmatrix} 0 \\ u_j\left(1 + \dfrac{\mathrm{Re}h^2}{6\Delta t}\right)u_i + P^*\big|_{n+1}^{p}\delta_{ij} \\ u_j\left(1 + \dfrac{\mathrm{RePr}\,h^2}{6\Delta t}\right)\Theta \\ u_j\left(1 + \dfrac{\mathrm{ReSc}\,h^2}{6\Delta t}\right)Y \end{Bmatrix}$$

https://doi.org/10.1515/9783112206430-018

$$\{f_j^d(\{q\})\} = \begin{Bmatrix} \dfrac{\partial \varphi}{\partial x_j} \\ \left(\dfrac{1 + \mathrm{Re}^T}{\mathrm{Re}} + \dfrac{h^2}{6\Delta t}\right)\left(\dfrac{\partial u_i}{\partial x_j} + \dfrac{\partial u_j}{\partial x_i}\right) + \dfrac{h^2\mathrm{Re}}{12(1+\mathrm{Re}^T)} u_j u_k \dfrac{\partial u_i}{\partial x_k} \\ \left(\dfrac{1}{\mathrm{Re}}\left(\dfrac{1}{\mathrm{Pr}} + \dfrac{\mathrm{Re}^T}{\mathrm{Pr}^{\mathrm{T}}}\right) + \dfrac{h^2}{6\Delta t}\right)\dfrac{\partial \Theta}{\partial x_j} + \dfrac{h^2\mathrm{RePr}}{12(1+\mathrm{Re}^T)} u_j u_k \dfrac{\partial \Theta}{\partial x_k} \\ \left(\dfrac{1}{\mathrm{Re}}\left(\dfrac{1}{\mathrm{Sc}} + \dfrac{\mathrm{Re}^T}{\mathrm{Sc}^{\mathrm{T}}}\right) + \dfrac{h^2}{6\Delta t}\right)\dfrac{\partial Y}{\partial x_j} + \dfrac{h^2\mathrm{ReSc}}{12(1+\mathrm{Re}^T)} u_j u_k \dfrac{\partial Y}{\partial x_k} \end{Bmatrix}$$

$$\{s(\{q\})\} = \left\{ \frac{\partial u_j}{\partial x_j},\ -\frac{\mathrm{Gr}}{\mathrm{Re}^2}\Theta \widehat{g}_i,\ \frac{\mathrm{Ec}}{\mathrm{Re}}\frac{\partial}{\partial x_j}\left(u_i \frac{\partial u_j}{\partial x_i}\right),\ s(Y) \right\}^T$$

These definitions are rendered specific to NS by defining $\mathrm{Re}^T \equiv 0$.

Rigorous incompressible DM differential constraint enforcement requires iterative coupling of NS/RaNS $O(h^4,\Delta t^4)$ $\mathrm{D}^m\mathbf{P}$ PDE with Chapter 4 derived DM^h

$$\mathrm{DM}^h\!: \frac{\partial^2 \varphi}{\partial x_k^2} + \frac{\partial u_j^h}{\partial x_j} = 0\ ,\quad \mathrm{NBC}\!: \left(\frac{\partial \varphi}{\partial x_j} - \left(u_j - u_j^h\right)\right)\widehat{n}_j = 0$$

for *any* approximation $u_j^h(x_i, t_{n+1})$ to divergence-free $u_j(x_i, t_{n+1})$. During time t_{n+1} NS/RaNS $\mathrm{D}^m\mathbf{P}$:DM^h + NBC iteration $p+1$, the pressure kinetic impact variable $\{f_j(\{q\})\}$ is

$$P^*(x_i, t_{n+1})|^p \equiv P^h(x_i, t_n) + \frac{1}{\theta\Delta t}\sum_{\alpha=1}^{p} \delta\varphi^h(x_i, t_{n+1})|^{\alpha}$$

for $\theta \geq 0.5$, time step Δt, prior time kinematic pressure $P^h(x_i, t_n)$ and for $\sum_{\alpha=1}^{p}\delta\varphi^h(x_i, t_{n+1})|^{\alpha}$ the accumulation of DM^h algorithm iterates

At NS/RaNS $\mathrm{D}^m\mathbf{P}$:DM^h + NBC t_{n+1} convergence, the significance of DM^h intrinsic energy norm $\|\varphi^h(t_{n+1})\|_E$ is equal to the iterative convergence requirement, usually $\varepsilon \geq$ E-04. Thus quantified is that $\mathrm{D}^m\mathbf{P}$:DM^h + NBC algorithm converged $u_j^h(x_i, t_{n+1})$ is $O(\varepsilon)$ *divergence-free*. Following GWS^h $\mathrm{D}^m\mathbf{P}$:DM^h + NBC algorithm time t_{n+1} convergence, the genuine kinematic pressure $P^h(x_i, t_{n+1})$ is post-processed via

$$\mathrm{DP}\!: \frac{\partial^2 P}{\partial x_k^2} + \frac{\partial}{\partial x_j}\left[u_i \frac{\partial u_j}{\partial x_i} + \frac{\mathrm{Gr}}{\mathrm{Re}^2}\bar{\Theta}\hat{g}_j\right] = 0\ ,\quad \mathrm{NBC}\!: \left(\frac{\partial P}{\partial x_j} - \left(\frac{1+\mathrm{Re}^T}{\mathrm{Re}}\right)\frac{\partial^2 u_j}{\partial x_k^2}\right)\widehat{n}_j = 0$$

Alteration of these identifications for compressible NS/RaNS TEA-altered $O(h^4,m^4,\Delta t^4)$ PDE system resides principally in $\mathrm{D}^m\mathrm{M}$ and superficially in $\mathrm{D}^m\mathbf{P}$ and $\mathrm{D}^m\mathrm{E}$. Incompressible DM^h replacement is TEA-altered $O(h^4)$ $\mathrm{D}^m\mathrm{M}$

$$\mathrm{D}^m\mathrm{M}\!: \frac{\partial \rho}{\partial t} + \frac{\partial}{\partial x_j}\left[m_j - \frac{h^2\,\mathrm{Re}}{12}u_k \hat{u}_j \frac{\partial \rho}{\partial x_k}\right] = 0 + O(h^4)$$

D^m**P** kinetic, dissipative flux vector $\{f_j\}$, $\{f_j^d\}$ compressible alterations exchange momentum m_i for velocity u_i with thermodynamic pressure p replacing P^*

$$\{f_j(m_i)\} = \left\{ u_j \left(1 + \frac{\mathrm{Re}h^2}{6\Delta t}\right) m_i + \mathrm{Eu}\, p\delta_{ij} \right\}$$

$$\{f_j^d(m_i)\} = \left\{ \begin{aligned} &\left(\frac{\mu^* + \mathrm{Re}^T}{\mathrm{Re}} + \frac{h^2}{6\Delta t}\right)\left(\frac{\partial u_i}{\partial x_j} + \frac{\partial u_j}{\partial x_i} - \frac{\partial u_k}{\partial x_k}\left(1 + \left(\frac{2}{3} + \rho k\right)\delta_{ij}\right)\right) \\ &\qquad + \frac{h^2\mathrm{Re}}{12(\mu^* + \mathrm{Re}^T)} u_j u_k \frac{\partial m_i}{\partial x_k} \end{aligned} \right\}$$

For supersonic/hypersonic onset Mach number, the D^{m}E total energy state variable E is often replaced with enthalpy H or stagnation enthalpy H_0. Either choice replaces the incompressible fluid Θ and inserts p into NS/RaNS D^{m}E PDE flux vector $\{f_j(E, H)\}$.

Dynamic pressure $p(\gamma,E,H)$ is the polytropic equation of state or equilibrium real gas variable γ data base predicted algebraically. Incompressible RaNS(k,ε) TEA-altered $O(h^4,\Delta t^4)$ PDE GWS + θTS algorithm for Re^T prediction is fully detailed in Chapter 6, hence not repeated here.

C.2 Continuous GWS^{h}enabled Newton Jacobian nonlinear identifications

Current practice CFD codes typically utilize state variable segregated, hence at best a *linearly convergent* linear algebra iteration package that can admit iteration *stall* which prevents algebraic convergence to *the* defined significance. To wit, Sajben transonic diffuser GWS^h and non-GWS^h RaNS:SA PDE system solutions were generated using the COMSOL code segregated linear algebra solver that decouples SA closure model update from the RaNS PDE iteration. Net is the *just barely* (!) convergent linear algebra process illustrated in Figure C.1 for *every* mesh M_n in the Chapter 6 documented regular mesh refinement 10-stage sequence. While this performance is tolerable for modest $n = 2$ discretizations, *high-fidelity* CFD prediction cannot mature without focused attention on linear algebra robustness!

The 1990s AIAA/ASME VBV protocol, Chapter 5, inadvertently recognized this linear algebra CFD issue in stating that the *measure* of CFD data publication *acceptability* was the *mesh-independent* solution. From continuous Galerkin weak formulation theory, this is a *highly unlikely* (!) occurrence as this solution exists *iff* CFD generated state variable DOF distributions correspond one-to-one with knots of pertinent order piecewise interpolation of NS/RaNS PDE exact solutions.

Open-source highly parallel nonsymmetric sparse matrix solver packages are widely available, e.g., GMRES [3], Bi-CGSTAB, van der Vorst [2], to upgrade the segregated linear algebra jacobian practice of current codes. Recall that this type solver is

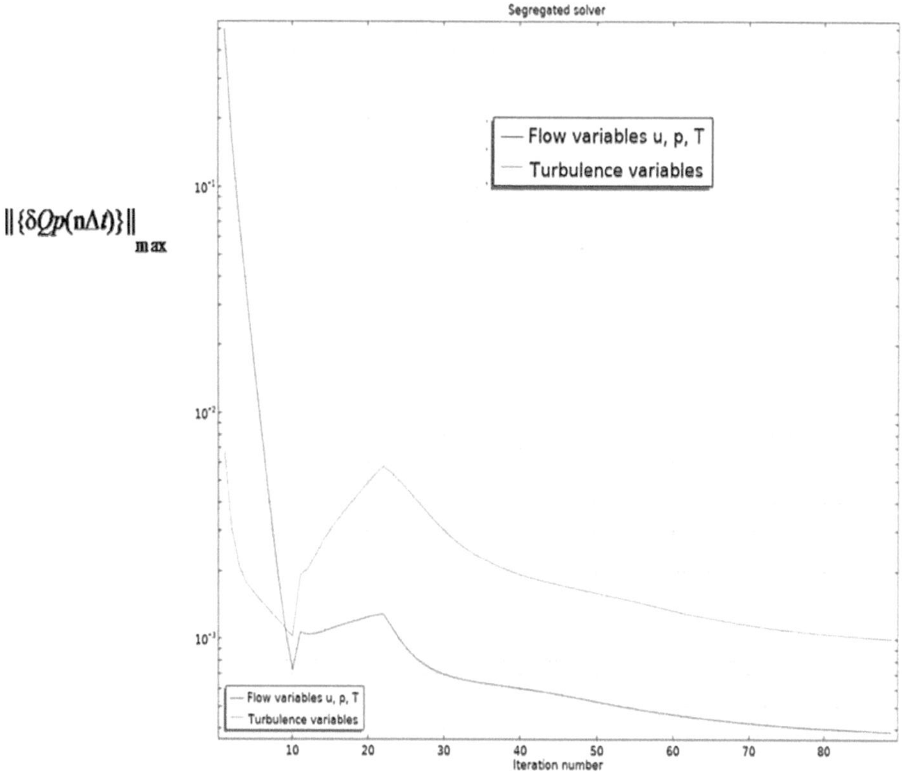

Figure C.1: Sajben transonic diffuser strong shock, TEA-altered compressible RaNS:SA $O(h^4)$ PDE GWS^h + θTS FE algorithm, COMSOL segregated linear algebra solver convergence history, mesh M_9.

absolutely required to achieve verified convergence of incompressible NS/RaNS rigorous conservation of mass.

GWS weak form FE basis matrix statements enable *precise* Identification of Newton algorithm *quadratic* convergent Jacobian *nonlinearities* via matrix differential calculus. For TEA-altered NS/RaNS $O(h^4, m^4, \Delta t^4)$ EBV PDE system flux vector statement

$$\text{NS/RaNS: } [m]\frac{\partial\{q\}}{\partial t} + \frac{\partial}{\partial x_j}\left(\{f_j\} - \{f_j^d\}\right) - \{s\} = \{0\} + O(h^4, m^4, \Delta t^4)$$

GWS^h + θTS algorithm iteration stage p global matrix statement is

$$\begin{aligned}\text{GWS}^h + \theta\text{TS: } \{F(\{Q_{n+1}\}^p\})\} &= [\mathrm{M}(\theta\{Q_{n+1}\}^p + (1-\theta)\{Q_n\})](\{Q^p_{n+1}\} - \{Q_n\}) \\ &\quad + \Delta t\left(\theta\{\mathrm{RES}(\{Q_{n+1}\})\}^p + (1-\theta)\{\mathrm{RES}(\{Q_n\})\}\right) = \{0\}\end{aligned}$$

for time stations $n+1$, n the DOF subscripts. Mass matrix [M(•)] holds GWS^h $[\mathrm{m}(y^h)]$ term while {RES({Q})} contains everything else.

The Newton iteration algebraic statement for solution of the global GWS^h + θTS matrix is

$$[\mathrm{JAC}(\{Q_{n+1}\}^p)]\ \{\delta Q\}^{p+1} = -\{F(\{Q_{n+1}\}^p)\}$$

$$[\mathrm{JAC}(\{Q_{n+1}\}^p)] \equiv \frac{\partial\{F(\{Q_{n+1}\}^p)\}}{\partial\{Q_{n+1}\}^p}$$

$$\{Q_{n+1}\}^{p+1} = \{Q_{n+1}\}^p + \{\delta Q\}^{p+1} = \{Q_n\} + \sum_{\alpha=0}^{p}\{\delta Q\}^{\alpha+1}$$

The Newton global Jacobian is formed by *assembly* of pertinent GWS^h + θTS element matrices. Therefore, the Newton Jacobian element matrix for time t_{n+1} terms in $\{F(\{Q_{n+1}\}^p)\}$ is element-level matrix calculus predictable as

$$[\mathrm{JAC}(\{Q\}^p_{n+1})]_e = \frac{\partial([\mathrm{M}(\theta\{Q_{n+1}\}^p_e)]_e\{Q_{n+1}\}^p_e + \theta\Delta t\{\mathrm{RES}(\{Q_{n+1}\}^p_e\})}{\partial\{Q_{n+1}\}^p_e}$$

$$= [\mathrm{M}(\theta\{Q_{n+1}\}^p_e)]_e + \frac{\partial[\mathrm{M}(\theta\{Q_{n+1}\}^p_e)]_e}{\partial\{Q_{n+1}\}^p_e}\{Q_{n+1}\}^p_e$$

$$+ \frac{\theta\Delta t\{\mathrm{RES}(\{Q_{n+1}\}^p_e)}{\partial\{Q_{n+1}\}^p_e}$$

The second term in the second line is due to the TEA γ term in diag[m].

C.3 Fidelity, GWS^h calculus statements translation to algebraic, Newton Jacobian

The GWS^h $\{\mathrm{RES}(\{Q\})\}_e$ element matrix for incompressible NS/RaNS TEA-altered $O(h^4,m^4,\Delta t^4)$ $\mathrm{D}^m\mathbf{P}$ nonlinear convection term generates the time- and DOF-*independent* (compute efficiency!) hypermatrix set $[\mathrm{M300}J]_e$, $1 \le J \le n$, which is readily Gauss quadrature evaluated for FE $p = 1,2,3$ bases. Accounting for incompressible DM, GWS^h element matrix contribution to $\{\mathrm{RES}(\{\mathrm{U}I\})\}_e$ is

$$\{\mathrm{RES}(\{\mathrm{U}I\})\}_e = \int_{\Omega_e}\{N\}\frac{\partial}{\partial x_j}\left(u_j\left(1+\frac{\mathrm{Re}h^2}{6\Delta t}\right)u_i\right)^h d\tau = \left(1+\frac{\mathrm{Re}h^2}{6\Delta t}\right)\int_{\Omega_e}\{N\}\frac{\partial}{\partial x_j}(u_j u_i)^h d\tau$$

$$= -\left(1+\frac{\mathrm{Re}h^2}{6\Delta t}\right)\{\mathrm{U}J\}^T_e\int_{\Omega_e}\{N\}\{N\}\frac{\partial\{N\}^T}{\partial x_j}\{\mathrm{U}I\}_e\, d\tau$$

$$= \left(1+\frac{\mathrm{Re}h^2}{6\Delta t}\right)\{\mathrm{U}J\}^T_e[\mathrm{M300}J]_e\{\mathrm{U}I\}_e$$

Matrix differential calculus operations on trailing DOF matrix $\mathrm{RES}(\{\mathrm{U}I_{n+1}\}^p_e)$, scalar matrix product, rearranged as needed, precisely identify *all* nonlinear element matrices that populate $[\text{JAC}(\{Q\}^{p}_{\text{n+1}})]_e$. The $\text{D}^m\mathbf{P}$ convection term contribution to Newton Jacobian state variable *self-coupling* element matrix is

$$[\text{JAC\_}\{\text{U}I,\text{U}I\}]_e = \theta\Delta t\left(1+\frac{\text{Re}h^2}{6\Delta t}\right)\left(\{\text{U}J\}^T_e[\text{M300}J]_e+\{\text{U}I\}^T_e[\text{M3}J\text{00}]_e\,\text{diag}[\delta_{\underline{IJ}}]\right)$$

which generates the new hypermatrix set $[\text{M3}J\text{00}]_e$, $1\le J\le n$. Underbar Kronecker deltas in $\text{diag}[\delta_{\underline{IJ}}]$ are indicators *only* (no summation). As well identified is $\text{D}^m\mathbf{P}$ state variable *off-diagonal coupling* Newton Jacobian element matrix contribution

$$[\text{JAC\_}\{\text{U}I,\text{U}J\}]_e \equiv \frac{\theta\Delta t\,\partial\{\text{RES}(\{\text{U}I_{\text{n+1}}\}^p_e)}{\partial\{\text{U}J_{\text{n+1}}\}^p_e} = \theta\,\Delta t\left(1+\frac{\text{Re}h^2}{6\Delta t}\right)\left(\{\text{U}I\}^T_e[\text{M3}J\text{00}]_e\right)$$

The TEA theory cubic/quartic nonlinearity resides in the NS/RaNS PDE dissipative flux vector. Altered to DOF rational tensor product interpolation, Green-Gauss divergence theorem imposition generates the TEA theory element matrix contribution to $\text{RES}(\{\text{U}I_{\text{n+1}}\}^p_e)$

$$\begin{aligned}
\{\text{RES}(\{\text{U}I\})\}_e &= -\theta\Delta t\int_{\Omega_e}\{N\}\frac{\partial}{\partial x_j}\left(\frac{h^2\text{Re}}{12(1+\text{Re}^T)}u_ju_k\frac{\partial u_i}{\partial x_k}\right)^h d\tau\\
&= \frac{\theta\Delta t\,h_e^2\text{Re}}{12}\int_{\Omega_e}\frac{\partial\{N\}}{\partial x_j}\left(\frac{\{\text{U}J\}^T_e\{N\}\{N\}^T\{\text{U}K\}_e}{\{1+\text{RET}\}^T_e\{N\}}\right)\frac{\partial\{N\}^T}{\partial x_k}\{\text{U}I\}_e d\tau\\
&\approx \frac{\theta\Delta t\,h_e^2\text{Re}}{12}\left\{\frac{\text{U}J{*}\text{U}K}{1+\text{RET}}\right\}^T_e\int_{\Omega_e}\{N\}\frac{\partial\{N\}}{\partial x_j}\frac{\partial\{N\}^T}{\partial x_k}\{\text{U}I\}_e d\tau\\
&\approx \frac{\theta\Delta t\,h_e^2\text{Re}}{12}\left\{\frac{\text{U}J{*}\text{U}K}{1+\text{RET}}\right\}^T_e[\text{M30}JK]_e\{\text{U}I\}_e,\ 1\le I\le n
\end{aligned}$$

The FE $p = 1,2,3$ basis-dependent *hypermatrix* set $[\text{M30}JK]_e$, $1\le J,K\le n$, *nonlinearly* distributes the DOF tensor product $\left\{\frac{\text{U}J{*}\,\text{U}K}{1+\text{RET}}\right\}^T_e$ to the nodes of Ω_e. Alteration of this term to NS simply involves defining $\text{RET}\equiv 0$.

The CFD code linear algebra *stability-critical* NS state variable *self-coupling* TEA term contribution to the Newton Jacobian $[\text{JAC\_}\{\text{U}I,\text{U}I\}]_e$ is

$$[\text{JAC\_}\{\text{U}I,\text{U}I\}]_e = \frac{\theta\Delta t\,h_e^2\text{Re}}{12}\left(\begin{aligned}\{\text{U}J{*}\text{U}K\}^T_e[\text{M30}JK]_e &+ \{\text{U}I\}^T_e[\text{M3}JK0]_e\text{diag}[\text{U}J\,\delta_{\underline{IK}}]\\ &+ \{\text{U}I\}^T_e[\text{M3}JK0]_e\text{diag}[\text{U}K\,\delta_{\underline{IJ}}]\end{aligned}\right)$$

no summation on underbar Kronecker deltas. $[\text{M30}JK]_e$ pre- and post-column matrix multiplier interchange enables the second and third matrix identifications, which generate tensor index-altered hypermatrix set $[\text{M3}JK0]_e$, $1\le(J,K)\le n$.

The Newton Jacobian matrix element contribution for incompressible RaNS $\text{RES}(\{\text{U}I_{\text{n+1}}\}^p_e)$ is the modest alteration

$$[\text{JAC\_}\{\text{U}I,\text{U}I\}]_e = \frac{\Theta\Delta t\, h_e^2\text{Re}}{12}\left(\left\{\frac{\text{U}J*\text{U}K}{1+\text{RET}}\right\}_e^T[\text{M30}JK]_e + \{\text{U}I\}_e^T[\text{M3}KJ0]_e\text{diag}\left[\frac{\text{U}K\,\delta_{\underline{IJ}}}{1+\text{RET}}\right]_e + \{\text{U}I\}_e^T[\text{M3}KJ0]_e\text{diag}\left[\frac{\text{U}J\,\delta_{\underline{IK}}}{1+\text{RET}}\right]_e\right)$$

Incompressible RaNS TEA-altered D^m**P** state variable Newton Jacobian off-diagonal coupling element matrix state variable *off-diagonal* coupling contributions are

$$[\text{JAC\_}\{\text{U}I,\text{U}J\}]_e = \frac{\Theta\Delta t\, h_e^2\text{Re}}{12}\left(\{\text{U}I\}_e^T[\text{M3}KJ0]_e\text{diag}\left[\frac{\text{U}K\,\delta_{\underline{IJ}}}{1+\text{RET}}\right]_e\right)$$

$$[\text{JAC\_}\{\text{U}I,\text{U}K\}]_e = \frac{\Theta\Delta t\, h_e^2\text{Re}}{12}\left(\{\text{U}I\}_e^T[\text{M3}KJ0]_e\text{diag}\left[\frac{\text{U}J\,\delta_{\underline{IK}}}{1+\text{RET}}\right]_e\right)$$

and RET $\equiv 0$ again defines the NS form.

Compressible NS/RaNS TEA-altered $O(h^4,m^4,\Delta t^4)$ D^m**P** PDE nonlinear convection term spatial differentiation is siimiplfied by Green-Gauss divergence theorem imposition

$$\begin{aligned}
\{\text{RES}(\{\text{M}I\})\}_e &= \left(1+\frac{\text{Re}h^2}{6\Delta t}\right)\int_{\Omega_e}\{N\}\frac{\partial}{\partial x_j}\left(\{\text{U}J\}_e^T\{N\}\{N\}^T\{\text{M}I\}_e\right)d\tau \\
&= -\left(1+\frac{\text{Re}h^2}{6\Delta t}\right)\int_{\Omega_e}\frac{\partial\{N\}}{\partial x_j}\{\text{U}J\}_e^T\{N\}\{N\}^T\{\text{M}I\}_e\,d\tau \\
&\quad+\left(1+\frac{\text{Re}h^2}{6\Delta t}\right)\oint_{\partial\Omega_e}\{N\}\left(\{\text{U}J\}_e^T\{N\}\{N\}^T\{\text{M}I\}_e\right)n_j\,d\sigma \\
&= -\left(1+\frac{\text{Re}h^2}{6\Delta t}\right)\{\text{U}J\}_e^T\int_{\Omega_e}\{N\}\frac{\partial\{N\}}{\partial x_j}\{N\}^T d\tau\{\text{M}I\}_e \\
&\quad+\left(1+\frac{\text{Re}h^2}{6\Delta t}\right)\{\text{U}J\}_e^T\oint_{\partial\Omega_e}\{N\}\{N\}\{N\}^T d\sigma\, n_j\,\{\text{M}I\}_e \\
&= \left(1+\frac{\text{Re}h^2}{6\Delta t}\right)\left(-\{\text{U}J\}_e^T[\text{M30}J0]_e\{\text{M}I\}_e + \{\text{U}J\}_e^T[\text{m3000}]_e\{\text{M}I\}_e\hat{n}_j\right)
\end{aligned}$$

which generates the $n = 2$ $\partial\Omega_e$ FE $p = 1,2,3$ basis “mass” hypermatrix $[\text{m3000}]_e$. Differentiating on trailing DOF matrices, repositioned as necessary, identifies compressible NS state variable *self-coupling* element matrix Newton Jacobian contribution

$$[\text{JAC\_}\{\text{M}I,\text{M}I\}]_e = \theta\Delta t\left(1+\frac{\text{Re}h^2}{6\Delta t}\right)\begin{pmatrix}-\{\text{U}J\}_e^T[\text{M30}J0]_e - \{\text{M}I\}_e^T[\text{M30}J0]_e\text{diag}\left[\delta_{\underline{IJ}}/\rho\right]_e \\ +\{\text{U}J\}_e^T[\text{m3000}]_e n_j + \{\text{M}I\}_e^T[\text{m3000}]_e\text{diag}\left[\delta_{\underline{IJ}}/\rho\right]_e\hat{n}_j\end{pmatrix}$$

as well as the compressible state variable *off-diagonal* coupling Jacobian contribution

$$[\text{JAC\_}\{\text{M}I, \text{M}J\}]_e = \theta\Delta t\left(1+\frac{\text{Re}h^2}{6\Delta t}\right)\{\text{M}I\}_e^T(-[\text{M30}J0]_e + [\text{m3000}]\,\hat{n}_j)\,\text{diag}\left[\delta_{\underline{IJ}}/\rho\right]$$

The NS TEA-altered $\text{D}^m\mathbf{P}$ cubic nonlinearity in $\{\text{RES}\{\text{M}I\}\}_e$ is

$$\begin{aligned}\{\text{RES}(\{\text{M}I\})\}_e &= \frac{h_e^2\text{Re}}{12}\int_{\Omega_e}\frac{\partial\{N\}}{\partial x_j}\left(\{\text{U}J\}_e^T - \{\text{DU}J\}_e^T\right)\{N\}\{N\}^T\{\text{U}K\}\frac{\partial\{N\}^T}{\partial x_k}\,d\tau\{\text{M}I\}_e \\ &\approx \frac{h_e^2\text{Re}}{12}\left(\{\text{U}J{*}\text{U}K\}_e^T[\text{M30}JK]_e - \{\text{DU}J{*}\text{U}K\}_e^T[\text{M30}JK]_e\right)\{\text{M}I\}_e\end{aligned}$$

The contribution of linear algebra *stability-critical* TEA term to compressible NS state variable self-coupling element Newton Jacobian matrix is

$$[\text{JAC\_}\{\text{M}I, \text{M}I\}]_e = \frac{\theta\Delta t\, h_e^2\text{Re}}{12}\begin{pmatrix}\{\text{U}J{*}\text{U}K\}_e^T[\text{M30}JK]_e - \{\text{DU}J{*}\text{U}K\}_e^T[\text{M30}JK]_e \\ +\ \{\text{M}I\}_e^T[\text{M3}KJ0]_e\text{diag}[\,\text{U}J\,\delta_{\underline{IK}}/\rho\,] \\ -\ \{\text{M}I\}_e^T[\text{M3}KJ0]_e\text{diag}[\,\text{DU}J\,\delta_{\underline{IK}}/\rho\,] \\ +\ \{\text{M}I\}_e^T[\text{M3}KJ0]_e\text{diag}[\,\text{U}K\,\delta_{\underline{IJ}}/\rho\,]\end{pmatrix}$$

and for compressible RaNS

$$[\text{JAC\_}\{\text{M}I, \text{M}I\}]_e = \frac{\theta\Delta t\, h_e^2\text{Re}}{12}\begin{pmatrix}\left\{\dfrac{\text{U}J{*}\text{U}K}{1+\text{RET}}\right\}_e^T[\text{M30}JK]_e - \left\{\dfrac{\text{DU}J{*}\text{U}K}{1+\text{RET}}\right\}_e^T[\text{M30}JK]_e \\ +\ \{\text{M}I\}_e^T[\text{M3}KJ0]_e\text{diag}\left[\dfrac{\text{U}J\,\delta_{\underline{IK}}/\rho}{1+\text{RET}}\right]_e \\ -\ \{\text{M}I\}_e^T[\text{M3}KJ0]_e\text{diag}\left[\dfrac{\text{DU}J\,\delta_{\underline{IK}}/\rho}{1+\text{RET}}\right]_e \\ +\ \{\text{M}I\}_e^T[\text{M3}KJ0]_e\text{diag}\left[\dfrac{\text{U}K\,\delta_{\underline{IJ}}/\rho}{1+\text{RET}}\right]_e\end{pmatrix}$$

To conclude, the *sparse matrix attribute* of NS D^mE PDE radiosity theory Neumann BC Newton Jacobian (Chapter 7)

$$\left[\text{JAC\_}\{Q_{n+1}^p\}\right]_e = \frac{4\text{det}_e}{\rho c_p}\frac{\sigma\varepsilon_e}{1-\varepsilon}[\text{M200}]\,\text{diag}\left[Q_{n+1}^3\right]_e$$

is retained in global $\left[\text{JAC\_}\{Q_{n+1}^p\}\right]$, [M200] the $\partial\Omega_e$ mass matrix.

Throughout, tensor index consistency serves as the check on $GWS^h + \theta TS$ algorithm residual/Jacobian element matrix veracity. The illustrated continuous weak formulation intrinsic capability, coupled with available highly parallel sparse matrix solver packages, portends total elimination of linear algebra *stall*-generated *non-converged* solutions at times admitted by current practice CFD codes.

References

[1] Van der vorst, H. (1992). “BiCGSTAB: a fast and smoothly converging variant of BiCG for solution of non-symmetric linear systems,” SIAM *J Sci & Stat Comp*,13.

Subject index

https://doi.org/10.1515/9783112206430-19